Network+

Targeted Courseware – 10 Steps To Certification

04 03 02 01 00   5 4 3 2 1

ISBN: 1-58619-011-3
Library of Congress Card Catalogue Number: 99-68909

First printing, January 2000
Printed in the United States

Published by Elton-Wolf Publishing
2505 Second Avenue, Suite 515
Seattle, Washington 98121
Tel 206.748.0345
E-mail: info@elton-wolf.com
Web Site: http://www.elton-wolf.com

Cover Design by Bassett and Brush Design

Text Conversion Services
Rave Communications
http://www.intergete.bc.ca/rave
780-998-4066
206-849-7966
Contact: Domhnall CGN Adams

Acknowledgments

Creating a study guide like this is a huge undertaking that requires the technical and creative assistance from many individuals. Several very important people must be recognized personally for their contributions in this project.

Courseware Development Manager—Candy Paape

Technical Writers—Anita Crocus, Greg Gille, Candy Paape, and Loral Pritchett

Editor—Anita Crocus

Technical Review—Wayne Cesaro, MCSE, MCT

Graphic Designer—Loral Pritchett

Photographer—Santo Roman

President—Michael Pastore, MSCE, MCT, A+

Vice President, New Business Development—Michael Holliday

Vice President, Learning Center Operations—Randall Thomas, MCSE, MCT, MCP+I, A+

Vice President, Online Systems—John Glassman

Vice President, Development—Don Drumheller

Administrative Director—Nancy DeChenne

Course Prerequisites

The study guide content targets the computer service technician with at least 18 to 24 months on-the-job experience.

In addition, we recommend that you have a working knowledge of the English language, so that you are able to understand the technical words and concepts this study guide presents.

To feel confident about using this study guide, you should have the following know-ledge or ability:

- The desire and drive to become a Network+ certified technician through our instructions, activities, quizzes, and study guide content
- Basic computer skills, which include using a mouse, keyboard, and viewing a monitor
- Basic networking knowledge including the fundamentals of the OSI model, TCP/IP, remote connectivity and security

Hardware and Software Requirements

To apply the knowledge presented in this study guide, you will need the following minimum hardware:

- Intel-based computer 486/66 MHz processor

- 16 MB RAM

- 500 MB of available hard disk space

- CD-ROM drive

- Mouse

- VGA monitor and graphics card

To apply the knowledge presented in this study guide, you will need the following minimum software installed on your computer:

- Microsoft Windows 95

- Microsoft DOS 6.0 or higher

Symbols Used in This Study Guide

To call your attention to various facts within our study guide content, we've included the following three symbols to help you prepare for the Network+ exam.

 Tip: The Tip identifies important information that you might see referenced in the certification exam.

 Warning: The Warning describes circumstances that could be harmful to you and your computer system.

 Note: The Note enhances your understanding of the topic content.

How to Use This Book

Although you'll develop and implement your own personal style of studying and preparing for the Network+ exam, we've taken the strategy of presenting the exam information in an easy-to-follow, ten-lesson format. Each lesson conforms to CompTIA's model for exam content preparation.

At the beginning of each lesson, we summarize the information that will be covered. At the end of each lesson, we round out your studying experience by providing the following four ways to test and challenge what you've learned.

Vocabulary—Helps you review all the important terms discussed in that lesson.

In Brief—Reinforces your knowledge by presenting you with a problem and a possible solution.

Activities—Further tests what you've learned in the lesson by presenting ten activities that often require you to do more reading or research to understand the activity. In addition, we've provided the answers to each activity.

Lesson Quiz—To round out the knowledge you'll gain after completing each lesson in this study guide, we've included ten sample exam questions and answers. This allows you to test your knowledge, and it gives you the reasons why the "answers" were either correct or incorrect. This, in itself, enhances your power to pass the exam.

You can also refer to the Glossary at the back of the book to review terminology. In addition, you can view the Index to find more content for individual terms.

Introduction to Network+ Certification

The Network+ certification exam is a testing program sponsored by the Computing Technology Industry Association (CompTIA) that certifies the competency of computer service technicians with at least 18 to 24 months on-the-job experience.

When you receive your Network+ certification, it proves your competence by having earned a nationally recognized credential as a computer service technician. In addition, major vendors, distributors, resellers, and publications back the Computing Technology Industry Association program, which gives this exam the credibility it deserves.

The Network+ exam covers a vast range of vendor-independent hardware and software technologies, as well as basic network knowledge, technical skills and practices, as defined by over 45 organizations in the information technology industry.

To help you bridge the gap between needing the knowledge and knowing the facts, this study guide presents networking technology and networking practices essential for passing the Network+ certification exam by CompTIA.

 Note: This study guide presents technical content that will enable you to pass the Network+ certification exam on the first try.

To become a certified Network+ Technician, you will need a 68% score to pass the Network+ exam. Ninety minutes is the allotted time to take the test, which consists of 65 questions.

**Tip:**   According to CompTIA, the Network+ certification is a
lifetime certification. Once you pass the exam, you will
not be required to ever take it again.

Network+ Study Guide Objectives

Successful completion of this study guide is realized when you can competently
understand, explain and implement networking technology and its recommended
practices. Your objectives are to:

- Demonstrate a basic understanding of a network's structure through its topology
options, and the characteristics of segments and backbones

- Identify major network operating systems, clients and directory services

- Define and associate IPX, IP, and NetBEUI with their functions

- Explain network terms encountered in a technician's daily tasks and define fault
tolerance technologies as applicable in a network environment

- Define the OSI model layers, including protocols, services and functions that
pertain to each layer

- Recognize and describe networking media and connectors and list the advantages and disadvantages of using each in a given network scenario

- Understand and apply the principles of basic network attributes, purpose and
function

- Explain and be able to apply the practices and concepts for each layer of the OSI
model

- Demonstrate knowledge, purpose and function of TCP/IP and the fundamental
concepts of TCP/IP addressing

- Identify TCP/IP configuration options and describe the utilities available in the
TCP/IP suite

- Explain methods for testing, validating and troubleshooting IP connectivity using TCP/IP utilities
- Explain remote connectivity concepts, including PPP, SLIP, PPTP, ISDN and PSTN
- Define the requirements for dial-up networking and a remote connection
- Identify strategies and good practices that can ensure network security
- Describe the planning and implementation methods required before installing a network, including environmental conditions that impact computer networks
- Recognize and demonstrate knowledge of common peripheral ports, external SCSI connectors, and common network components
- Demonstrate knowledge and understanding for maintaining, supporting and troubleshooting a network

Figures

Tables

Table of Contents

Network+ 10 Steps to Certification

Lesson 1: Basic Network Knowledge 1

Lesson 2: Network Architecture 111

Lesson 3: Network Operating Systems......161

Lesson 4: TCP/IP Fundamentals 195

Lesson 6: Connectivity..............................271

Lesson 8: Network Requirements 357

Lesson 9: Network Support and Maintenance

Lesson 10: Troubleshooting 425

Lesson 1: Basic Network Knowledge

A network consists of two or more computers connected together by cables with the purpose of sharing tasks and information. The complexities of modern networks have evolved from these fundamental premises. Currently, the use of networks as a business tool is increasing as more organizations discover the ease with which they can reliably communicate and share data, messages, graphics, printers, and other hardware resources.

To understand network functions, it is necessary to understand the physical and logical components comprising a network. This includes how physical components connect as well as the electrical and logical interactions between them. Included are descriptions of network models, topologies, cabling, and network expansion components.

After completing this lesson, you should have a better understanding of the following topics:

- Network Basics

- Network Structure

- Network Components

Network Basics

The concept of networking emerged from the need of stand-alone computers to share information with other users in a timely manner. Without a network, information must be either printed out or copied to floppy disks so others can copy information to their computers. With a network, sharing data and online communication among users is simplified. To accomplish this, a network utilizes a group of physically and logically connected computers, printers, and other devices to enable sharing files, printers, communication, and other resources.

Additional advantages of networks include:

- Communication and scheduling efficiencies are maximized
- Applications can be standardized so all users have the same application and version
- Costs are cut through timely data and peripheral sharing

Local Area Network (LAN)

A LAN consists of a limited number of computers connected together in a common area within a limited physical distance. For example, a small company or office located on a single floor of a building might use a LAN's logical environment. Earlier LANs allowed about 30 users to connect on a cable not exceeding 600 feet. Current LAN technologies support many more users and have fewer cable distance restrictions.

A LAN combines hardware and software technologies to allow users to share resources such as data, programs, storage devices, printers, and other peripherals. A LAN also enables users to collaborate and interact by sending messages and data to each other. Collaboration is enabled by groupware applications that run on both the server computer and client workstations.

LAN Communication

A LAN enables communication among network nodes consisting of workstations, hubs, repeaters, bridges, switches, routers, and servers. To enable communication, the networking hardware, and software on LAN workstations, servers, and other devices must be compatible. This includes hardware such as network adapter cards, hubs (for linking network devices), and transmission media, as well as the software for server and workstation operating systems, application servers, and shared network devices. An example of LAN connectivity between a server and clients is shown in Figure 1.1.

Figure 1.1 LAN Servers and Client Connectivity

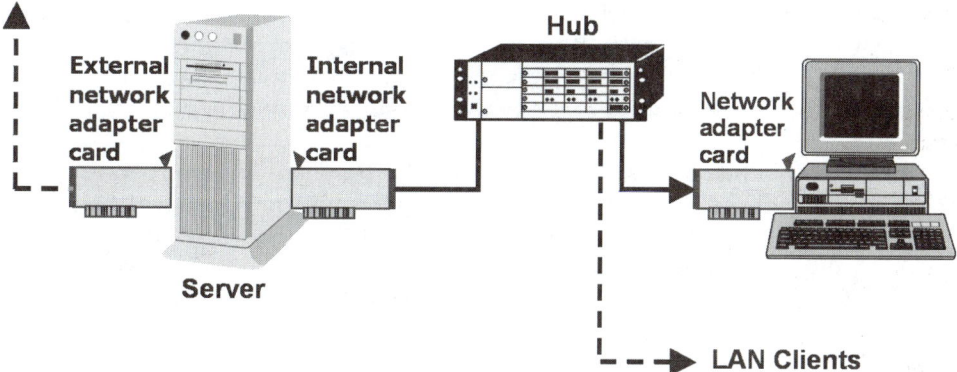

Topology

LAN topology describes the way a network is organized both physically and logically. The most widely used LAN topology implementations are the star, linear bus, and ring configurations. For example, Ethernet network architecture utilizes star and linear bus configurations. LANs typically use twisted-pair, coaxial, or fiber-optic cable.

Network Access

When a LAN computer has data to send to another computer, it must have a way to access the network without interfering with other transmitting computers on the network. Different LAN topologies require different methods for computers to gain network access, but the goal of all access methods is to avoid data collisions by managing when and how computers transmit their data. The most common access methods are:

Carrier Sense Multiple Access with Collision Detection (CSMA/CD)—With CSMA/CD, each network computer that is ready to send data, including the server, listens to the network cable for traffic. When the network is busy, a neutral carrier signal is detected which signifies to the listening computer that it must wait until data reaches its destination and the network clears before transmitting. This is considered a "contention" access method since computers contend for network access and must wait for transmission clearance to avoid data collisions.

Carrier Sense Multiple Access with Collision Avoidance (CSMA/CA)—In this system, computers signal their intent to transmit data on the network in advance of that occurrence. Other computers listen for these signals so they can sense when a collision might occur and thus delay their data transmissions to avoid collisions. However, CSMA/CA places additional traffic on the network when signaling the intent to transmit, causing this access method to be slower than CSMA/CD.

Token passing—In a token passing network, a packet known as a free token circulates around a ring of computers. When a computer has data to send, it takes control of the free token, which enables it to transmit. While the token is in use by a computer, no other computers can access the network for data transmission. For this reason, token passing is considered a "non-contentional" network access method that provides collision free access.

Demand priority—A method for the 100 megabytes per second (Mbps) Ethernet standard known as 100VG-AnyLAN. In this type of network, repeaters manage network access by polling network nodes for data send requests. The repeater maintains a table of node addresses and verifies that the nodes are functioning. If two computers cause contention by simultaneous transmissions, demand priority implements a system that grants access by servicing the request with the highest priority level. If two contending computers have the same priority level, both requests are serviced simultaneously by alternating between the two.
The advantages of using demand priority are that it uses four pairs of wires, two for transmitting and two for receiving, which allows computers to transmit and receive at the same time. In addition, transmissions are under the centralized control of the hub and not broadcast to all network computers. Therefore, computers never contend independently for network access.

Note: In demand priority networks, computers can transmit and
                  receive at the same time since they use four pairs of wires
                  and quartet signaling (a 25 Mhz signal is transmitted on all
                  four wire pairs to regulate transmissions).

Data Transmission

LAN architectures accommodate various transmission rates. For example, current
Ethernet architecture can handle data transmission rates up to 10 Mbps. Another
commonly used architecture is token ring, which can handle either 4 or 16 Mbps
transmission rates. Arcnet architecture uses star bus topology with data
transmission rates of 2.5 Mbps and Arcnet+ supports up to 20 Mbps. In all these
architectures, the media in use must support the data transmission rate or network
communications can fail.

Duplexing

Duplexing refers to the directional criteria for the network's send and receive
signals. Networks support either half-duplexing or full-duplexing.

Half-duplexing—Allows data to flow in one direction. Transmit signals sent in one
direction must reach their destination before response signals can be sent on the
same path in the opposite direction.

Full-duplexing—Increases network speed by allowing transmits and response
signals to travel simultaneously in both directions. For this to work, the signal must
connect point-to-point between two pieces of equipment and cannot traverse a
shared access device, such as a hub. In full-duplexing, each device uses two
different wire pairs (one for transmit, the other for receive) to allow bi-directional
signals on the network.

Transmission Types

The two basic types of signal transmissions used in network communications are
baseband and broadband.

Baseband transmissions, shown in Figure 1.2, utilize a single digital signal to send data over a single transmission channel between devices connected by coaxial cable or twisted-pair cables. Baseband network devices only transmit when there is no channel traffic, although a technique called time division multiplexing shares channels. Baseband network transmissions travel as packets, which contain source and destination addresses and data. Baseband signals use a detection scheme called Manchester encoding, where the positive and negative edges of the data pulses register as the signal's logic levels.

Figure 1.2 Baseband Transmission

Data stream   **Time** ⟶

Baseband networks typically support data transmission rates from 56 up to 100BaseT or 100BaseTX but are limited to relatively short distances because of signal attenuation. Baseband technology has strict requirements for distance, cable types, shielding, and topology to ensure proper network implementation. Typical baseband configurations include Ethernet and token-ring networks.

Broadband transmissions use analog signals at high frequencies and large bandwidths. A broadband transmission medium carries multiple simultaneous messages, as shown in Figure 1.3, with each message modulated by a separate carrier frequency using a device, such as a broadband modem.

Figure 1.3 Broadband Transmissions

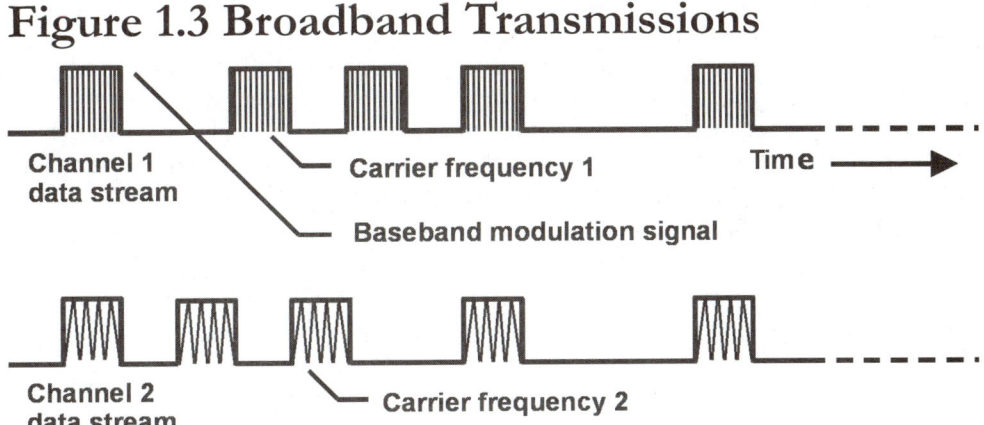

Broadband technology allows several networks to coexist on a single cable without transmission interference since each network has a unique carrier frequency, as shown in Figure 1.3. Broadband networks utilize radio-frequency signals that travel on separate outbound and inbound channels to deliver and receive data, voice, and video information. You can use broadband technology in WAN communications because it supports data rates that extend into the gigabit range.

Media

The transmission media used on a LAN depends on several factors including the required data transmission rate, security needs, type of network adapter cards, and the physical deployment environment. LAN media must be compatible with network adapters, hubs, and other network devices or LAN communication can fail. Several types of transmission media include unshielded twisted pair (UTP) cables, shielded twisted pair (STP), coaxial cable (Thinnet and Thicknet), and fiber-optic cable.

Wide Area Network (WAN)

When the infrastructure of a small business LAN expands to encompass users in different cities and states, it is a WAN. A WAN can serve thousands of users and is often referred to as an enterprise network. In contrast, a LAN cannot support the needs of a large business with multiple offices spanning geographically separate locations.

WANs consist of multiple LANs interconnected by routers, channel service unit/data service unit (CSU/DSUs), and leased lines from telephone carrier service providers. Some organizations create WANs using satellite links, microwave transceivers, and packet radio, while others build virtual private networks using encrypted communication over the Internet.

Routers connect LANs to the WAN and act as internetwork gateways. The router's WAN port connects serially to a CSU/DSU, which in turn connects to a digital line, such as T1, provided by a local telephone carrier. A LAN-to-LAN connection across the WAN can be a dedicated point-to-point circuit or can terminate in a frame relay cloud (a packet-switching network that breaks up packet routings to achieve the fastest data delivery time).

Network Structure

Network structures vary depending on the specific needs of organizations. These needs define the physical hardware, connection media, interconnection devices, software components, and protocols used in the network configuration. Basic network structure consists of a networking model, topology, and network connection media.

Network Models

The networking model refers to a hierarchical relationship between network computers and the way they interact within that hierarchy. The three types of hierarchies commonly used in networks today are mainframe, peer-to-peer and client/server.

Mainframe

A network with one central computer (server) that contains all the programs, performs all tasks, and maintains the databases while workstations having minimal computing power access these resources only as needed. Distributed processing is a variation of this network model where independent workstations perform some of the processing, but the central computer handles most tasks. In some cases, data is contained on a central server that distributes information to a series of smaller servers, called a "staging system."

Peer-to-Peer

A peer-to-peer network does not rely on a central server, but functions with workstations acting as equal peers for handling tasks, data storage, and security matters.

A peer-to-peer network can usually meet the needs of small organizations or workgroups. A peer-to-peer network does not have any dedicated servers or computer hierarchy, as shown in Figure 1.4. Instead, each computer is an equal peer that functions as both a client and server, thus decentralizing data access and security control. In peer-to-peer networking, individual users perform certain administrative functions and decide what data to share on the network. Peer-to-peer networks are less secure than server-based networks since access control tools on local computers are limited.

Figure 1.4 Peer-to-Peer Network Model

Client-server stations

You can use a peer-to-peer network when:

- Organizational growth is limited to 10 or less network users

- Users are centrally located

- Security and central administration is not important

- Users are capable of performing administrative tasks such as making resources available, data backup, and installing, maintaining, or upgrading application software

- The cost of a high-capacity network server is prohibitive

Before implementing a peer-to-peer network, you should consider the following requirements:

- Peer-to-peer network users must access resources such as directories, printers, and modems
- When acting as a server, the workstation must support network resource access.

 Note: In Microsoft's Windows NT Workstation, Windows for Workgroups, and Windows 95/98, full peer-to-peer networking software is included in the operating system.

- The workstation must devote most of its resources to supporting its users
- Peer-to-peer network users set and track passwords for shared network resources. Since users set their own passwords and shares are not located on a dedicated server, centralized share access control is not possible, which results in minimal network security.
- Before users can manage both user and administrator tasks, you should consider offering a training program for managing administrative tasks.

Client/Server

A client/server network has a central server, which manages users, data, security, hardware, and shared resource access. Workstations can share the processing load with the central server. The client/server model is the primary focus of this course. When a user environment exceeds more than 10 workstations, peer-to-peer networking is difficult to manage and quick access to network resources deteriorates. In this situation, most organizations deploy the client/server model, as shown in Figure 1.5, to utilize the speed, functionality, and security of a server-based network.

A central server manages the tasks of storing data, implementing network security, managing users, backing up critical data, sharing files, managing hardware, and providing access to shared resources. The server responds quickly to service requests from network clients while also ensures file and directory security. With a centrally managed server, it is easy to deploy fault tolerance systems and plan for disaster recovery strategies.

Clients (workstations) share the processing burden with a server such as Microsoft Windows NT, Novell NetWare, or UNIX. Workstations connecting to the server receive a security logon prompt to enter user credentials (name and password). The logon defines their level of access to server/network resources based on assigned user rights and profile restrictions.

Figure 1.5 Client/Server Networking Model

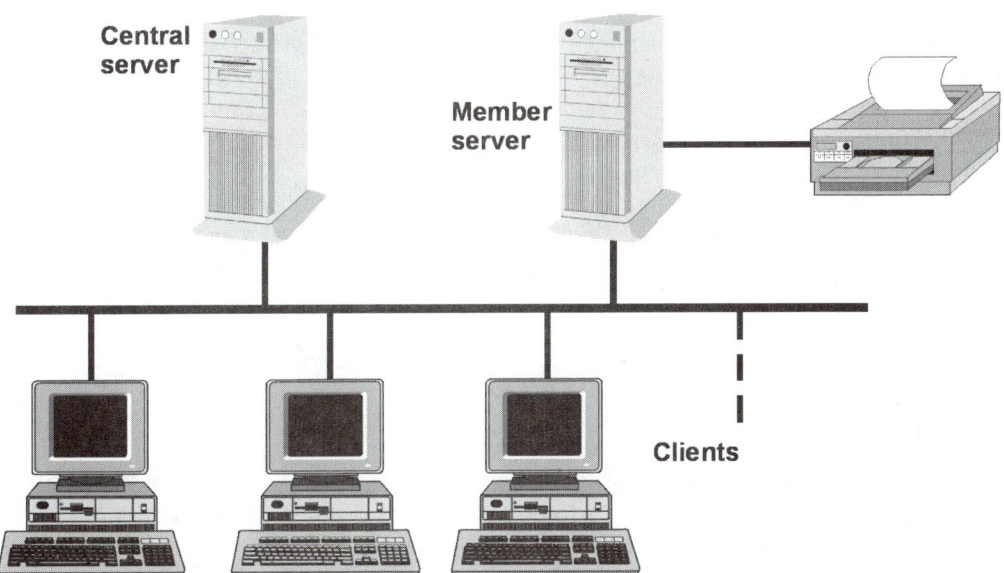

If the network becomes large enough, additional stand-alone network servers, sometimes referred to as member or application servers, may be required. A member server does not exercise the full capability of the central server, but is a network member that helps redistribute the processing load for specific tasks. The following describes several types of member servers, which utilize the client/server model:

File and print server—Manages user access and use of file resources. For example, when clients need to access a document file, they download the file to their workstations from the file server where it is stored. You can edit the document locally using a word processor application running on the workstation.

The print server manages print jobs sent from the client. In large organizations, a dedicated print server can expedite quick turnarounds for multiple simultaneous print jobs.

Application server—Maintains a database that can be accessed by network client requests. The database always resides on the application server and only specific information related to the client request is sent to the client (instead of the entire database). The client sends requests with a local application that accesses a database such as Microsoft SQL Server.

Mail server—Manages e-mail services for network users. Mail servers carry a heavy processing load, especially if they support groupware applications as Microsoft Exchange Server does. In large organizations with many users communicating via e-mail, a dedicated server is needed to handle the heavy traffic generated.

Communication server—Handles the exchange of information between the central server network and remote networks utilizing dial-up or dedicated connectivity access methods. An example of a communication server is Microsoft Windows NT Remote Access Services (RAS).

Client/Server Advantages

Following are the major advantages of the client/server environment:

Resource sharing—Central location of resources on a single server allows central management and control for administrative support and a centralized location for user share creation and access.

 Note: Most networks utilize a combination of server-based and peer-to-peer networks with simultaneous client access to both server resources and other client workstation shares.

Security—A security policy can be centrally administered and applied from the server to every network user. The superior security offered by the client/server networking model is the primary reason it is most often chosen.

Data backup—When critical company data is stored in one central location, the backup process is made simpler and more reliable.

Fault tolerance—With company data on a single server, a fault tolerance system utilizing data redundancy techniques can be easily implemented. An online backup copy of data is replicated from the server and can be retrieved if the server crashes.

User numbers—A server-based network can support many more users than a peer-to-peer network.

Client hardware minimized—Since clients do not perform server functions, the requirement for client hardware resources such as RAM and hard disk space are reduced compared to server requirements.

Network Topology

The structure of a network depends on the type of topology. Network topology refers to the way cables, computers, and other components are laid out on the network. Network topology, or basic network design, can be described in terms of both physical and logical (conceptual) elements.

Physical Topology

Physical topology is the physical arrangement of computers, printers, cables, and other devices. To a certain degree, the physical layout defines the logical configuration of the network.

Logical Topology

Logical topology defines data routes from source to destination nodes, message-passing methods, and logical interaction paths between network devices.

Network topology impacts network requirements in the following areas:

Equipment type—Includes capabilities and configuration of network equipment such as the interconnection of different cable types, network adapter cards, and other devices.

Capacity planning—The number of users assigned to a network can determine the type of topology required, however, you must plan for network expansion as your organization grows.

Communication—Network topology impacts computer communication requirements such as the data transmission rate. Various topologies support specific data transmission rates and network access methods.

Network management—Plan for ease of use since some network topologies are easier to maintain, such as a bus network.

Network performance—Fast access methods and more reliable (collision-free) data transmissions, such as a token-ring network guarantee better network performance than other network types.

 Tip: Network engineers define network topology before installing and configuring network elements and components. Mapping out the physical and logical environment in advance can save time and effort when it comes time to implement the design in a real working environment.

Most network designs are based on five fundamental topologies: bus, star, ring, mesh, and combined.

Bus Topology

A bus topology is a simple design that uses a single cable to connect network devices. The single cable, sometimes called a trunk or backbone, connects devices along a linear segment, shown in Figure 1.6.

Figure 1.6 Bus Topology Network

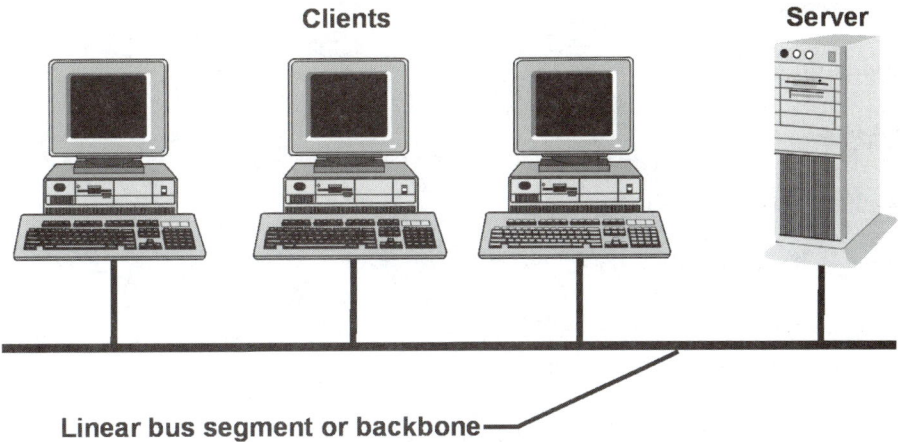

Linear bus segment or backbone

Communication in a bus topology involves sending data across the backbone from the source to destination computer. The data consists of streams of encoded electronic pulses that in both directions on the backbone to all connected computers. However, only the computer with an address that matches the data address accepts the message, shown in Figure 1.7. The other computers and devices ignore the message.

The data transmission rate in bus topology is determined in part by the speed the backbone segment and device connection cabling can support. A typical Ethernet network using baseband architecture and bus topology transmits at 10 Mbps.

In a bus topology network, only one computer at a time transmits data, while the other computers wait for the network to clear. When there are many computers connected in a bus topology configuration, network traffic can quickly be congested. The more computers waiting to place data on the bus, the slower the network.

Figure 1.7 Sending Data in a Bus Topology

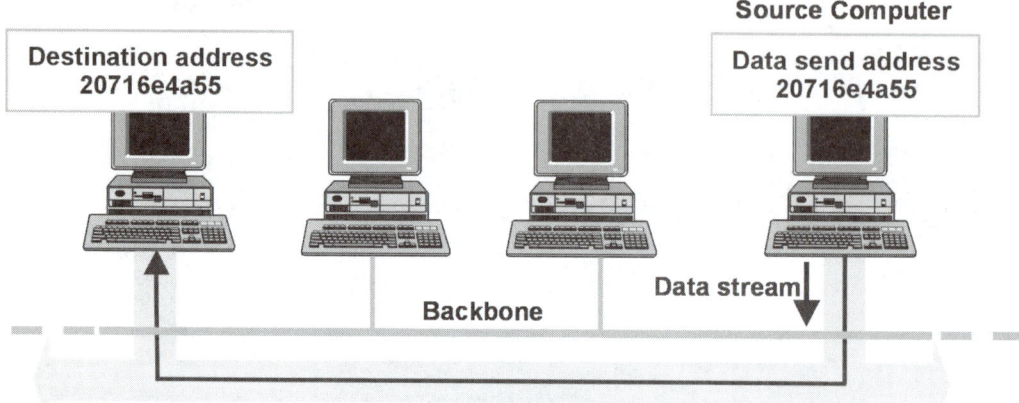

Data signal propagation

CSMA/CD manages network access for computers in a bus topology configuration such as Ethernet.

Bus topology network performance is affected by computer hardware capabilities, applications running on the network, and cable types and the distance between computers.

While the destination computer receives the data, the original data transmission continues to propagate on the backbone. If a device known as a terminator is not installed on the bus ends, the open circuit cable end causes signal reflections to interfere with concurrent data transmissions from other computers. Signal reflections create a standing wave pattern that can distort and degrade subsequent data transmissions, which causes the network to go down because distorted data cannot be recognized and processed.

The terminator, shown in Figure 1.8, is a device installed at the ends of the bus segment in a typical Ethernet bus network. The terminator absorbs any energy from the data transmission still propagating on the network after the destination computer receives the original message. To do this without creating signal reflections, the terminator must match the cable impedance.

For example, a cable with a 50-ohm characteristic impedance requires a 50-ohm terminator for optimum signal absorption. In most applications, a passive terminator uses a static resistive element to absorb energy.

However, cabling components (including terminators) contain a frequency-sensitive or reactive impedance element that contributes to signal reflections. This is not a problem since terminator reactance values are tuned out by the manufacturer and, therefore, have no significant affect on signal reflection levels.

 Note: Some manufacturers make active terminators that automatically adjust their impedance value to match the impedance of the attaching cable segment.

Figure 1.8 Terminators on a Bus Network

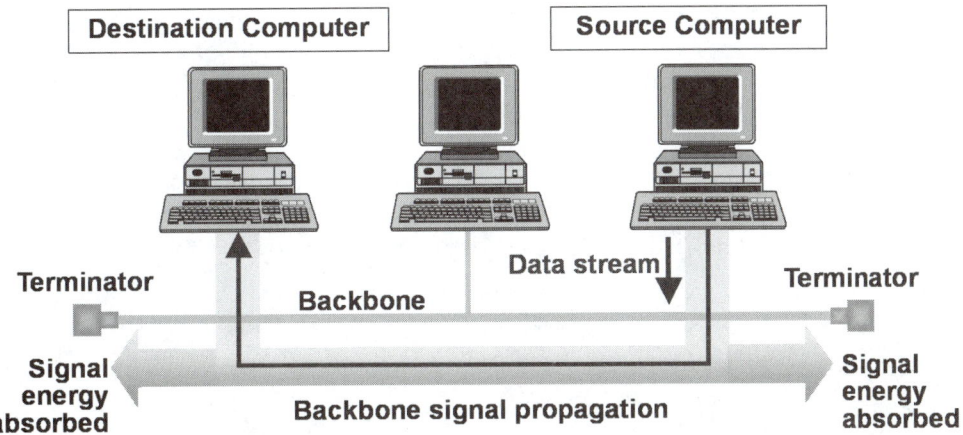

Bus topology is passive because all computers listen for data but do not regenerate the signal and pass it along to the next computer, as in an active topology. If one computer in the bus topology configuration fails, other computers are not affected and the network stays up.

Note: If a cable breaks or is disconnected on the backbone
segment in a bus topology network, the network stops
functioning although computers can still work on a
stand-alone basis.

As the bus topology network expands, cable distances can be extended to
accommodate growth by using longer cables, and interconnecting cables with barrel
connectors. However, limit your usage of barrel connectors since mismatches can
occur at connection interfaces causing signal loss.

You can use repeaters (signal-restoring devices) to interconnect cables and boost
the signal for longer cable distances.

Star Topology

A star topology supports network computers connected to a centralized mainframe
computer; however, the basic aspects of this design are implemented in the
client/server environment. The star topology network consists of computers,
printers, and other devices connected by point-to-point cable segments to a central
hub in a configuration that resembles a star.

In a star topology network, data is transmitted from the sending computer to a hub
where the signal is transmitted to all computers on the network as shown in Figure
1.9. Only the computer with an address that matches the data address accepts the
message.

 **Tip:**   A hub is a device that provides a central location where
computers and devices can connect together. Passive hubs
organize cabling and distribute signals while active hubs
provide signal regeneration and retransmission to their
distribution points.

Figure 1.9 Star Topology Network

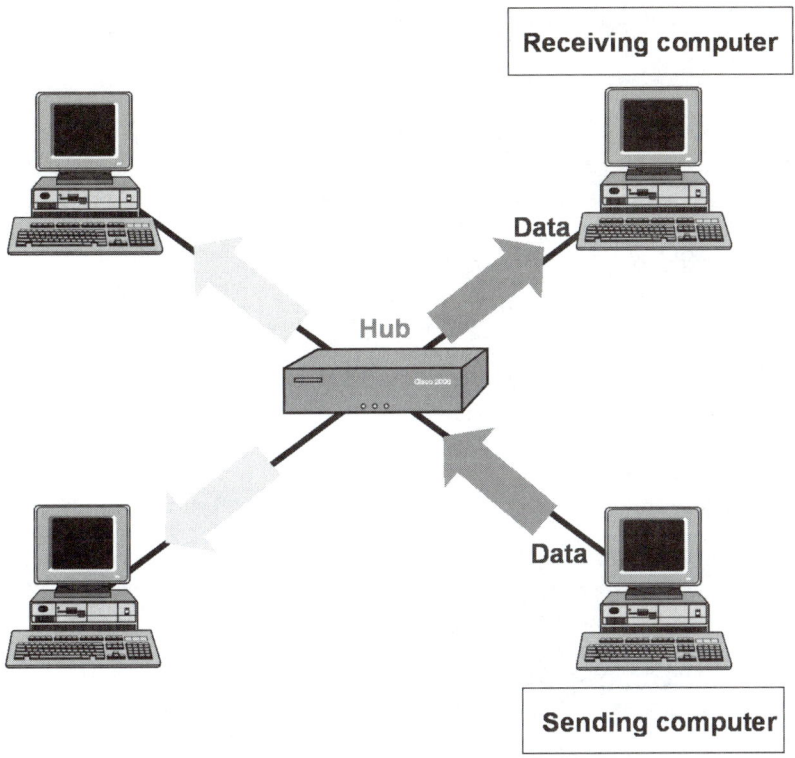

The star topology advantages are:

■ It provides centralized resources and management

■ If a network device malfunctions, the device can be shut down and repaired without affecting other devices in the network, and that can send and receive network data

■ It supports centralized multiple network access polling from the hub to outlying nodes.

The star topology disadvantages are:

■ If the hub fails, the network is effectively down since the devices connected to the hub cannot network with other computers.

■   For large network installations, star topologies require a lot of cable since each
    device must be separately connected to the hub. This negatively impacts
    budget in terms of installation time and materials.

Ring Topology

The ring topology connects all computers on a single circle of cable with no
terminated ends, as shown in Figure 1.10. The data signal travels around the loop in
one direction and passes through each computer. Since each computer regenerates
the data signal, ring topology is considered an active network. In a ring topology
network, if one computer fails, the entire network goes down.

Figure 1.10 Ring Topology Network

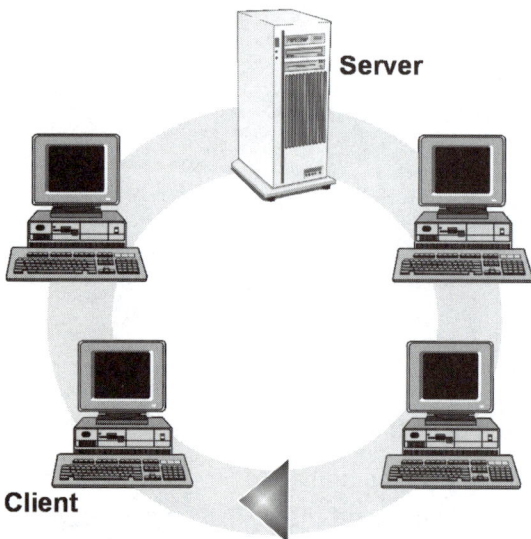

The most common implementation of the ring topology is the token-ring network.
Token-ring was originally introduced in 1984 by IBM as a solution for PCs, mid-range
computers, and mainframe SNA environments. It utilizes an electronic token

to manage client computer network access. The token is a predefined 24-bit code that enables only the computer possessing it to put data on the network.

In a token-ring network, the most common method of controlling data transmission around the ring is called token passing, as shown in Figure 1.11. In token passing, a free token circulating around the ring manages network access for all client computers. Only a single token can be active on the network at any given time, traveling in one direction around the ring.

Figure 1.11 Token Passing Network

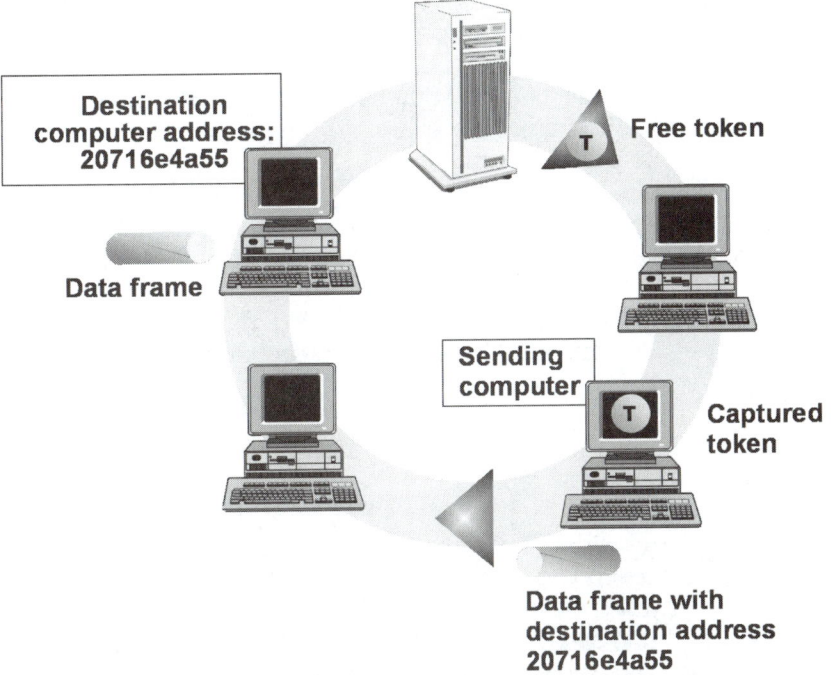

With the token-passing network access method, the data transmission process is as follows.

1. The first active computer coming online in a token-ring network generates the token, which travels around the ring in one direction.

Note: The first active computer online in the token-ring network is assigned the network monitor to make sure only one token at a time circles on the ring. The assigned computer verifies accurate data delivery functions by detecting frames that circulate the ring more than once.

2. The token travels around the ring polling each computer until one computer signals it has data to transmit. At that time, the sending computer captures the token while changing token status from "free token" to "busy." Computers without data to send regenerate the token and pass it on.

3. The sending computer assigns a destination address to all data frames and passes them on to the network. The frames travel around the network until reaching the destination address specified in the frame.

4. The destination computer copies the frames into a receive buffer and marks the frame status field to indicate the information was received.

5. The destination computer places the frame back on the ring, destined for the sending computer. When received, the data transmission/reception process is acknowledged as successful.

6. The sending computer removes the frame from the ring, creates, and sends a new token on the ring, and the process repeats.

When computers are ready to send data on a token-ring network, access to the network is managed so that interference with other computer transmissions does not occur. In a token-ring network, token passing is a collision-free access method since only one computer at a time can transmit data. Token passing is known as a "deterministic" access method since computers without the token cannot coerce network access as they can in an Ethernet CSMA/CD environment.

Tip: If two computers attempt to transmit data at the same time, data collisions can occur which distort and degrade the data signals, making them unreadable.

If an error occurs on a token-ring network, the token returns to the sending computer and creates an error. When the sending computer detects the error, a process called "beaconing" is launched.

All computers on a token-ring network are aware of the token passing process. When an error is detected by a ring computer, it sends out beacon signals onto the network to notify other computers that token passing was interrupted by an error. This continues until the ring computer detects a beacon from its upstream neighbor, and then it stops. This process repeats until the only computer sending a beacon is the one directly downstream of the failure, isolating the failure location.

The specification for data transmission speed on a token-ring network includes 4 and 16 Mbps rates. Current standards for token-ring utilize the 16 Mbps transmission rate. The transmission media in use must support the data transmission speed.

Tip: Token-ring computers must have a network adapter card that supports token-ring passing and the data transmission rate in use. Most network adapter cards currently designed for 16-Mbps token-ring networks are backward compatible to 4 Mbps. However, the inverse is not true.

Token-ring networks use a hub that contains the internal wiring to form the logical ring. The hub in a token-ring network is known as a multistation access unit (MSAU). MSAUs also have the ability to sense when a ring computer's network adapter card is faulty and bypass that computer (disconnect from it) to prevent the entire network from going down.

When a computer in a token-ring network comes online, the token-ring system initializes it through the computer's physical address configuration is checked for conflict with existing computer addresses. In addition, the computer's presence on the network is registered with existing ring computers.

Mesh Topology

Mesh topology is characterized by intermediate nodes and multiple redundant paths that interconnect network devices, shown in Figure 1.12. Multiple communication paths to a destination allows for selecting the best network traffic route at any particular time. It is common to use mesh topology in a WAN to connect remote sites across telecommunication links by interconnecting multiple routers, as in the largest mesh topology network, the Internet.

Figure 1.12 Mesh Topology Network

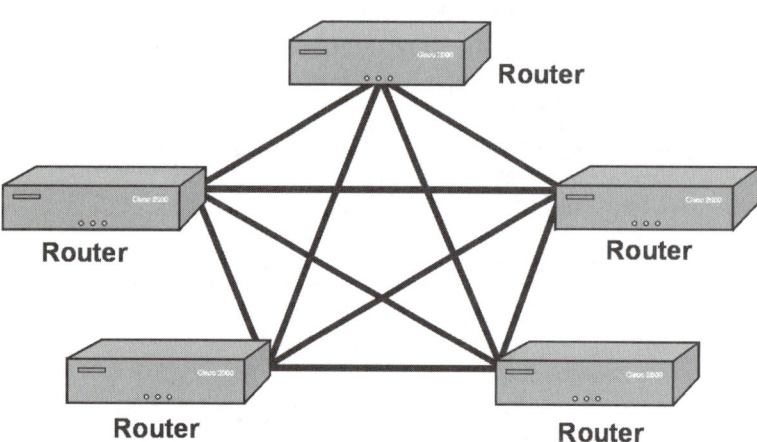

Routers, shown in Figure 1.12, search for multiple active paths in the mesh network to determine the best data delivery path at a given moment. All routers maintain a routing table that provides addressing information used to locate all other accessible routers across multiple communication paths.

As mentioned earlier, current networks utilize combinations of the basic bus, star, and ring topologies to gain certain layout and performance advantages. The sections that follow discuss the star bus and star ring topologies.

Star Bus Topology

Star bus topology is one of the most commonly used networks. It combines the simplicity of bus topology with the point-to-point connections of star topology to form a reliable high-speed network. The star bus topology has several star networks linked together with linear bus trunks as shown in Figure 1.13.

Figure 1.13 Star Bus Topology

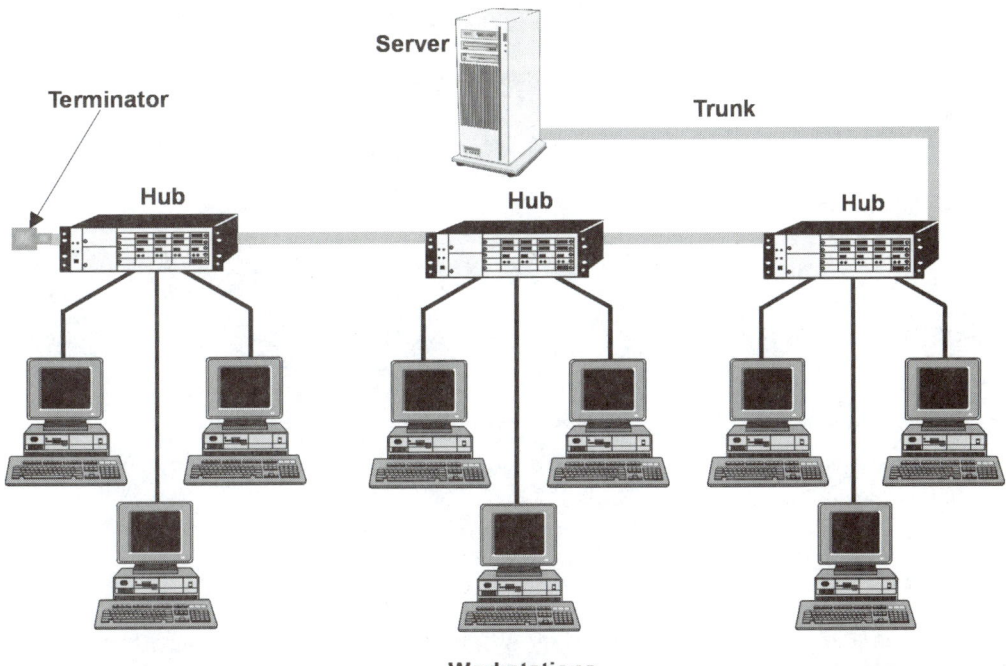

If one computer on a hub goes down in the star bus configuration, other network computers are unaffected and continue to transmit and receive data. If one hub goes down, the computers on that hub can no longer communicate on the network. If the faulty hub is linked to others, all downstream hubs are also disconnected from the network.

Star Ring Topology

Star ring topology uses a main hub to star wire the ring hubs that attach the ring to computer networks, shown in Figure 1.14. The ring hub contains the internal wiring to form the logical ring for network computers. If a ring hub goes down, all the computers in that ring network go down, but the remaining star-wired ring hubs are unaffected. If the main hub goes down, all ring hubs and attached computers go down.

Figure 1.14 Star-Wired Ring Topology

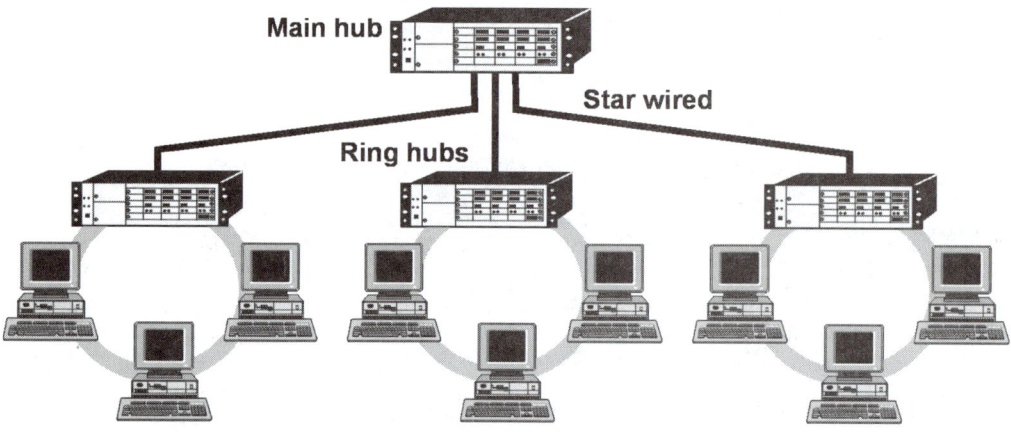

Network Segments

A segment is a LAN logically broken up by a bridge or router, yet they are still connected physically to the same network. To improve network performance, a heavily populated segment can be divided into two logically separate segments and then physically joined with a bridging device. This improves network access time since the total number of computers in each segment contending for network access is reduced. Segmentation is useful when many new users are added to a network or if high-bandwidth programs are introduced.

Note: Another process for segmenting a network is
                      microsegmentation using a switch.

Segmenting with a bridge also reduces network traffic by separating and controlling
the flow of traffic on connected segments, shown in Figure 1.15. The bridge does this
by using a routing table to determine whether traffic is destined for computers on
one segment or another.

Figure 1.15 Network Segmenting with a Bridge

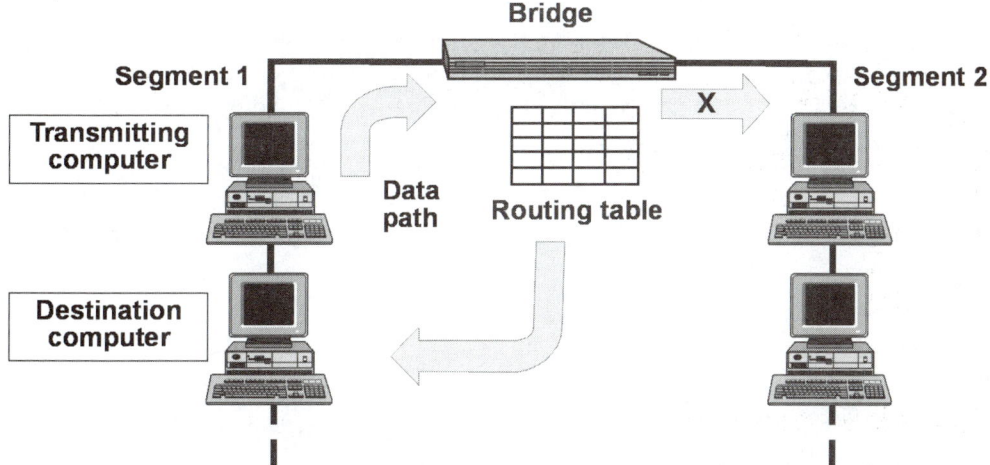

In Figure 1.15, when network traffic passes across the bridge, the bridge updates a
routing table in its RAM by capturing transmitting computer MAC addresses. This
enables the bridge to know each computer's location on its connected segments. For
example, if the source and destination nodes for a particular transmission are both
located on segment 1, the bridge uses its routing table to prevent the broadcast
from reaching segment 2 reducing segment traffic by limiting its flow.

A disadvantage of segmenting with a bridge is that it can create excess traffic. For
example, if the bridge does not have a destination address in its routing table, it

broadcasts the packet to all bridged segments (except the segment of the transmitting node), thus increasing segment traffic.

Network Backbones

A backbone is the main cable or trunk segment that connects computers, bridges, routers, and other devices in a network. A hierarchical topology might have a high-speed backbone connecting servers, LANs, and workstations as shown in Figure 1.16. This configuration is useful when servers are remotely located and high-speed backbone access is needed.

Hierarchical networking isolates local LAN traffic on network architectures such as Ethernet or token-ring while transmitting internetwork traffic over a high-speed backbone architecture like Fast Ethernet. LAN workstations access the backbone through routers as needed.

For example, some Internet backbone implementations use Fiber Distributed Data Interface (FDDI) with Asynchronous Transfer Mode (ATM) switching. A linear bus network might use a Thicknet coaxial cable backbone with Thinnet coaxial cables connecting workstations to it.

Figure 1.16 Hierarchical Backbone Topology

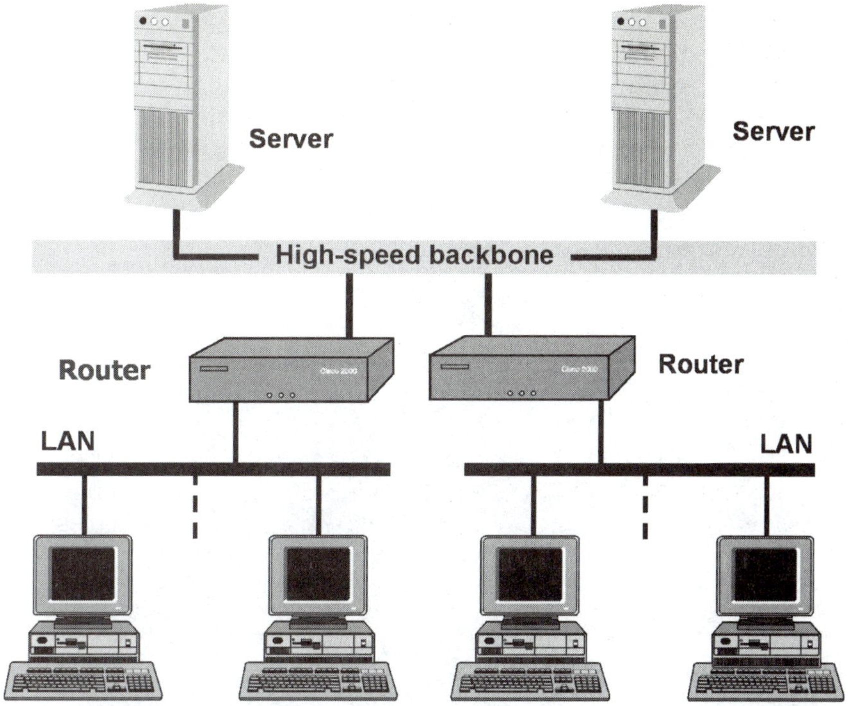

Network Connection Media

Network connection media are the cabling components that provide a physical path over which data travels to reach specified destinations. Certain characteristics of the network media have an impact on how fast and efficiently the data can travel from source to destination. Improper connection media can actually prevent the data from reaching its destination in a recognizable form and impede network communication.

The properties of the connection media should be completely understood with respect to the characteristics of the proposed network to ensure compatibility before implementation. The connection media that support the requirements of LANs are discussed in the following sections.

Coaxial Cable

Coaxial cable is a widely used network cabling that transmits voice, data, and video information over relatively long distance with reasonable security. Coaxial cable is constructed of a solid copper inner conductor surrounded by a dielectric Teflon layer, a layer of braided wire-mesh shielding, all covered by an outer layer of nonconductive material such as rubber (see Figure 1.17). Coaxial cable with multiple shielding layers, such as dual or quad shielding, is available for environments subject to high levels of electromagnetic interference (EMI).

Figure 1.17 Coaxial Cable Layers

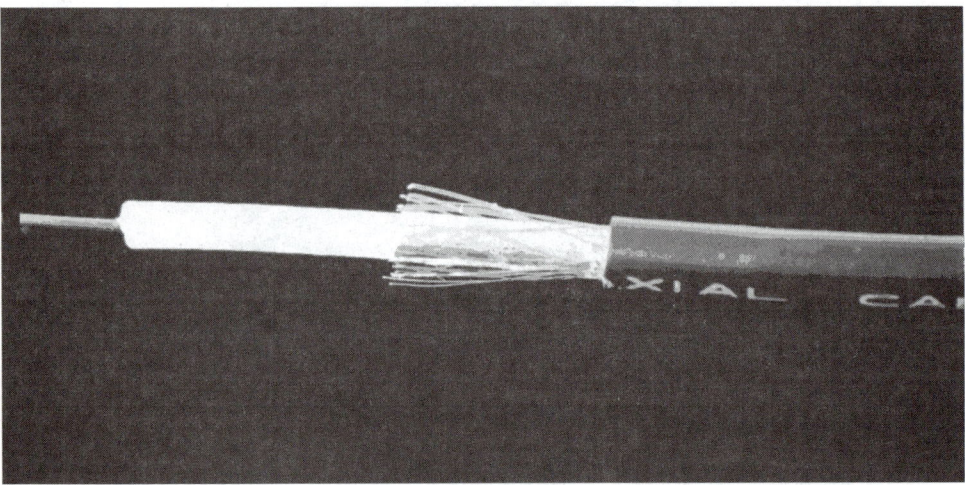

In coaxial cable, data signals travel on the copper inner conductor while the shielding, which is normally grounded, prevents outside noise, crosstalk, or EMI from reaching the inner conduction path and interfering with the data.

 **Note:**   With unshielded cables, crosstalk and EMI can be a problem. Crosstalk occurs through electromagnetic induction when data on adjacent cables radiates across unshielded cable runs and induces baseband frequency energy onto other data paths.

For coaxial cable to function properly, the inner conductor and braided shielding must not make contact, or data signals will short to ground and never reach their destination. As cable lengths increase, coaxial cable introduces signal loss resulting from impedance variations. Signal transmission loss is known as attenuation. A typical coaxial cable impedance value is 50 ohms.

 **Tip:**   A cable tester is an essential piece of test equipment programmed with different media specifications to enable testing various types of network media. During installations, all cable runs should be tested and certified.

The following are types of coaxial cables and mating connectors that can be used as network connection media:

Thinnet—A thin, flexible coaxial cable belonging to the RG-58 family of cables. It has a thickness of ¼-inch and a 50-ohm characteristic impedance. The following describes the main difference among RG-58 family cables:

- RG-58/U inner conductor is solid

- RG-58 A/U inner conductor is stranded wire

- RG-58 C/U is the military specification of RG-58/U

Other RG family cable types include RG-59 for broadband transmission and RG-62 used in ARCNet® networks. With Thinnet, attenuation starts to occur around 185 meters (607 feet). Networks using Thinnet connect coaxial cables directly to network adapter cards.

Tip: The conversion factor between meters and feet is 3.28. For
                example, 185 meters x 3.28 = 606.8 feet.

Thicknet—A thick, rigid coaxial cable sometimes referred to as standard Ethernet.
It is similar to Thinnet coaxial cable, but it has a thickness of ½-inch. Since
Thicknet's inner conductor is larger than Thinnet, it can carry signals further before
attenuation occurs—the typical cable distance where attenuation occurs is 500
meters (1640 feet).

Since Thicknet can transfer data over longer distances, it is often used as a LAN
backbone segment with smaller Thinnet-based networks connecting to the
backbone through transceivers. The transceiver to network adapter card
connections is made with a transceiver cable attaching to an attachment unit
interface (AUI) port or DB-15 connector on the card.

Warning: PVC grade coaxial cable should not be used in
                overhead spaces because it emits extremely harmful
                poisons when burning. To meet fire inspection codes,
                you must use plenum grade fire-resistant coaxial
                cable in overhead spaces near electricity and lighting.

Coax connectors—Thinnet and Thicknet both use standard coaxial cable
connectors, called British Naval Connector (BNC), to make connections between
cables and computers. The following are four types of BNC connection components:

BNC cable connector—Soldered or crimped to the end of coaxial cables, shown in Figure 1.18.

Figure 1.18 BNC Cable Connector

BNC barrel—Joins two cable segments to make a single long cable length, as shown in Figure 1.19.

Figure 1.19 BNC Barrel Connector

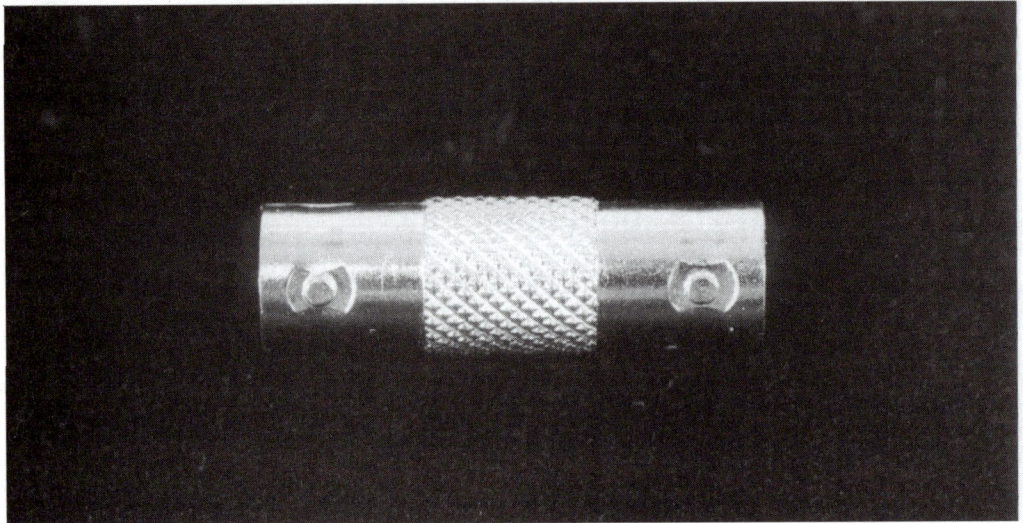

BNC-T connector—Connects network adapter cards to the network, as shown in Figure 1.20.

Figure 1.20 BNC-T Connector

BNC terminator—Absorbs data signals still propagating on the network after the destination computer receives the data (Figure 1.21). For example, this component is used in a bus network to terminate the bus cable ends. A bus network cannot operate without terminators.

Figure 1.21 BNC Terminator

Twisted-Pair Cable

Twisted-pair cable consists of two insulated copper wires twisted around each other. Groups of twisted-pair cables are often placed together in a common sheath to form a cable. The cable twists create an out-of-phase condition for electrical noise and unwanted signals from other twisted pairs, thus canceling them out to minimize crosstalk and other interference. To further reduce crosstalk, the lay length of the twists is different for each pair.

Tip: You can use twisted-pair cable if budget constraints and
            installation ease are an issue. If you are concerned with
            data integrity over long distances at high speeds, do not use
            twisted pair.

The following describes unshielded and shielded twisted pair cable characteristics:

Unshielded twisted pair (UTP)—UTP consists of two wires twisted around each other with no shielding, as shown in Figure 1.22. UTP is commonly used for telephone installations and already exists in many buildings, making it advantageous for networks that can utilize the characteristics of UTP. UTP specifications define how many twists per foot of cable can be used, depending on the purpose.

Figure 1.22 UTP Cable

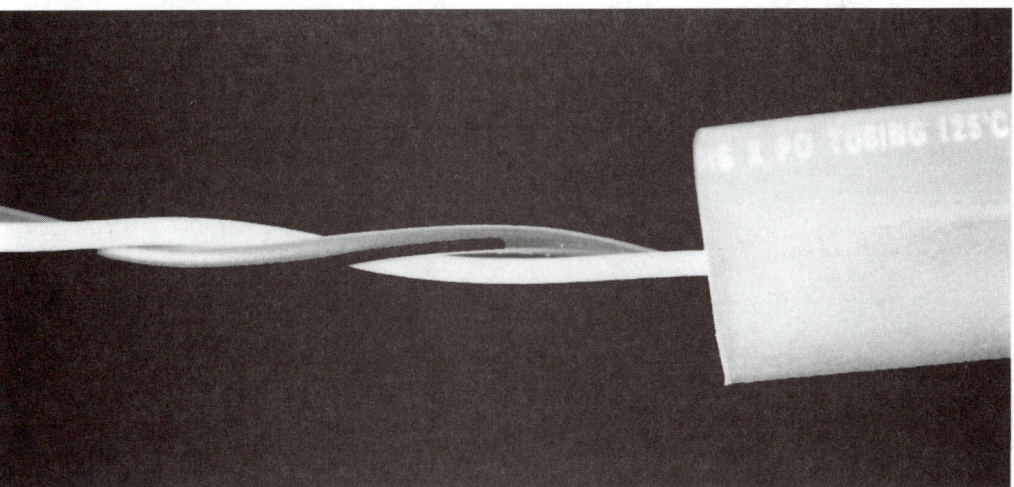

Note: Normal telephone UTP may not have the twisting and
                      other electrical characteristics for clean and secure
                      computer data transmissions.

UTP is a commonly used cable configuration for LANs having media guidelines
defined by the 10BaseT specification for Ethernet. Other popular network
installations use a star bus topology with Category 5 UTP to support data
transmission rates up to 100 Mbps.

Under 10BaseT definitions, UTP uses RJ-45 connectors and can have a maximum
cable length of 100 meters (328 feet) before significant attenuation occurs. Design
features of UTP are intended to minimize crosstalk.

UTP standards include five categories as follows:

Category 1—Standard telephone cable that supports voice only.

Category 2—Four twisted pairs that handle data transmission rates up to 4 Mbps.

Category 3—Four twisted pairs with three twists per foot to support transmission
rates up to 10 Mbps.

Category 4—Four twisted pairs that support data transmission rates up to
16 Mbps.

Category 5—Four twisted pairs that support data rates up to 100 Mbps. This is the
most popular version of UTP.

**Tip:**   To plan for future network upgrades, Category 5 UTP can be
                      installed to accommodate enhanced data transmission rates
                      up to 100 Mbps.

The following summarizes UTP characteristics:

Cost—UTP is the least expensive cabling type for a network. Since many buildings are pre-wired with UTP, it is cost effective for an organization setting up a network.

Bandwidth—UTP supports data transmission rates between 4 and 10 Mbps with some implementations reaching 100 Mbps.

Attenuation—UTP cable lengths are restricted to 100 meters, beyond which significant attenuation occurs.

Node capacity—UTP supports networks with a maximum of 1,024 nodes.

Connectors—UTP utilizes RJ-45 connectors, which are similar to RJ-11 telephone connectors, except they are larger and contain eight wires instead of four. RJ-45 and RJ-11 connectors are not compatible.

Components—Components such as distribution racks and patch panels are available to make UTP installations easier. Patch panels come in versions that can handle up to 96 ports with speeds of 100 Mbps. Other components include jack couplers (single or double RJ-45 jacks with 100 Mbps capacity) and wall plates.

EMI resistance—UTP has no shield and is therefore susceptible to EMI. Installing UTP too close to other cable runs or overhead fluorescent lighting should be avoided. UTP is also susceptible to electronic eavesdropping by wire-tapping.

Shielded twisted pair (STP)—STP and UTP both use internally twisted-wire pairs, but STP uses foil and braided mesh for ground shielding, as shown in Figure 1.23. The shielding captures any unwanted noise or electro-magnetic energy radiating near the STP cable run and grounds it to prevent interference. Shielding also prevents the signals on the cable from radiating out to other cables and creating crosstalk. The shielding not only reduces the susceptibility to EMI, but also allows STP to support a higher data transmission rate over longer distances.

Figure 1.23 STP Cable

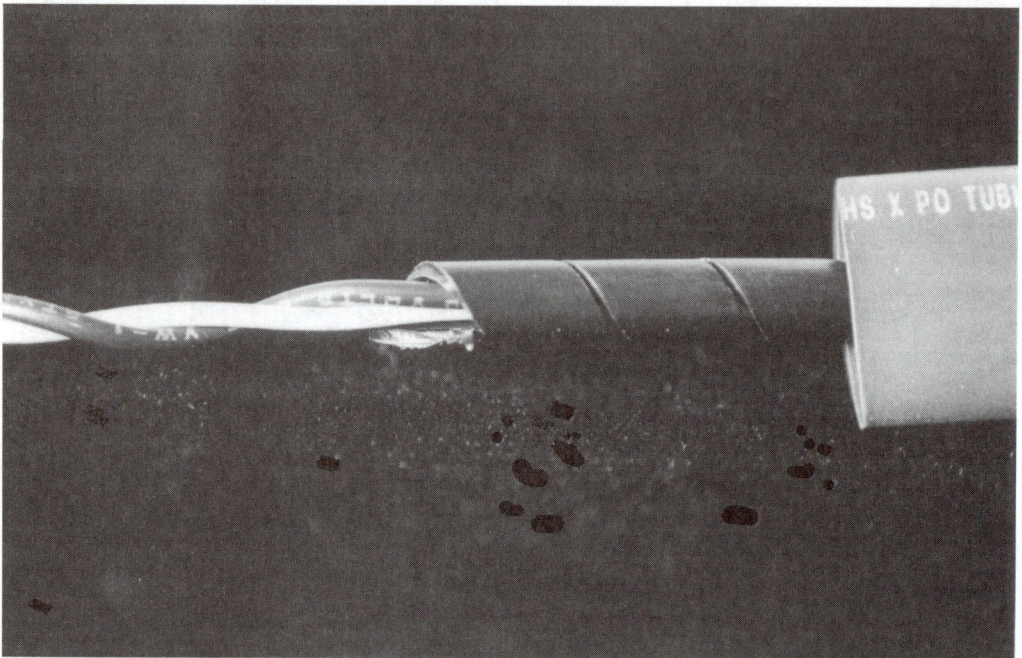

STP is available in a variety of cable types. Some have shielded individual cable twists while others shield groups of twists in multiple-pair cables. IBM categorizes STP according to specific characteristics, the most common of which is IBM Type 1 cabling.

The following summarizes STP characteristics:

Cost—STP is more expensive than UTP and Thinnet, but less expensive than Thicknet and fiber-optic cable.

Bandwidth—STP supports data transmission rates as high as 500 Mbps, although most implementations do not exceed 155 Mbps. IBM token-ring networks use STP at a 16-Mbps transmission rate.

Attenuation—STP cable length specification is approximately the same as UTP; only it supports higher transmission rates over the same cable segment with an equivalent attenuation factor.

Node capacity—STP supports networks with a maximum of 260 nodes and is commonly used in token ring topologies.

Connectors—STP can utilize RJ-45 connectors in token ring and AppleTalk networks.

Installation—STP is substantially heavier and less flexible than UTP cable, which extends to installation time, but is more flexible than coaxial cable.

EMI resistance—The shielding of STP reduces susceptibility to EMI and is the preferred choice in high EMI environments.

Fiber-Optic Cable

In this type of cable, optical fibers carry digital data signals using light pulses modulated at baseband frequencies. Since only light waves traverse the fiber-optic medium instead of electrical signals, the cable is highly efficient, resistant to extraneous noise and EMI, and cannot be tapped like a copper-based cable. This means fiber-optic is a very secure transmission media suitable for high-speed, high-capacity data transmissions over long distance with little attenuation.

Fiber-optic transmission rate is very fast, typically handling data rates of 100 Mbps with a distance capability of 2 kilometers (6,562 feet).

A fiber-optic cable assembly is constructed of a central optical fiber (a thin cylindrical glass core) surrounded by a concentric layer of glass cladding with an outer reinforced layer of plastic, as shown in Figure 1.24. Sometimes the core is made of plastic. However, this material supports a shorter transmission distance than glass. A typical fiber-optic cable core has a size of 62.5 microns with a 125-micron cladding. Optical fibers carry light signals in one direction only. Therefore, a fiber-optic cable assembly must have two fibers—one for the transmit signal path and one for receive.

When a light pulse enters the fiber-optic medium, transmission mode varies depending on the physical properties of the core and cladding. The following are three different types of fiber-optic cable, each with increasing distance capabilities:

Figure 1.24 Fiber-Optic Cable

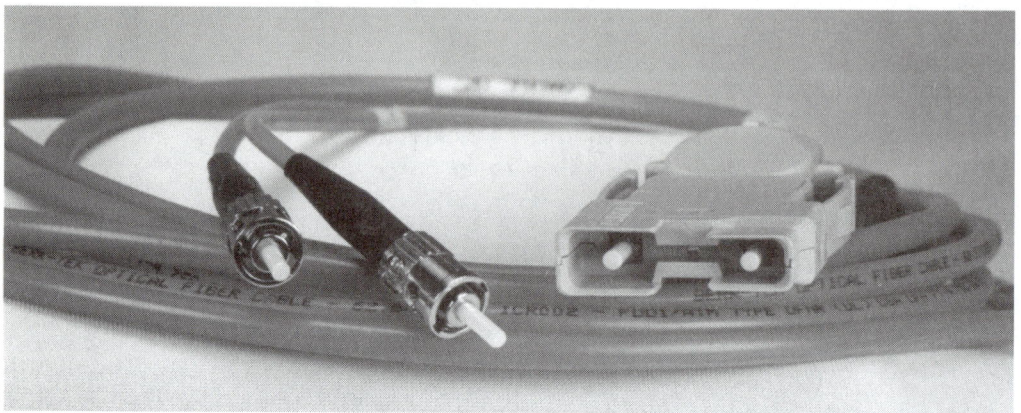

Multimode—This fiber-optic cable uses light emitting diodes (LEDs) to transmit data as pulse-modulated light signals. In the transmission media, some light waves bounce off the cladding at different angles while others are absorbed. This sets up interference patterns that distort and limit the transmission rate to approximately 200 Mbps over a distance of 1 kilometer.

Multimode graded index fiber—This fiber-optic cable uses LEDs to transmit data as pulse-modulated signals. The design also decreases the core refractive index to focus light waves more efficiently, yielding bandwidths as high as 3 Gbps with a distance capability of several kilometers.

Single mode—This fiber-optic cable uses lasers to focus light waves so that only a single wavelength at a time can pass through the medium. This technique reduces multiple reflections and eliminates distortion, resulting in maximized bandwidth and distance capabilities. This is the highest performing and most expensive type of fiber-optic cable.

 Note: Fiber-optic cable should not be used if you have budget constraints or if expertise for connecting fiber-optic cables is not available.

The following summarizes fiber-optic cable characteristics:

Cost—Fiber-optic cable is more expensive than copper-based cable. However, the network equipment required to support fiber-optic cable is very expensive.

Bandwidth—Fiber-optic cable supports data transmission rates from 100 to 622 Mbps.

Attenuation—Since light signals transmitting in a fiber-optic environment do not disperse the way that electromagnetic signals radiate in copper cables, fiber-optic cables have a weaker attenuation rate. This gives fiber-optic cable the ability to handle transmission distances of 2 kilometers or longer without signal attenuation problems. However, fiber-optic transmissions can deteriorate from chromatic dispersion.

Node capacity—Fiber-optic cable in an Ethernet network can support up to 75 nodes.

Uses—Fiber-optic cable can be used in Ethernet networks as the backbone segment of a Fiber Distributed Data Interface (FDDI) network connecting slower LANs, or as the transmission media in an FDDI token-passing ring network. It is also used with ATM switching networks supporting data transmission rates up to 622 Mbps and to connect back-end computers such as mainframes and minicomputers.

Connections—The integrity of fiber-optic cable connections can be a critical factor during installation. For example, a maximum bend radius specification cannot be exceeded. In addition, fiber-optic cabling components require special expertise, such as terminating and polishing techniques for proper installation.

EMI resistance—Fiber-optic cable is unaffected by EMI, noise, or high-voltage environments, and is not susceptible to electronic eavesdropping or wire-tapping. In addition, fiber-optic cable does not use electrical grounds, therefore it is unaffected by shifts in ground potential at the equipment site.

When planning for connection media in a network, consider the following:

- Network traffic level

- Shielding and security needs

- Transmission speed and distance (attenuation factors)

- Budget and installation logistics

Table 1.1 summarizes and compares connection media characteristics.

Table 1.1 Connection Media Characteristics

Specification	Thinnet	Thicknet	Twisted Pair	Fiber-Optic
Distance	185 meters	500 meters	100 meters	2 kilometers
Relative cost	2	3	1 (cheapest)	4
Transmission rate	10 Mbps	10 Mbps	4 to 100 Mbps	100 Mbps to 1 Gbps
Noise and EMI susceptibility	Good	Good	Poor	None
Preferred uses	Medium to large networks with high security		UTP: smaller sites on budget. STP: any size sight in token ring	Any size installation requiring high speed, data integrity, and security.

Network Components

Networks contain a variety of components crucial to network operation and connectivity. Network components are devices that facilitate and manipulate data transfer from its source to its destination. During data transfer, network components perform the following crucial functions:

- Provide signal distribution and connectivity within the LAN or out to other networks
- Modify, condition, or convert data
- Route data over the best available path
- Expand the network or segment traffic

To accomplish these functions on a typical network, you'll need the following devices: network adapter, hub, modem, repeater, switch, bridge, router, brouter, and a gateway.

The following section describes these devices and explains how you can use them in your network environment.

Network Adapter Card

A network adapter card, otherwise known as a network interface card (NIC), is a device that connects a computer to a LAN by accepting the physical connection to the network media, as shown in Figure 1.25. The network adapter mediates the demands of a server or client on one side and the network architecture, with its rules for accessing the shared network media, on the other.

Figure 1.25 Network Adapter Card

Components

A network adapter card is installed in an expansion slot of each computer requiring network access. The network adapter circuit card assembly contains various chips including an on-board microprocessor, a UART, and ROM. ROM contains firmware (embedded software routines) that implements network adapter functions using the microprocessor, UART, and other on-board devices.

Tip: A computer with two network adapters, servicing internal and external networks is called a "multihomed" computer.

When handling data, the network adapter performs the following basic functions:

- Converts the data format for transmission across the physical network media

- Controls data flow onto the physical media

- Sends the data to other devices

- Accepts incoming data and converts it to a format recognized by the receiving computer's CPU

Data Conversion

Internal computer circuits move data on buses in a "parallel" fashion to accommodate rapid processing. A bus is an internal signal path on a circuit card assembly, such as a NIC or motherboard that passes binary information. Early computers used 8 bits of data moving on 8 parallel-running buses. Today's computers are constructed using 16- and 32-bit parallel buses.

Data travels through the shared network media in a single baseband stream, known as serial transmission. To transmit data from a computer onto the network media, a single coaxial cable converts data from a parallel to serial data format. The network adapter component UART receives data from the network and converts the serial data stream to the parallel format the computer recognizes.

 Note: Data format conversion can also be done for an optical interface, if the network adapter card supports that kind of connection.

Since the network adapter card interfaces with the shared physical network media, it must have the ability to implement the network access method of the existing topology. For example, the CSMA/CD method. The standards that control access to the shared network media are known as media access control (MAC) protocols. These protocols reside at the MAC sublayer of the OSI model's Data Link Layer. The address of a network adapter must be uniquely identified to other devices on the network. All network adapters contain a unique physical address, known as a MAC that is hard-coded by the manufacturer into the network adapter's firmware. The Institute of Electrical and Electronic Engineers (IEEE) assigns blocks of addresses to each network adapter card manufacturer, allowing each computer containing a network adapter card to have a unique address on a network. An example of a network adapter address is the following hexadecimal identification number: 685E40237F1A.

When data is moved from the computer to the network adapter card, the following events occur:

1.   The network adapter signals the node requesting the computer's data.

2.   The computer data is moved from the CPU/memory across internal bussing to the network adapter. On network adapters that can utilize direct memory access (DMA), the computer assigns the adapter some memory space to reduce data access time.

3.   If communication rates are too fast, the data is stored in RAM during transmit and receive.

Network Adapter Communication Protocol Parameters

Before sending data from its source to its destination, network adapter communication protocols establish the following parameters:

■   Network adapter data capacities, including the data-send group size

■   Data amount sent before confirmation signals respond

■   Time to wait before confirmation

■   Time intervals between sending data blocks

■   Data transmission rates

Communication begins when all parameter adjustments are made for both cards.

 Note: Newer network adapters accommodate older adapter data transmission rates by falling back to the slower speed to enable the communication.

Software Configuration Parameters

Some network adapters are software configurable, some are set with dual inline package (DIP) switches, and others are plug and play (PnP). Configurable options include the following:

The interrupt request (IRQ) setting is the interrupt priority level assigned to the network adapter. In most cases, NICs use IRQ10 (the default). You can use the WinMSD utility to display the computer's existing IRQ designations to check for assignment conflicts.

 Note: IRQs are also assigned to various internal computer components and exist in a hierarchy from IRQ0 to IRQ15, with IRQ0 being the highest priority. This allows the CPU to recognize the order in which processing requests should be carried out.

The base I/O port address designates a channel for information to flow between the network adapter card and the CPU. This channel is identified as an address that the CPU recognizes. Typical base I/O port addresses for network adapters are in the 300 to 30F and 310 to 31F range.

 Note: All hardware devices have a different base I/O port address in hexadecimal format.

The base memory address identifies a starting location in RAM for the network adapter's buffer area. A typical value for a network adapter is D8000 or D800. In addition, some network adapters allow for specifying the amount of RAM, such as 16K or 32K.

In addition, the network adapter card may have a jumper configuration to define whether an onboard or external transceiver is used.

Note: Network adapter cards must also be set up with a TCP/IP configuration so that computers can be identified on the network. Parameters include the IP address, default gateway, and subnet mask. On Windows 95/98 and Windows NT computers, the network adapter TCP/IP configuration is set up using the Network utility.

Network adapter cards are not compatible across platforms. For example, a network adapter card for a Macintosh computer could not be interchanged with a network adapter for an IBM computer since they both use different access methods. In addition, certain topologies, such as bus and ring, require different network adapter card types.

Data Bus Architecture

A network adapter card must match the data bus architecture in the computer in which it is installed. The following are four different types of data bus architectures:

Industry Standard Architecture (ISA)—ISA refers to the size of the computer's expansion slot where cards are plugged in. ISA architecture has either an 8-bit or 16-bit expansion slot and is used on IBM PCs.

Extended Industry Standard Architecture (EISA)—An extension of ISA architecture that utilizes a 32-bit expansion slot. EISA architecture was introduced in 1988 and maintains compatibility with ISA architecture.

Micro Channel Architecture (MCA)—An IBM architecture physically and electrically incompatible with ISA buses, functioning as either a 16- or 32-bit bus.

Peripheral Component Interconnect (PCI)—A 32-bit architecture used in most Pentium and Macintosh computers. Many current implementations of PCI are PnP compatible, meaning that devices installed in this architecture require minimal user assistance to configure computer changes.

The network adapter card connectors must support the physical characteristics of the network media in use, such as coaxial, twisted-pair, or fiber-optic cable. Some typical network adapter connector types are BNC, AUI, and RJ-45. Some network adapter cards have a combination of connector types, requiring you to use either software or DIP switches to configure the card with the appropriate connector type used in your application.

Network Adapter Performance

Since network adapters are directly involved with data transmission, their performance can have a significant impact on network speed and client/server performance. If a network adapter is in a bottleneck, send and receive network traffic flow can be critically impacted, especially in a bus network where computers must wait for bus traffic to clear before transmitting. Some network adapters support enhancements that increase data transmission speed, as follows:

Direct memory access—If the network adapter card has DMA capability, data can be moved directly from the adapter's buffer space in RAM to computer memory without the use of the computer's CPU.

Shared system memory—A section of computer memory is used by the network adapter's CPU to process data. This is faster than I/O or DMA methods.

Shared adapter memory—A section of the network adapter card RAM is shared with the computer. The address space is identified as part of the computer's memory.

RAM buffers—Enhances adapter performance by using RAM buffers. When network traffic outpaces the network adapter, it uses its RAM to store data until it can be processed, which prevents bottlenecks on the card.

Bus mastering—Moves data directly to computer memory by bypassing the computer's CPU and taking control of its bus, releasing the microprocessor to focus on other tasks. MCA and EISA architectures both use bus mastering to improve network performance up to 75 percent.

Additional Network Adapter Types

The following special network adapters are available for applications:

Wireless—Wireless network adapters create an all-wireless LAN or interfaces wireless workstations to a cabled LAN.

Fiber-optic—Allows direct connections from the desktop to high-speed fiber-optic networks to accommodate high data rate, high-bandwidth applications.

Remote-booting PROMS—Provides an on-board programmable read only memory (PROM) chip from which the workstation boots up. These are used in a high security environment where workstations are required to have no disk or floppy drives.

Hub

Hubs are standard equipment in most networks. A hub is a central point of connectivity for network devices that distributes computer signals and data. A hub works at the Physical Layer of the OSI model and repeats its input signal to all devices or segments connected to the output ports. Hubs can also be used to expand a network and increase the number of computers in a LAN or a particular segment.

Hubs typically have anywhere from 4 to 24 ports using female RJ-45 connectors on the front panel of the unit for connecting with, for example, Category 5 twisted-pair cables. In addition, link lights on the front panel indicate proper connectivity with other devices. If two hubs are connected using front panel RJ-45 ports, a crossover cable must be used. Otherwise, hybrid hubs equipped with multiple connectors, such as RJ-45, BNC, and AUI, can be used to allow connecting hubs without the use of crossover cables.

Following are three types of hubs available today:

Passive hubs—Organizes cabling and distributes signals without regenerating the signal. They function as wiring distribution panels that pass the signal and require no power to operate.

Active hubs—Multiple ports that transmit distribution signals to attached devices while providing signal regeneration like a repeater. Active hubs are referred to as multiport repeaters. Because an active hub regenerates the signal, it can be used to extend the distance between network nodes. Active hubs require electrical source power to run.

Intelligent hubs—Contains the software-based intelligence to perform certain network functions. An example of an intelligent hub is an MSAU used in token ring networks. If a computer in a token-passing network fails, the token stops circulating and the network goes down. The MSAU senses when a network adapter card fails and bypasses it to enable the token to continue circulating. This prevents faulty computers or connections to the MSAU from affecting network operation.

Figure 1.26 shows a LAN segment using active and passive hubs.

Figure 1.26 Active and Passive Hubs

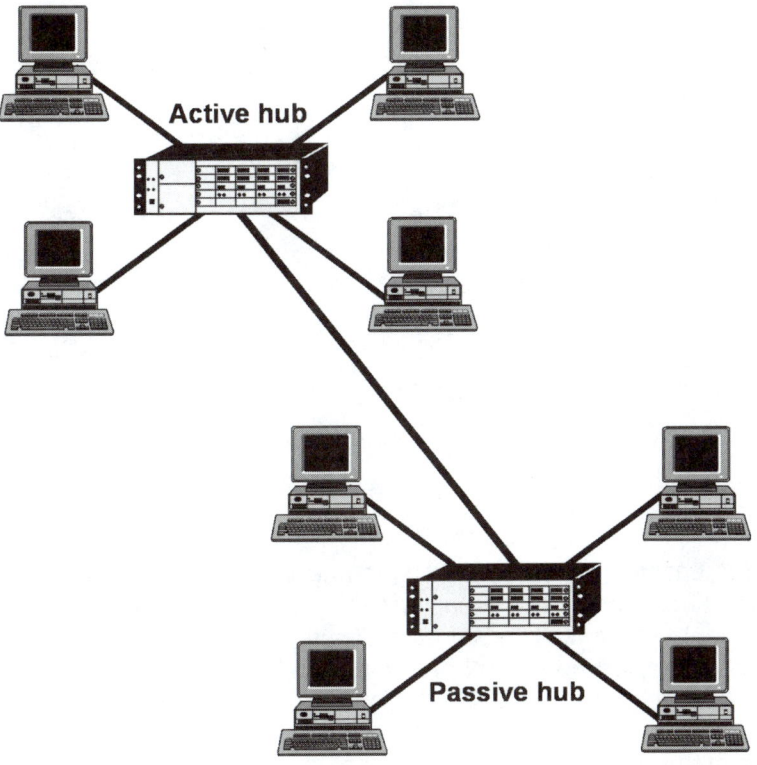

Note: Intelligent hubs control network access in demand priority 100VG-AnyLAN Ethernet networks. In these networks, the intelligent hub maintains all addresses, links, and end node identification. This enables the intelligent hub to manage network access by performing round-robin searches for data-send requests on all network nodes. A node can be a computer, router, bridge, or switch.

Some intelligent hubs provide management features that centralize network diagnostics and traffic monitoring.

The advantages of using a hub network:

- Isolates failing computers from the network. If one computer on a hub goes down or its connection fails, only that computer is affected

- Provides easy expansion of wiring to additional computers

- Provides a simple way to increase cable distances

- Accommodates different network cable types

- Offers diagnostic capabilities for network activity and traffic

Modem

A modem is a device that enables two computers operating on different networks to communicate over normal telephone lines, expanding the network. The term "modem" stands for "modulate/demodulate." Since computer data signals do not transmit efficiently over Public Switched Telephone Network (PSTN) wires, they must be converted to lower frequency signals.

Modems convert digital data signals to analog audio signals using carrier modulation techniques to transmit over normal analog telephone lines. At the destination node, the analog signals are converted back to digital signals by the demodulation process of another connected modem so the data format is recognizable by the receiving computer.

A modem can be a stand-alone external device connecting to the computer's serial RS-232 link or installed internally as a circuit card. A modem is considered Data Communications Equipment (DCE) and uses a serial RS-232 communications interface and a RJ-11 telephone line (four-wire).

Modem Standards

In the early 1980s, Hayes Microcomputer Products introduced a smart modem that could dial out when a phone was on hook. Since then, Hayes-compatible modems have become the adopted standard with current speeds of 28 Kbps or more. Other standards developed by the International Telecommunications Union (ITU) include the following:

V.22bis—An early standard modem.

V.32bis—The current standard modem with a 14,400-bps transmission rate.

V.34bis—A relatively fast and backward-compatible modem with 28,800-bps transmission rate.

V.42bis—A current standard with data transmission rates of 57,600 bps using error correction and compression. Modems with the V.42bis standard are backward compatible with other V. modems. This means that during modem negotiation signaling, the V.42bis protocols cause the modem to fall back to the data communication rate of the handshaking modem. A current industry standard is V.42bis/MNP5.

Transmission Rates

Modem transmission rates are described using two sometimes-confusing terms known as the baud and bit rate. A baud rate is the number of signal events per second transmitted. Bit rate is the number of bits transmitted per second. Baud and bit rates are not the same parameter. For example, with compression and encoding, a 2,400-baud modem could transmit 2,400 signal events, each with 4 bits of data. The actual number of bits transmitted is 4 x 2,400 signal events/sec or 9,600 bps.

Communication Performance

Asynchronous modem communication performance depends on the following two parameters:

Signaling or channel speed—Describes how fast bits are encoded onto the communication channel.

Throughput—Measures the amount of useful information transmitted across the channel. By using compression, data transmission throughput can be doubled.

Modem Types

There are two basic types of modems used in computer network communications: asynchronous and synchronous.

Asynchronous modems—Transmits data in a serial stream. This is the most common form of modem connectivity used in applications such as dial-up sessions for stand-alone computers accessing the Internet or remote access to a LAN. In asynchronous communications, the data stream looks like the illustration in Figure 1.27.

Figure 1.27 Asynchronous Modem Communication

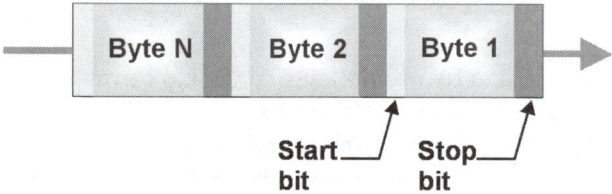

With asynchronous modem communications, each transmitted character converts to a string of bits (a byte). A start-of-character bit separates each byte and a stop bit. Both sending and receiving computers are configured to agree on the start and stop bit sequence, so the timing is organized for reception of subsequent bytes.

Asynchronous communication traffic consists of 25% traffic control and data match checking. Data communication rates for asynchronous devices are between 28,800 and 115,200 bps. Other features of asynchronous communication include error control and compression, as follows:

Error control—Asynchronous communication includes a parity bit for error checking to make sure the number of bits sent equals the number received. Two error correction protocols in use today are Microcom Network Protocol (MNP) and V.42 link access procedure for modems (LAPM).

Compression—Removes redundant elements (repeating characters) and empty space. Transmission times can be cut in half using a data compression protocol such as Microcom's MNP Class 5. Even greater performance can be achieved using V.42bis standard using data compression. For example, a 9,600 bps modem using V.42bis can achieve a throughput rate of 38,400 bps. A good combination of protocols for a LAN-to-LAN link with fast reliable service is the following:

- V.32bis signaling

- V.42 error control

- V.42bis compression

Synchronous modems—Use a clock to coordinate transmissions between sender and receiver. The timing scheme organized by the clock separates groups of bits and transmits them into blocks called frames. Synchronization bits at the beginning and end of a frame coordinates the transmission, shown in Figure 1.28.

Figure 1.28 Synchronous Modem Transmission

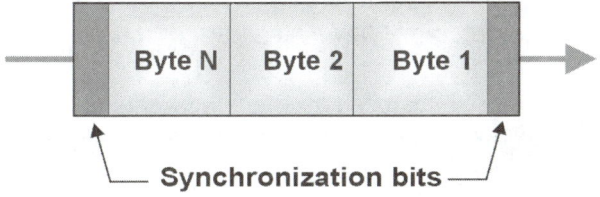

Some features of synchronous modem communications include error detection and the use of several different communication protocols.

Error detection—If an error occurs in synchronous transmissions, the error detection and correction mechanism implements a frame retransmission.

Synchronous protocols perform several tasks that asynchronous do not, as follows:

■ Data is formatted into blocks

■ Control information is added

■ Error control is provided by checking the data

Protocols—The primary synchronous protocols in use are the following:

■ Synchronous Data Link Control (SDLC)

■ High-level Data Link Control (HDLC)

■ Binary synchronous communication (BISYNC)

 Note: A synchronous modem is the first choice in almost all digital/network communication applications because synchronous transmissions are more efficient than asynchronous.

The functions of components are discussed within the context of how they are used for network expansion in addition to performing other functions. Some indications that a LAN should be expanded are the following:

■ The cable gets crowded with network traffic

■ Print jobs are slow

■ Traffic-generating applications such as databases have increased response times

The components described in this section can be installed by a network engineer to expand the size of a network within its existing environment. Expansion can be implemented in the following ways:

■ Segmenting existing LANs so each segment becomes its own LAN

■ Joining two separate LANs

■ Connecting to other LANs and computing environments to join them into one larger network

The following sections describe the repeater, switch, bridge, router, brouter, and gateway components.

Repeater

A repeater is a network device that restores and repeats an analog or digital signal. LAN data traffic consists of digital signals with discrete voltages and durations. As a digital signal traverses the network medium, it can sustain transmission losses and distortion that impairs the ability of a destination node to recognize it as valid digital data (Figure 1.29).

Figure 1.29 Restoring Digital Signals with a Repeater

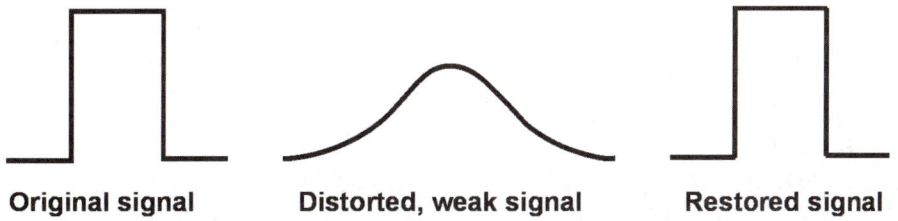

Original signal   **Distorted, weak signal**   **Restored signal**

An application might use a repeater to regenerate data signals traveling long distances or to take data signals from one LAN segment to restore and pass them on to another. Network engineers use workgroup repeaters and local repeaters to accommodate connections and translations between different media types. These are known as workgroup repeaters. Repeaters also provide for LAN expansion by allowing signals to be repeated to multiple LAN segments over extended distances, as shown in Figure 1.30.

Figure 1.30 LAN Expansion with a Repeater

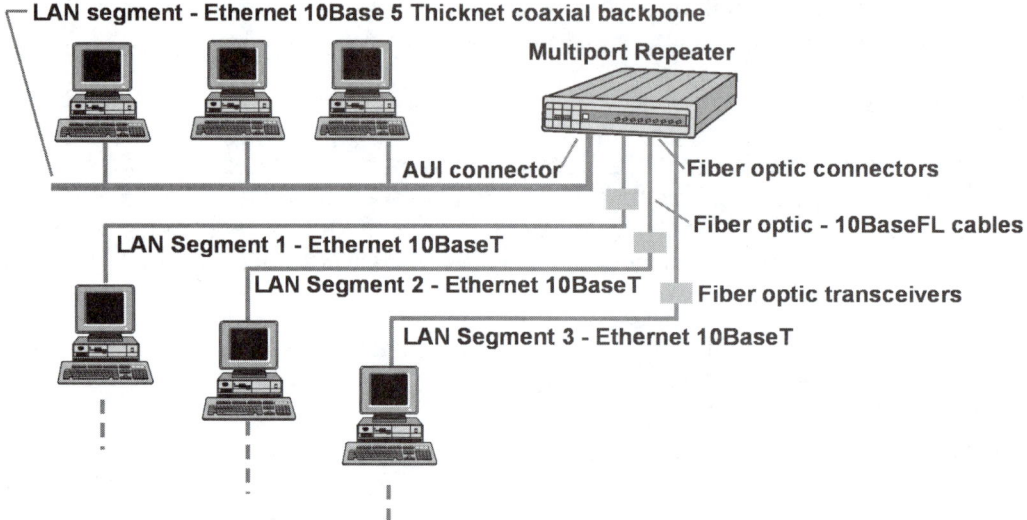

Data Packets

To pass data packets through a repeater, the following must exist:

Identical packet frames—Packet frames on attaching segments must be the same. For example, a repeater does not enable communication between an 802.3 Ethernet LAN and an 802.5 token-ring LAN.

Similar access methods—Both segments joined by a repeater must have the same access method, such as token passing or CSMA/CD.

Different media—A repeater must have the appropriate matching connectors to move packets from one physical media to another, for example, when connecting a Thicknet coaxial segment to a fiber-optic cable segment.

Note: Some multiport repeaters act as multiport hubs connecting different media types.

Logical Link Protocol (LLC)—The LLC protocols on attaching segments must be identical to enable proper communication.

Repeaters operate at the Physical Layer of the OSI model, work with the signaling protocols at this level to pass all data packets between attached segments without discriminating between packet types, and ignore packet destination addresses.

When extending a physical network beyond distance and node limitations, use a repeater to link segments having low traffic volumes and when budgetary constraints are a factor. Do not use repeaters when there is heavy network traffic, segments use different access methods, or packet filtering is required.

 Note: Repeaters do not provide filtering since they pass all packets across their interface including broadcast storms, malformed packets, and packets with erroneous destinations.

Switch

A switch is a multiport device that can microsegment (separate) a large network, such as an Ethernet network, into smaller segments to reduce network traffic, increase bandwidth, and reduce the occurrence of collisions. Each port on a switch services a separate network (or workstation) since the switch uses MAC addresses to repeat data only to the port servicing the destination node. By contrast, a hub repeats data to all its ports. Two implementations of a switch are shown in Figure 1.31.

Figure 1.31 Switch Implementations

Isolating Servers

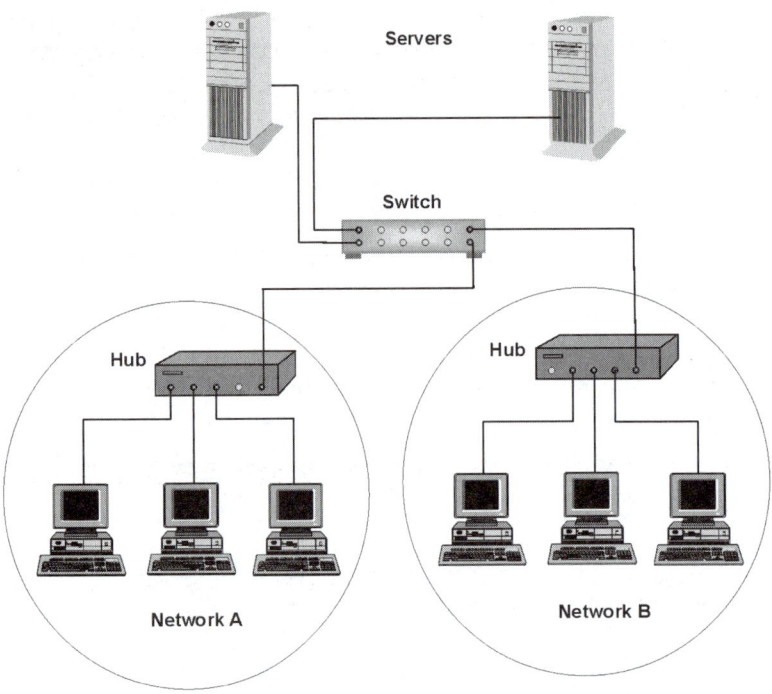

Microsegmentation

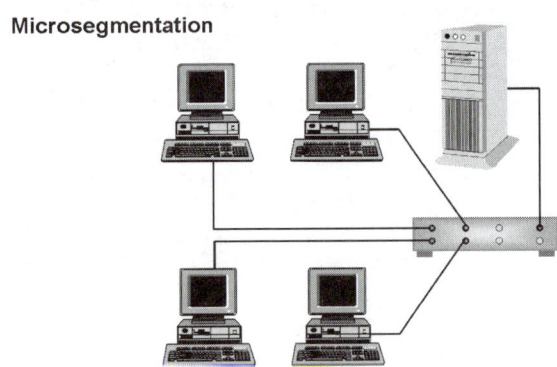

Hardware Switching

Like a bridge, a switch operates at the Data Link Layer of the OSI model. A switch functions similar to a bridge, except that switching is managed through hardware with application specific integrated circuits (ASICs). A bridge uses MAC addresses in a routing table to identify the path to the destination node. Hardware switching is an extremely fast technique compared to the methods used in bridging. A switch can increase bandwidth dramatically if workstations and servers are organized properly on LAN switch segments.

A switch operates in the following manner:

■ The first Data Link Layer frame transmission to be directed to a destination node or LAN segment arrives at the switch.

■ The switch reads the MAC address on the incoming frame.

■ The switch rapidly builds a dedicated switched connection (in hardware) to the LAN segment or destination node.

■ The newly dedicated circuit provides a routing path for all subsequent packets directed to that destination.

Switch ports handling different LAN segments discriminate between packets destined for one segment or another. If a switch receives non-local traffic, it builds a switched connection to a port where a router is connected to access the remote address.

To segment a network, review the following advantages and disadvantages of a switch and a router.

■ Switches obtain bandwidth increases while routing is optimal for filtering packets and protocols

■ Since switches cannot recognize network layer protocols, they cannot perform sophisticated filtering or security functions

■ Switches do not discriminate multiple paths or best route criteria as compared to a router

■ Switches offer minimal management system information as compared to a router

■ Since switched connections exist for only a few microseconds, monitoring switch traffic can be very difficult.

A switch can be controlled from a management port requiring a null modem connection for access. After the initial configuration, the switch can be managed with a Telnet session.

 Note: A null modem connection is made with a cable that crosses the transmit and receive pins from end-to-end so that connected devices can communicate. The cable wiring allows the transmit pin on one device to access the receive pin on the other, and vice-versa.

Bridge

A bridge and a repeater share the ability to connect network segments, but a bridge can also divide a network and isolate traffic. When using a bridge to separate an overburdened LAN into two linked segments (Figure 1.32), some guidelines must be followed to avoid inter-LAN bottlenecks.

Figure 1.32 Bridge LAN Segmentation

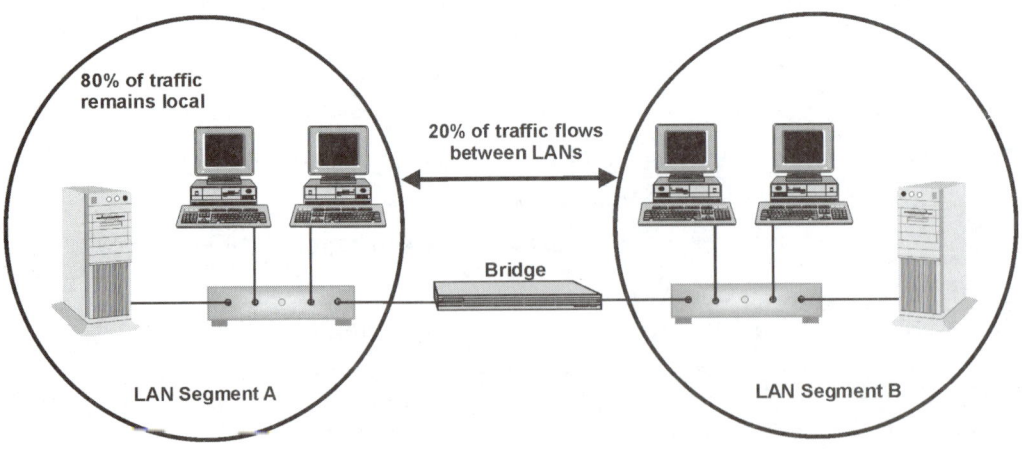

This requires deciding what devices to place on each side of the bridge so that 80 percent of the traffic on each LAN segment remains local, and 20 percent or less of the traffic requires bridge processing and forwarding. Segmenting a network with a bridge is a good strategy for expanding a LAN and relieving traffic congestion.

The following are situations where using a bridge might provide solutions:

Dividing a network to isolate and filter traffic—Bridges can alleviate traffic bottlenecks from excessive numbers of computers on a network by segmenting it into separate networks with reduced traffic on each.

Joining LAN segments—Bridges allow connecting two or more LANs to form one larger network while managing and filtering traffic to each segment.

Expanding segment distances—Bridges provide a convenient way to add a group of computers onto an isolated segment.

Linking media—Bridges allow linking different media, such as coaxial and UTP cable.

Connecting networks—Bridges allow connecting different frame types, such as Ethernet and token ring, and forwarding packets between them (depending on the software).

A bridge works at the Media Access Control sublayer of the OSI model's Data Link Layer and regenerates the signal at the packet level. For this reason, a bridge is sometimes referred to as a MAC layer bridge. A bridge cannot access other information embedded at the higher OSI layers. This causes the bridge to pass all network protocols since it cannot distinguish among them.

 Note: A MAC address is the unique hexadecimal address hard-coded into the network adapter card of a computer.

Segmentation

Using a bridge to divide a network reduces the traffic level on each new segment. A bridge works on the premise that each network node has a unique address and forwards packets based on the destination node address. A bridge reads the destination address of data frames, determines if the destination is local or remote (another attached segment), and allows only the data frames with non-local destinations to bridge to the remote segment.

If a workstation group generates excessive traffic, a bridge can isolate the group to prevent excess traffic from inundating other parts of the network. The bridge keeps the group traffic in the local segment from which a routing table filters the group traffic, redirecting it back to the workstation group's segment. This process occurs on all segments serviced by the bridge.

Traffic generated in one segment and destined for another segment, is routed as long as the bridge routing table contains the necessary MAC addresses. If the bridge routing table does not know the location of a destination node, it broadcasts the packet to all connected segments, except the segment where it originated.

Bridge Routing Table

Bridges build their routing tables based on the MAC addresses of computers that have transmitted data on the network. When a bridge receives a packet, it compares the packet's source address to the routing table record. The routing table contains information that the router uses to determine the best path to a specific destination. This information is contained in the routing table fields and if the address is not there, it is added. In addition, it compares the packet's destination address to the routing table record according to the following criteria:

- If the destination address is in the routing table, and the destination is on the same segment as the source address, then the packet is not forwarded, thus isolating and reducing network traffic.

- If the destination address is in the routing table, and the destination is not on the same segment as the source address, then the packet is forwarded to the appropriate bridge port for routing to the destination address.

- If the destination address is not in the routing table, and the destination is not on the same segment as the source address, then the packet is forwarded to all ports except the one handling the source computer transmission.

All known network addresses—IP addresses that identify hosts on internetworks (using the host network adapter IP address). This is the key look-up field for finding destination address records along with other best-path information.

Methods to connect to other networks—Specifies the MAC address of the next router along the path to the target network. It also includes the number of hops (intermediate routers) to the target network and possible paths between other routers with the use of routing algorithms.

Cost—The cost of sending data on a specific path. Based on the paths available and the cost, the router selects the best route.

Router sending port—Sends the re-created Data Link Layer frame.

Entry age—To avoid choosing a route based on outdated information, the age of the entry containing the above information is determined and ignored if dated.

Bridges and WANS

WANS can have more than one bridge. Multiple bridges can combine several smaller LANs into one WAN. Remote network segments can be connected across telecommunication lines using synchronous modems and a dedicated data-grade telephone line.

If LANs are joined by more than one path across telephone lines, the data transmission can become trapped in a continuous loop. To handle this situation, the CCITT implemented the 802.1 standard, and the spanning tree algorithm (SPA) to detect more than one route, calculate the best alternative, and configure the bridge routing table to use that path.

Consider the following characteristics when you want to use a bridge or repeater in a network:

- Bridges have all the features of repeaters but can support more nodes.

- Bridges improve network performance by generating fewer packets, resulting in fewer collisions. However, a repeater cannot reduce packet traffic.

- Bridges divide busy networks into separate segments to reduce the number of computers contending for resources, whereas, a repeater only regenerates the signal and passes it from one segment to another.

Additional considerations for using a bridge include:

- Bridges are not suited to WANs slower than 56 Kbps, nor can they take advantage of simultaneous multiple paths
- Bridges pass packets of unknown destinations and all other broadcasts, contributing to broadcast storms
- Bridges can be a stand-alone external device or installed in a server as a network adapter card (an internal bridge), if the NOS supports it
- Bridges are inexpensive, installation is simple and transparent at the workstation

Router

A router is a sophisticated network component that allows you to logically interconnect complex network environments having several network segments with differing protocols and architectures. A router can:

- Connect networks using multiple redundant paths
- Separate administrative domains
- Limit unnecessary traffic
- Prioritize packet processing according to the Network Layer protocol
- Provide security by acting as a firewall to isolate one network from another and can act as a gateway from one network to another

Basic router tasks include the following:

- Identify and store the address of connected networks
- Determine the best data route between network destinations
- Limit local broadcast traffic to local segments
- Provide better traffic management than bridges

An implementation of hierarchical networking using routers is shown in Figure 1.33.

Figure 1.33 Router Implementation

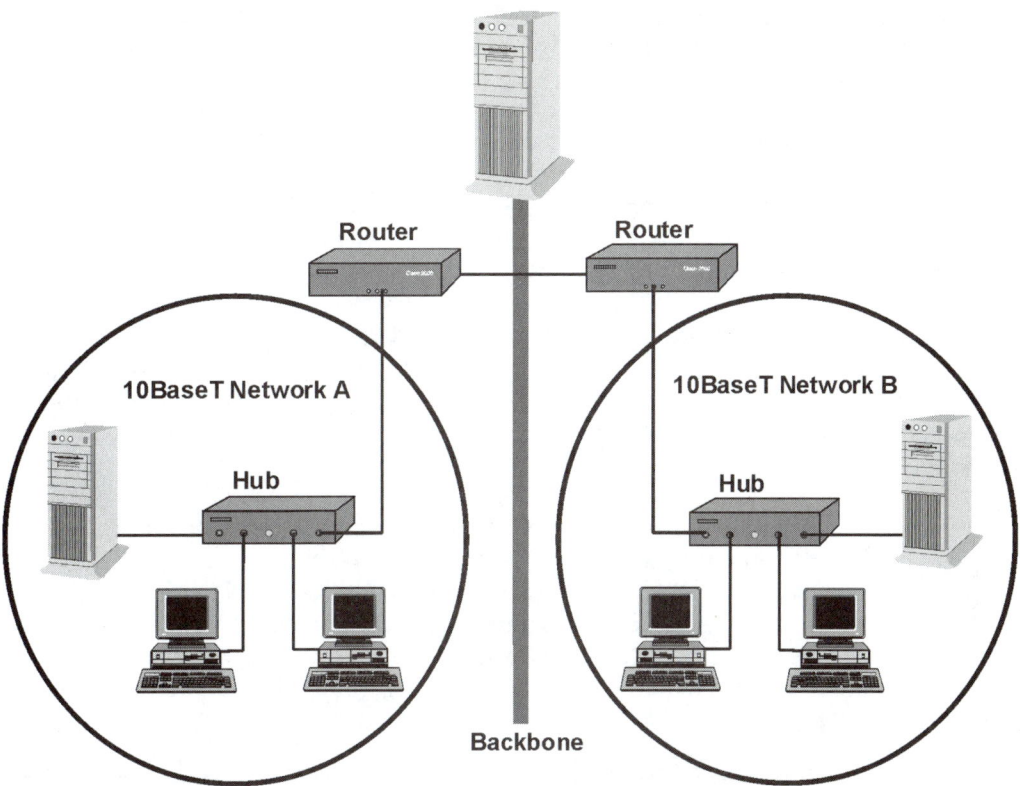

Routers work at the OSI model's Network Layer, allowing them to switch and route packets across multiple networks. Since routers function at a higher layer than a bridge, they are able to read complex network addressing information that a bridge cannot and use this information to improve packet deliveries. Routing exchanges protocol-specific information between separate networks. Routers share status with other routers and determine when to bypass slow or faulty connections.

Before a router forwards a data packet, it confirms the validity of the destination address and analyzes the destination paths. It then chooses the best available path for the timeframe based on certain criteria and then passes the packet along.

Router Addressing

A router forwards packets to a destination based on the Network Layer addresses embedded in the data field of Data Link Layer frames. A bridge uses the MAC layer addresses in the header of Data Link Layer frames to make its routing decisions.

A router is located on the network by its physical or MAC address. Packets arriving from a network that are to be forwarded to connecting internetworks contain the router's MAC address embedded in the destination address field of the Data Link Layer frame. The router's MAC address acts as the destination address/gateway for a network, such as Ethernet or token ring.

The router discards the MAC sublayer envelope containing its own address and then reads the data field of the Data Link Layer frame to access the embedded Network Layer destination address. After reading this final destination address, the router consults its routing table to determine the best path on which to reach the destination.

When destination address validation and best-path analysis are complete, the router repackages the data packet for that particular route. This involves the router creating a new Data Link Layer frame with the source address field containing its own MAC address and the destination address field containing the MAC address of the next router along the best path. However, in this new frame, the Network Layer destination address remains the same and the router then releases the data packet onto the network.

 Note: A router can route packets from networks such as TCP/IP Ethernet to TCP/IP token ring because it re-creates the Data Link Layer frame.

Router Paths

Routers can accommodate multiple paths between LAN segments or internetworks and choose among alternate redundant routes. Since routers can link networks using different data packaging and media access schemes, many redundant paths are available to a router.

A router monitors a network, identifies the busiest parts, and uses this information to determine the best packet route. The router also identifies the best route by determining the number of hops between internetwork segments. A routing table containing best-path information is created using the following routing algorithms:

OSPF—Open Shortest Path First is a link-state routing algorithm that controls the routing process. Link-state routing uses the Dijkstra algorithm to calculate the route based on number of hops, line speed, traffic, and cost. Link-state algorithms are more efficient and create less network traffic than distance vector algorithms such as RIP in the TCP/IP environment.

RIP—Routing Information Protocol uses distance vector algorithms to determine routes. TCP/IP and IPX support RIP. Used in NetWare, XNS, and TCP/IP environments.

NLSP—NetWare Link Services Protocol is a link-state algorithm used with IPX. Used in the NetWare 4.1 environment.

RTMP—Routing Table Maintenance Protocol is used in the AppleTalk network environment.

Internetwork Routers

Routing tables contain IP addresses for internetwork identification of hosts, while bridge routing tables contain MAC addresses for routings to workstation/devices among segments. Routers do not look at destination node addresses, only network addresses, such as that of another internetwork router.

Internetwork routers do not communicate directly with remote hosts. If a packet is destined for a remote host, the router will send it to the internetwork router that manages the destination network.

Routers can perform the following tasks:

- Segment larger networks into smaller networks by using the router's addressing scheme

- Control traffic flow by only passing information when the network address is known, reducing traffic between networks

- Act as a segment barrier by denying passage of certain packets such as broadcast storms. Since routers only read addressed network packets, this also prevents bad data from passing onto a network

Routable and Non-Routable Protocols

One of the most important features of a router is that it can distinguish between multiple Network Layer protocols. Since multiple protocols can exist inside a Data Link Layer envelope, a router can be programmed to process them independently.

For example, when the router opens the envelope (frame), it can forward NetWare traffic using IPX protocol to one network and AppleTalk (AFP) traffic to another. To accommodate different Network Layer protocols with varying packet structures and destination address lengths, multiprotocol routers are used to interpret, process, and forward the data packets of multiple protocols.

Table 1.2 lists common routable Network Layer protocols handled by multiprotocol routers along with their associated network environments.

Table 1.2 Routable Network Protocols

Protocol	Network Environment
Internet Protocol (IP)	TCP/IP
Internetwork Packet Exchange	(IPX) Novell NetWare
Open Systems Interconnection (OSI)	Open systems
XNS	3Com
DDP and AFP	AppleTalk
VIP	Vines

Certain protocols are Data Link Layer protocols without a Network Layer addressing scheme. These protocols are considered non-routable. Table 1.3 lists some non-routable protocols along with their associated network environments.

Table 1.3 Non-Routable Network Protocols

Protocol	Network Environment
Local Area Transport (LAT)	Digital DecNet
Network Basic Enhanced User Interface (NETBEUI)	Microsoft
System Network Architecture/ System Data Link Control (SNA/SDLC)	IBM

Router Types

The two types of routers you'll want to know for the exam are static routers and dynamic routers.

Static routers require manual set up and configuration of the routing tables. Static routers always use the same route specified by the hard-coded routing table entry. They are considered more secure since the administrator controls the routes.

Dynamic routers perform automatic discovery of routes using link-state or distance vector algorithms. They examine information about other routers and decide packet-by-packet how to send network data. The first route is manually configured. Thereafter, automatic discovery updates the routing table.

 Note: You can improve dynamic router security by manually configuring discovered network address filtering.

Bridges Versus Routers

Although there are distinct differences between bridging and routing, similarities can lead to confusion regarding the appropriate applications. When deciding whether to use a router or bridge in a network, consider the information presented in Table 1.4.

Table 1.4 Bridge and Router Comparison

Feature	Bridge	Router
Address recognition	Recognizes only MAC addresses in packets. Examines the destination address of every packet on attached LAN segments.	Recognizes IP network addresses and protocols. Examines only the packets specifically addressed to it.
Packet forwarding	Forwards (broadcasts) unrecognized packets to all segments except the transmitting segment. Forwards recognized packets to appropriate segments.	Identifies specific addresses of other routers and determines which packets to forward and where to send them (to which routers).
Path recognition	Recognizes only one path between networks.	Searches multiple active paths and determine the best path.
Protocols	Does not recognize network protocols.	Works only with routable network protocols.
Filtering	Does not offer filtering, but indiscriminately forwards all packets with destinations outside the bridge routing table.	Filters addresses by forwarding specific protocols to specific addresses (other routers).
Speed	Passes traffic faster than a router.	Works slower than a bridge due to the complex packet functions performed.

Brouter

A brouter combines the best features of the bridge and router. A brouter provides more cost effective and manageable internetworking than separate bridges and routers. A brouter acts as a router for one protocol and bridges all others. A brouter handles routable and non-routable protocols by routing selected routable protocols, and bridging non-routable protocols.

 Note: Most routers today have bridging capabilities.

Gateway

A gateway converts data transmitted from one environment to another to make communication between the environments possible. It repackages data sent from the source network in one format or protocol to the protocol of the destination network. For example, an X.400 gateway for e-mail receives messages in one format, translates it, and forwards it in X.400 format to a receiving system compatible with that format. Gateways can link two systems that have incompatible communication protocols, data formatting structures, languages, and architectures.

In addition, a gateway can connect a Windows NT Server heterogeneous network to the IBM SNA environment.

Gateway Tasks

A gateway takes data from one environment, strips off the protocol stack, and replaces it with the protocol stack of the destination network. Gateways work at the Application Layer of the OSI model, but some gateways work at all seven layers. Gateways are task specific and dedicated to a particular type of data transfer. Therefore, they are often identified by the task they perform. For example, the Windows NT Server-to-SNA gateway.

Router Gateways

You can use a router as a gateway between networks having different Physical and Data Link Layers. In Figure 1.34, the router accesses data at Layers 1 through 7 of packets from the Ethernet network and then places it in a newly created token-ring packet destined for the token-ring network.

Figure 1.34 Router-Gateway Implementation

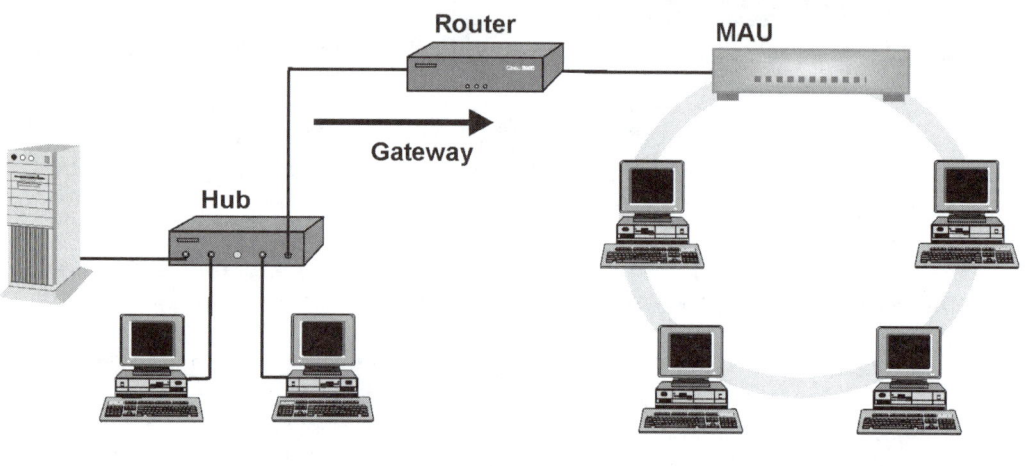

Ethernet Network A   **Token Ring Network B**

Server Gateways

Gateways allow communication from one type of server to communicate with another. For example, Microsoft Gateway Services for NetWare (GSNW) allows a Windows NT Server client to have to a NetWare server. When a client sends a request to the Windows NT Server running GSNW, the request converts into its NetWare equivalent. The NetWare server then converts its response to its Microsoft equivalent.

Mainframe Gateways

A mainframe gateway translates among PCs and minicomputers or mainframe computers. A host gateway connects intelligent LAN computers with mainframes that do not recognize them. In the LAN environment, one computer is usually designated as the gateway computer. A desktop application on other LAN computers accesses the mainframe through the gateway computer and has full use of mainframe resources from their desktops.

Gateways are typically dedicated servers on a network, using a significant percentage of the server's bandwidth for the resource-intensive task of protocol conversion. Gateway servers should have adequate RAM and CPU bandwidth to perform multiple tasks.

Vocabulary

Review the following terms in preparation for the certification exam.

Term	Description
100VG-AnyLAN	A 100 Mbps Ethernet standard using the demand priority network access method
10BaseT	An Ethernet standard for bus type LANs using a 10 Mbps transmission rate, baseband signals, and twisted-pair cable
active hub	A hub that regenerates data signals before distributing them to attached devices
active terminator	A terminator that automatically adjusts its impedance value to the network cable where it is installed
AppleTalk	An Apple LAN for communication and resource sharing using a layered set of protocols similar to the OSI model, transferring information in frame format
application server	In a client/server environment, a member server dedicated to a specific task or application such as a database
ARCNet	Attached Resource Computer Network is a network architecture loosely mapping to the 802.4 specification, used for workgroup sized token-passing bus networks having broadband cable and a 2.5-Mbps data rate
ASIC	Application specific integrated circuits are used to build switched circuits that route data on a dedicated path

asynchronous	Refers to data transmissions that rely upon start and stop bits to pace the exchange of information, instead of using a timing mechanism such as a clock
ATM Asynchronous	Transfer Mode is used to transmit data, voice, and frame relay traffic in real time by breaking data into packets containing 53 bytes each and transmitting between 1.5 and 622 Mbps
AUI Attachment	Unit Interface is a DB-15 connector found on some network adapter cards that accepts a mating transceiver cable connector
backbone	The trunk or main segment of a network that carries its major traffic and to which computers, devices, or LANs attach
base I/O port address	A channel for information to flow between a network adapter card and the CPU, identified as an address recognized by the CPU
base memory address	Identifies a starting location in RAM for a network adapter's buffer area, used when the network adapter needs to share CPU memory space for temporary data storage
baseband	Refers to networks, such as Ethernet or Token Ring, that utilize a transmission media handling one message stream at a time in serial (digital) data format
baud rate	In modem communications, baud rate is the number of signal events occurring per second, not to be confused with bit rate
bit rate	In modem communications, bit rate is the actual number of bits transmitted per second across a telecommunications link

BNC	British Naval Connector(s) are used to interconnect Thinnet and Thicknet coaxial cabling using barrel, T-connectors, terminator, and connector components
bridge	A network device that connects segments
broadband	Refers to the technology used in WAN communications that transmits multiple analog, high-frequency radio signals carrying multiple simultaneous messages and supporting data rates extending into the gigabit range
brouter	A network device that combines the best features of a bridge and router, including the ability to bridge certain protocols and route others
bus topology	A simple network topology utilizing a single trunk or backbone segment to daisy chain devices together, such as a typical Ethernet network does
carrier frequency	An analog high-frequency signal used to carry data signals across telecommunication links using WAN technology
characteristic impedance	The base impedance of a coaxial cable calculated on the basis of physical cable elements, without the reactive or frequency-sensitive components included
client/server model	A networking model utilizing a central server to manage users, security, resource sharing, data storage, and fault tolerance where the client workstation participates in sharing the processing load
coaxial cable	A transmission media used in Thinnet and Thicknet networks consisting of a copper center conductor surrounded by a Teflon dialelectric layer and a rubber outer sheathing

collision	When data signals simultaneously transmitted by two computers run into each other and create distorted or garbled data
communication server	A server that handles the exchange of information between the central server network and remote networks utilizing dial-up or dedicated connectivity access methods
contentional access	A network access method where computers contend for transmit clearance
crosstalk	Cross talk occurs when data on adjacent cables radiates across unshielded cable runs and induces baseband frequency energy onto other data paths
CSMA/CA	Carrier Sense Multiple Access with Collision Avoidance is a network access method where computers ready to send data signal their intent to transmit in advance to avoid data collisions
CSMA/CD	Carrier Sense Multiple Access with Collision Detection is a network access method used in most Ethernet networks where computers ready to send data wait for traffic to clear on the network cable before transmitting their data, avoiding data collisions
CSU/DSU	Channel Service Unit/Data Service Unit provides an interface between a LAN gateway router and a digital line, such as T1, in a WAN internetworking scenario
data bus architecture	Describes the physical and logical layout of the internal bus upon which data travels in a computer, such as ISA, EISA, or PCI architecture
data transmission rate	The rate at which data is transmitted across a network in kilobits or megabits per second

DCE	Data Communication Equipment is an intermediate device that modifies data sent from Data Terminal Equipment (DTE) in RS-232 format, such as a modem
demand priority	A network access method using a repeater to poll network nodes for data-send requests, where contending computers ready to send data are handled on a priority basis or with alternate processing if priority levels are equal
demodulation	The process of converting a modulated data signal back to its original form as a baseband signal, as in modem communications
deterministic access	Describes a type of network access, such as token passing, which determines that only one computer at a time in possession of the token can transmit its data
distributed processing	Distributed processing is a type of mainframe computing environment where independent workstations perform some of the processing, but the central computer handles most tasks
DMA	Direct Memory Access allows a device to directly access the computers memory without using the CPU for data transfer
dynamic router	A type of router that automatically updates its routing table using automatic discovery of routes and link-state or distance vector algorithms, as it interacts with other internetwork routers
EISA	Expanded Industry Standard Architecture is an extended version of ISA architecture that utilizes a 32-bit expansion slot while also maintaining compatibility with ISA

electromagnetic induction	The process by which electromagnetic signals induce energy at the same frequency onto other conduction paths in close proximity
EMI	Electromagnetic interference consists of random noise, frequency bursts, and crosstalk, that in sufficient quantity, can obscure recognition of data transmitted across a network media
Ethernet	An IEEE 802.3 standard for networks utilizing bus or star topology, baseband signals at 10 Mbps, a contention access method such as CSMA/CD, and coaxial, fiber-optic, or twisted-pair cables
fault tolerance	Refers to various systems that implement physically redundant versions of computer data to accommodate for disk failures, but does not substitute for data backup
FDDI	Fiber Distributed Data Interface is a standard for 100-Mbps token-passing ring networks with fiber-optic media for high end computers needing more bandwidth than 10-Mbps Ethernet or 16-Mbps token-ring can provide
fiber-optic medium	A fiber-optic conductor having a central optical fiber (a thin cylindrical glass core) surrounded by a concentric layer of glass cladding with an outer reinforcing layer of plastic
fiber-optic cable	A glass cable media utilizing modulated light pulses to send baseband signals across a network and provide a very secure transmission media suitable for high-speed, high-capacity data transmissions over long distance with little attenuation
file and print server	A member server dedicated to serving files and printers to a group of workstations

frame	In data communications, a frame is a package of information for transmission across a network, containing source and destination addresses, data, and other control information
frame relay cloud	In X.25 packet switching networks, a large array of switches, circuits, and routers that form spontaneous best data paths according to the need of the moment are sometimes called a frame relay cloud since the path configurations change so rapidly
free token	An uncaptured token circulating in a ring network waiting to be captured and used by a computer ready to send data
full-duplexing	A network communication method that increases network speed by allowing transmit and response signals to travel simultaneously in both directions
gateway	A dedicated network conversion device that allows one network with different protocols, data formats, or architectures to communicate with another
gigabit per second (Gpbs)	Annotation describing very high data transmission rates, for example, 2 Gbps = 2×10^9 bits per second
half-duplexing	A network communication method that allows data to flow in one direction at a time, with transmit signals sent in one direction reaching their destination before response signals can be sent on the same path in the opposite direction
handshaking	The process by which two devices interchange control signals when negotiating a communications link

HDLC	High-level Data Link Control is a bit-oriented protocol used for information transfer in synchronous transmission systems, utilizing the Data Link Layer of the OSI model to transmit frames between computers
hierarchical networking	Hierarchical networking isolates local LAN traffic on networks such as Ethernet or token ring while transmitting internet-work traffic over a high-speed backbone
hub	A network device providing a central location for connecting computers and devices, while serving to organize cabling and pass signals to their distribution points
intelligent hub	A hub that contains the software intelligence to manage network access for connected nodes, senses when a faulty node exists and bypass it, or provides management features that centralize network diagnostics and traffic monitoring
IRQ	Interrupt request is a numerical setting for the interrupt priority levels of internal computer devices, allowing the CPU to recognize the order in which processing requests are carried out
ISA	Industry Standard Architecture refers to the size of a computer's expansion slot where cards are plugged in, having either an 8-bit or 16-bit slot size as used on IBM PCs
LAN	A LAN consists of a limited number of computers connected together in a common area within a limited physical space, combining hardware and software technologies that allow users to share resources such as data, programs, storage devices, printers, and other peripherals

laser	Light amplification by stimulated emission of radiation is an existing technology utilized in fiber-optic cable transmissions for high speed, large bandwidth, low attenuation, and high security computer networks
LED	Light emitting diodes are used in multimode grade fiber-optic cable transmissions to generate data-modulated light pulses
logon	The process that validates a user's credentials and retrieves the applicable user rights and user profile when signing onto a network
MAC	Media Access Control is a sublayer of the Data Link Layer of the OSI model, dealing with network access and collision detection
MAC address	Media Access Control addresses identify devices on a network with unique hexadecimal numbers
mail server	A member server dedicated to processing e-mail for a network
mainframe	A central computer (server) containing all programs, performing all tasks, and maintaining the databases while workstations having minimal computing power access these resources only as needed
Manchester encoding	A detection scheme for baseband signals where the positive and negative going edges of the data pulses register the signal's logic levels
media	The physical medium used to carry computer data signals across a network
megabit per second (Mbps)	Annotation describing high data transmission rates, for example, 4 Mbps = 4 million bits per second

member server	A stand-alone network server, sometimes referred to as an application server, or file/print server, dedicated to redistribute the processing load by performing specific tasks for which the network is designed
mesh topology	A network topology using routing devices to create multiple redundant paths to interconnect WAN segments
Micro Channel Architecture	A standard IBM bus architecture incompatible with ISA buses, functioning as either a 16- or 32-bit bus
microsegmenting	A method of reducing traffic by isolating LAN devices with switched access
modem	A device used to convert (modulate) digital signals to analog and back again (demodulate) to enable computer communications across telecommunication links
modulation	The process of superimposing lower frequency information signals on a higher frequency signal to carry information to its destination across a medium suitable only to the high-frequency signal
MSAU	Multi Station Access Unit is a multiport hub used in token-ring networks to sense when a ring computer's network adapter is faulty and bypasses it to keep the network up
multihomed computer	A host computer containing two network adapter cards, one for the internal network (LAN) and one connecting to an internetworking device such as a router
multiprotocol router	A router capable of independently processing data frames with multiple embedded protocols and sending them to their appropriate destination networks

network access	The control method used to manage the way computers access a network, such as CSMA/CD or token passing
network adapter	An internal device on a computer that interfaces the computer to a network by accepting the physical connection to the network media and mediates between the demands of a server or client on one side, and the network architecture, with its rules for accessing the shared network media on the other
network architecture	The standards that define the architectural components of a network such as the topology, data transmission rate, media, signal type, cable distances, number of nodes, and so on
NIC	Network interface card is a term used interchangeably with network adapter card and has the same functions
node	A computer or device on a network
non-contentional access	A network access method, such as token passing, where computers do not contend for access but are managed to reduce the chance of data collisions
non-routable protocol	A Data Link Layer protocol, such as DecNet's LAT or Microsoft's NetBEUI, that cannot be routed by a router because it contains no Network Layer addressing scheme
OSI	Open Systems Interconnection is a specification that defines the layered architecture of services and interactions for computers exchanging information across a network
packet	A transmission unit of fixed maximum size containing data and a header with source and destination addresses, as well as error control information

packet forwarding	The process by which packets are sent out to multiple network segments to locate their destination
parallel data	Data traveling on a bus with binary digit values on separate conduction paths running in parallel with each other
passive hub	A hub that repeats and distributes its input signal to the output ports where computers are attached, without regenerating the signal
passive topology	A topology, such as a linear bus network, in which data transmissions are not regenerated and passed along by computers on the network
PCI	Peripheral Component Interconnect is a 32-bit bus architecture used in most Pentium and Macintosh computers, with many implementations having Plug and Play compatibility
peer-to-peer	A network of 10 computers or less with no central server where network security and file sharing is managed by equal peers (users)
polling	A managed process by which a device searches for specific information from a group of network computers
protocol conversion	The translation of one protocol to another to accommodate communication between networks of differing protocols, formats, or architectures, as performed by a gateway
PSTN	The Public Switched Telephone Network is the standard telecommunication link used by ordinary phone calls and dial-up connections for two-way computer communications with a modem

reactance	A frequency-sensitive component of impedance, such as inductive or capacitive reactance, that can affect signal transmission loss in a coaxial cable segment
refractive index	The measure of light refraction in a fiber optic medium determining the transmission efficiency of the medium
repeater	A network device that restores data signals and passes them along to other devices to ensure data signal recognition
ring topology	A network topology using computers attached in an unterminated ring and a free token to facilitate network access and communication between nodes
RJ-11	An ordinary modular telephone jack containing four wires for connecting and interfacing devices such as modems to the PSTN
RJ-45	A modular jack similar in appearance to an RJ-11 connector, but containing eight wires for connecting UTP or STP cabling to various network devices
ROM	Read-only memory is a computer chip containing firmware that can only be read by a program and not altered
routable protocol	A protocol, such as TCP/IP or Novell's IPX, capable of being routed by a router
router	A network device used to discover and maintain multiple redundant paths from one network to another, sometimes used as a gateway or firewall that isolates a LAN from an internetwork

routing table	The table maintained within a router that is updated either manually by an administrator or automatically by discovered network addresses as the router communicates with other internetwork routers
SDLC	A data transmission protocol commonly used on IBM networks with SNA architecture
security	The means of protecting an organization's network from malicious intrusion or unwanted solicitation using server security systems and appropriate transmission media
segment	A section of cable to which computers are attached that forms a separate logical network
segmentation	The process of dividing a network into smaller segments to reduce traffic on each
serial data	A data format using a single stream of baseband signals across a single cable to transmit data between computers on a network
server	A central computer on a network dedicated to handling the processing burden of specific tasks and responds to requests by the client workstations it services
signal propagation	The manner in which a signal traverses the physical network media
signal reflections	Energy at the baseband frequency that bounces off poorly terminated or unterminated cable connections and interferes with data recognition
SNA	System Network Architecture is an IBM communications framework that defines network functions and standards that enable computer information exchange and processing

star topology	A network topology resembling a star configuration where computers are connected point-to-point with a central hub, offering centralized management and network fault tolerance
start bit	In asynchronous communications, a bit that signifies the beginning of a character
static router	A router that must have its routing table updated manually by an administrator, rendering routes more secure for that reason
stop bit	In asynchronous communications, the stop bit signifies the end of a character
STP	Shielded Twisted Pair is a network transmission media using internally twisted wire pairs with foil and braided mesh ground shielding that shunts noise and EMI to ground to prevent interference and provide a high data transmission rate over long distances
switch	A network device that can be used to segment (or microsegment) a network using hardware-based switched circuits, resulting in dramatic bandwidth increases
synchronous	Synchronous communication refers to data transmissions that rely upon the use of a timing mechanism such as a clock to pace the exchange of information between two digital systems
T1	A widely used digital line, also known as a T-carrier network, for point-to-point transmission over a wire pair with full duplexing at a 1.544 Mbps rate, handling voice, data, and video, and using multiplexing to place up to 24 voice channels per frame on a single cable

terminator	A cabling component used to absorb reflected data signal energy propagating on a network to prevent interference with the data streams of transmitting computers
Thicknet	A rigid coaxial cable similar to Thinnet in construction (sometimes referred to as standard Ethernet) having a thickness of ½-inch, and carrying signals up to 500 meters before attenuation starts to occur
Thinnet	A thin, flexible coaxial cable belonging to the RG-58 family of cables with a thickness of ¼-inch, a 50-ohm characteristic impedance, and carrying signals up to 185 meters before attenuation starts to occur
throughput	The measure of useful information transmitted across a communication channel
token passing	A non-contentional network access method free of collisions that uses a token circulating around a ring of computers to enable individual computers to transmit only when taking control of the token
token ring	A network topology connecting computers and devices in ring configuration and utilizing a token passing method to manage network access
topology	Network topology refers to the way cables, computers, and other components are physically and logically laid out and organized on a network
transceiver	A network device, such as a UART, that handles both transmission and reception of data signals and performs parallel to serial data format conversions when placing data on the network media
trunk	A linear segment also referred to as a backbone which handles the major traffic of a network

twisted-pair cable	A network transmission media consisting of two insulated copper wires twisted around each other to create an out-of-phase condition for electrical noise and unwanted signals from other twisted pairs, canceling them out to minimize crosstalk and other interference
UART	A universal asynchronous receiver/transmitter is a device residing on a computer's network adapter card that handles parallel to serial data format conversions when transmitting and receiving data from the network media
user profile	A computer-based record maintained for an authorized network user that defines the user's environment, configuration options, and preference settings such as installed applications, desktop settings, color options, and so on
user rights	Refers to the range of system-level rights applied to a user, such as the right to do backups or operate a printer, as distinguished from permissions which are assigned to users for accessing different resources, such as files, directories, and printers
UTP	Unshielded Twisted Pair is a network transmission media consisting of two wires twisted around each other without shielding, attached with RJ-45 connectors, and used in networks defined by the 10BaseT specification to allow cable lengths of 100 meters before significant attenuation occurs
WAN	A wide area network, often referred to as an enterprise network, can serve thousands of users in different cities and states, and consists of multiple LANs interconnected by routers, channel service unit/data service unit (CSU/DSUs), and leased lines from telephone carrier service providers

In Brief

If you want to...	Then do this...
Network a group of computers in a localized area within physical distance limitations	Implement a LAN
Manage network access in a linear bus topology	Use the CSMA/CD access method
Utilize non-contentional network access	Provide a token passing or demand priority access method
Transmit serial data across a network	Use baseband signal transmission
Transmit multiple data channels over a single transmission media with separate high-frequency carriers	Use broadband signal transmission
Link multiple LANs encompassing separated geographical areas such as cities or countries	Implement WAN technologies
Connect 10 workstations or less in a minimum security environment with decentralized data access	Use peer-to-peer networking
Utilize the speed, functionality, and security of a server-based network	Use the client/server model
Use a separate server to perform a dedicated task specific to your network application	Implement a member server
Connect computers on a linear segment known as a backbone	Set up a bus topology network

Extend a bus network	Use BNC interconnecting cable components or repeaters
Connect computers to a central hub with point-to-point cable segments	Set up a star topology network
Manage network access for multiple devices using an intelligent hub to poll outlying nodes for data-send requests	Use a star topology network
Have a collision free network with deterministic access	Use a token ring (or token passing) topology
Have a centralized connection point in a token ring network that can sense when a ring computer's NIC is faulty	Utilize an MSAU
Connect remote WAN sites across telecommunication links	Use a mesh topology network with routers
Segment a network to reduce traffic and improve performance	Use a bridge or router
Connect remote servers for high speed access by interconnecting LANs	Implement a Fast Ethernet backbone
Reduce EMI in a network	Attach network cable segments under 185 meters in length using coaxial cable
Set up a backbone segment 500 meters or less in length using coaxial cable	Use Thicknet coaxial cable to minimize attenuation problems
Interconnect Thinnet or Thicknet cable segments	Use BNC connector components

Accommodate future network upgrade to higher data rates in a twisted-pair media environment	Use Category 5 UTP cable to support data rates up to 100 Mbps
Connect devices in a 10BaseT Ethernet network with cable segments less than 100 meters in length	Use UTP with RJ-45 connectors
Support data transmission rates as high as 500 Mbps with attenuation factors equivalent to UTP	Use STP
Support data transmission rates in the 100 to 600 Mbps range with immunity to noise, crosstalk, and EMI over distances up to 2 kilometers	Use fiber-optic cable
Prepare a computer requiring network access	Install and configure a network adapter card
Reduce network adapter bottlenecks	Use a network adapter card that supports DMA, shared system memory, RAM buffering, or bus mastering
Provide a central point from which to connect computers and distribute restored data signals	Use an active hub
Connect two computers in different networks over normal telephone lines with data rates up to 56 Kbps	Install and configure a V.42bis asynchronous modem
Restore signals traveling over long cable lengths	Utilize a repeater
Divide a network into segments to isolate and reduce traffic	Utilize a bridge

| Connect networks using multiple redundant paths, separate administrative domains, or isolate a LAN from an internetwork | Utilize a router |
| Link networks with different protocols, data formats, or architectures | Utilize a gateway |

Lesson 1 Activities

Complete the following activities to better prepare you for the certification exam.

1. Describe how the CSMA/CD access method works.

2. Describe the circumstances where the client/server networking model is more desirable than the peer-to-peer model.

3. Discuss the functions of a bus network. Include the addressing scheme, a typical Ethernet data transmission rate, network access method, and the purpose of terminators in your description.

4. Explain some of the advantages the star topology network has over the bus topology network.

5. Describe how a token-ring network functions. Include the access method and addressing scheme in your discussion.

6. Describe the network media required for very long cable distances, extremely high data transmission rates, EMI immunity, and the highest security possible. Where appropriate, include the typical values or properties supported by this type of media.

7. Explain the basic functions of a network adapter card.

8. Show why a switch provides increases in network bandwidth and reduces traffic.

9. Discuss how a bridge can be used to reduce network traffic.

10. Describe three things a router is used for. In your descriptions, discuss some of the tasks a router performs to connect to remote networks.

Answers to Lesson 1 Activities

1.   CSMA/CD is a contention access method that prevents data collisions by forcing computers with data send requests to wait for transmission clearance. With this method, computers ready to send data on the network listen for traffic on the network cable. When traffic is present, a neutral carrier signal is detected which notifies the listening computer that it must wait for the network to clear before transmitting. When clear, the sending computer can transmit its data.

2.   Peer-to-peer networks support small organizations with up to 10 users in environments where high security and central data access are not required. In user environments with more than 10 workstations, peer-to-peer networks break down in the areas of user management and quick access to network resources. In this situation, a client/server environment is more desirable, since it provides a central, high-speed server dedicated to the following tasks: Network security implementation, User management, File and resource sharing, Hardware management, Data storage, Critical data backup, Fault-tolerance deployment

3.   A bus network is used to connect workstations, devices, and servers along a linear backbone in a simple, easy to implement topology. Data communication is accomplished by sending messages across the backbone from source to destination computer. The sending computer encodes the destination address in the data packets to cause only the receiving computer to respond to the transmission, while other computers ignore the message. A typical Ethernet network using baseband architecture and bus topology transmits at 10 Mbps and uses the CSMA/CD network access method. Since the signal continues to traverse the network after the receiving computer acquires the message, terminators are used at the bus ends to absorb reflective signal energy, which can create interference and corrupt the data.

4.   Since a star topology network utilizes a central hub for point-to-point cable connections, the star network is able to remain up even if a network device fails. The failed device can be removed and repaired while other network devices continue to send and receive data. In the linear bus topology, if a device fails, the entire network goes down. In addition, when the star topology uses multiple network access polling with a central intelligent hub, the chance of data collisions is greatly reduced in high network access contention scenarios. In a bus topology network, the access method is based on network clear/transmit criteria, which can greatly retard network access time when many computers are trying to gain access.

5.   All computers in a token-ring network are connected on a single circle of cable with no terminated ends. This allows the data to travel around the loop and regenerate as it is passed through each computer. An electronic token, which is a predefined 24-bit code, manages client computer network access by enabling only the computer that captures it to transmit data.

        When a computer ready to send data has the token, it generates token ring data packets that are tagged with the source and destination addresses. This allows the data to reach the proper destination node and affirm data reception back to the source node. The source node acknowledges this, creates another free token, and sends it out on the network to be captured by the next computer ready to send data.

6.   Fiber-optic cable supports data transmission rates in the 100 to 600 Mbps range utilizing data-modulated light signals to traverse the optical medium. The "multimode graded index fiber" type of cable allows for bandwidths as high as 3 Gbps. Since fiber-optic cable uses pulses of light to deliver data from source to destination, it operates in a frequency range that provides immunity to EMI, noise, crosstalk, and powerline interference. In addition, it does not suffer signal attenuation like copper cable, enabling it to cover distances up to 2 kilometers with little signal loss. In addition, the construction of fiber-optic cable makes it invulnerable to wire tapping and eavesdropping, and therefore provides a very high level of data security outside the immediate network.

7.   A network adapter is a device that interfaces a computer to a network by
      accepting the physical connection to the network media. The network adapter
      mediates between the demands of a server or client on one side and the
      network architecture, with its rules for accessing the shared network media,
      on the other. The basic functions of a network adapter are as follows: convert
      the data format for transmission across the physical network media, control
      data flow onto the physical media, communicate with other network adapters
      to facilitate sending data to other devices, and accept incoming data and
      convert it to a format recognized by the receiving computer's CPU.

8.   A switch operates similar to bridge segmenting, except that switching is done
      in hardware instead of a routing table. For this reason, a switch provides very
      high bandwidths. It does this by utilizing application specific integrated circuits
      (ASICs) to very rapidly build dedicated switched circuits in hardware when
      routing data to LAN segments or destination nodes. Since it only routes data
      through the switched port servicing the destination node, a switch also
      reduces traffic.

9.   A bridge can be used to alleviate network traffic bottlenecks from excessive
      numbers of computers by segmenting the network and isolating the traffic on
      each segment. A bridge builds a routing table using MAC addresses to identify
      the location of computers on each attached segment. The table is then used to
      prevent traffic generated on a local segment, destined for nodes on the same
      segment, from being broadcast to other attached segments.

10.   A router can be used for the following purposes: Connect remote networks
      using multiple redundant paths, filter unwanted broadcast packets, including
      bad data and broadcast storms, to prevent them from reaching a network,
      determine the priority for processing packets based on the Network Layer
      protocol, provide firewall security by isolating one network from another, act
      as a gateway from one network to another by passing packets between
      networks with differing protocols.

      Some of the basic tasks performed by a router when connecting to a remote
      network include: Identifying and storing connected network addresses,
      determining the best data path between network destinations, filtering local
      broadcast traffic to local segments

Lesson 1 Quiz

These questions test your knowledge of features, vocabulary, procedures, and syntax.

1. The goal of all network access methods is to do which of the following?

    A. Provide a way for data to be placed on the network media
    B. Avoid data collisions by managing when and how computers transmit their data
    C. Convert the transmission protocol of one network to enable access to another network with a different protocol
    D. Establish a gateway from one network to another

2. Which of the following selections best describes baseband signal transmission?

    A. Multiple data signals transmitted simultaneously on multiple channels across a WAN
    B. Multiple data signals transmitted simultaneously on a single channel
    C. A single data stream transmitted on a single channel between devices in an Ethernet or token-ring network
    D. A single message transmitted across low bandwidth circuits

3. Which topology utilizes a linear backbone segment? (Choose all that apply.)

    A. Star
    B. Bus
    C. Ring
    D. Mesh

4. Which type of network uses a predefined 24-bit data packet circulating in the network to enable computers capturing it to send data?

    A. Switched LAN
    B. Star-wired bus
    C. Demand priority
    D. Token ring

5.   Which topology combines the simplicity of a bus network with point-to-point hub-to-computer connections to form a reliable high-speed network?

      A.   Token bus
      B.   Segmented bus
      C.   Star-wired bus
      D.   High-speed backbone

6.   To segment a network, which of the following can be used?

      A.   Switch
      B.   Bridge
      C.   Router
      D.   All the above

7.   When designing a small Ethernet network having budget constraints, data transmission rates anywhere between 4 and 100 Mbps, cable distances less than 100 meters, and where EMI susceptibility is not an issue, what type of cable do you use?

      A.   STP
      B.   Thinnet
      C.   Thicknet
      D.   Category 5 UTP

8.   A MAC address is hard-coded on a computer's network adapter card to do what?

      A.   Identify the network adapter card to the CPU
      B.   Channel data from the MAC sublayer to the Physical Layer
      C.   Uniquely identify the computer on the network with a hexadecimal address
      D.   Implement MAC protocols and control access to the shared network media

9.   What parameter or method locates a router on a network?

       A.   Network Layer address
       B.   MAC address
       C.   Distance vector algorithms
       D.   Automatic route discovery

10.   From the following, choose which are required for a repeater to pass packets
       properly?

       A.   The same packet frame types on attaching segments
       B.   The same network access method on both attached segments
       C.   The appropriate matching connectors for connecting segments with
              different physical media
       D.   The ability to restore the original voltage and duration of digital signals to
              ensure proper recognition of the data

Answers to Lesson 1 Quiz

1.   Answer B is correct. The goal of all network access methods is to manage how computers ready to transmit data can gain access to the network so the chance of data collisions is reduced.

      Answer A is incorrect. A network adapter card is responsible for preparing data and placing it on the network media.

      Answer C is incorrect. Protocol conversion is a gateway property.

      Answer D is incorrect. Network access is associated with the way computers access the network media, not the way they access another network with a different protocol or architecture.

2.   Answer C is correct. Baseband transmission uses a single data stream at one baseband frequency transmitted across a single channel between devices in Ethernet or token ring networks.

      Answer A is incorrect. Multiple simultaneous data signal transmissions are not utilized in baseband networks.

      Answer B is incorrect. Multiple data signals transmitted simultaneously on a single channel are utilized in broadband transmissions.

      Answer D is incorrect. Baseband signals are not limited to a single message nor does a baseband signal of itself have bandwidth constraints.

3.   Answer B is correct. Bus topology uses a linear backbone segment to connect computers and devices.

      Answer A is incorrect. Star topology uses a central hub connecting computers with point-to-point cable connections.

      Answer C is incorrect. Ring topology attaches computers in a continuous loop configuration.

Answer D is incorrect. Mesh topology connects computers or networks with multiple redundant paths.

4.   Answer D is correct. A token-ring network uses a predefined 24-bit encoded packet, known as a free token, to circulate around the ring and enable computers to transmit their data only when in possession of the token.

Answer A is incorrect. Token passing is irrelevant to the operation of a switched LAN. In the switched LAN configuration, a switch is used to very rapidly build routing paths to the LAN with switch ASICs.

Answer B is incorrect. A star-wired bus does not use token passing to manage network access, but uses CSMA/CD or CSMA/CA instead.

Answer C is incorrect. Demand priority is a priority-driven access method for networks based on the 100-Mbps Ethernet standard and does not use a token.

5.   Answer C is correct. A star-wired bus combines the simplicity of a bus network with the convenience of a point-to-point connected star network.

Answer A is incorrect. A token bus network utilizes a single shared network media with token passing as the traffic management method.

Answer B is incorrect. A bus segment is merely a logically divided section of a network and does not combine any different topologies.

Answer D is incorrect. A star-wired ring network combines the star and ring topologies.

6.   Answer D is correct. A switch, bridge, or router can each be used to segment a network.

Answer A is correct. Since a switch builds routing paths to specific destination nodes, it filters traffic specific to those destinations and effectively segments the network.

Answer B is correct. The primary purpose of a bridge is to segment (divide) a network and reduce traffic.

Answer C is correct. A router can be used to isolate one network from another using its routing table to filter traffic between attached networks, effectively segmenting networks.

7.   Answer D is correct. UTP is the least expensive of the choices listed. Category 5 cable accommodates data transmission rates up to 100 Mbps and handles signal transmission without significant attenuation over cable lengths of 100 meters or less. It is not the best choice for EMI resistance, but the twisted-pair wiring configuration does offer some immunity to noise and electrical interference.

Answer A is incorrect. Although STP handles data transmission rates up to 100 Mbps and has an attenuation factor equivalent to UTP, it is more expensive since it offers high immunity to EMI and noise.

Answer B is incorrect. Although Thinnet coaxial cable handles transmission distances up to 185 meters and provides EMI resistance, it is more expensive than UTP and cannot support transmission rates as high as 100 Mbps without attenuation problems.

Answer C is incorrect. Although Thicknet coaxial cable handles transmission distances up to 500 meters and provides EMI resistance, it is more expensive than UTP and cannot support transmission rates as high as 100 Mbps without attenuation problems.

8.   Answer C is correct. The Media Access Control (MAC) address is hard-coded in the network adapter card's ROM to uniquely identify the computer on the network.

Answer A is incorrect. The network adapter card is identified to the CPU using the base I/O address assigned to the card.

Answer B is incorrect. Although the network adapter card is responsible for preparing data and placing it on the physical media, the only relevance the adapter's MAC address has in this process is that it is embedded in the source address field of Data Link Layer frames to inform the destination node of the transmitting computer's network location.

Answer D is incorrect. Although the MAC protocols controlling access to the shared network media reside at the MAC sublayer of the OSI model's Data Link Layer, the hard-coded MAC address is not related to this process.

9.   Answer B is correct. Routers use their MAC address to identify their locations to each other on an internetwork. When a router recreates a Data Link Layer frame, it embeds its MAC address in the destination address field of the frame and sends the frame to the next router.

Answer A is incorrect. The Network Layer address of a router is its IP address and is used to find the host of a destination network.

Answer C is incorrect. Distance vector algorithms are used to determine best routing paths at any given moment based on certain routing criteria.

Answer D is incorrect. Automatic route discovery occurs as a router interacts with other routers and the information is used to update the routing table.

10.   Answers A, B, D, and D are all correct. For a repeater to function properly, all the selections are applicable.

Lesson 2: Network Architecture

Network architecture refers to the way communication protocols, logical and physical topologies, access methods, and standards combine to create a fully functional network. Communication protocols are rules defining how data is formatted, transmitted, and recognized, as defined by the OSI model, and topology is the physical and logical network layout. Closely associated with these are access methods that characterize the way computers transmit data onto the network media and standards specifications identifying network media configurations and their supporting data transmission rates.

After completing this lesson, you should have a better understanding of the following topics:

- OSI Model

- Project 802 Specifications

- Packet Functions in Networks

- Network Protocols

- Network Architecture Models

OSI Model

The Open Systems Interconnection (OSI) model provides a basis for vendors to design their network products using layering techniques to facilitate open data communications. Transmitting data between network source and destination nodes is a complex process involving the following:

- Dividing data blocks into manageable portions called packets

- Adding source and destination node information to each packet

- Adding error correction and timing or control information to each packet

- Launching data packets onto the network

- Recognizing data packets at the destination node

The rules controlling these operations, known as protocols, are implemented very carefully by network operating systems. Because of the large number of differing vendor software and hardware products available today, standard protocol models were developed to handle communication incompatibility. These models are known as Open Systems Interconnection (OSI) and Project 802 (a variation of the OSI standard). These standards describe open network architecture for connecting dissimilar devices and operating system environments and provide a common basis for communication and information exchange.

OSI Model Architecture

The OSI model was originally introduced in 1978 by the International Standards Organization (ISO) and then revised and reissued in 1984.

The OSI model creates a hierarchical architecture that divides network communication into seven layers with each layer representing different network functions and services. Each layer has associated protocols that implement layer functions while communicating and working with the layers immediately above and below it. For example, the Transport Layer interacts with both the Session Layer above and the Network Layer below.

Note: The higher layers of the OSI model carry out the most complex tasks while task complexity diminishes with each subsequent lower layer.

Layer Relationships

Each layer builds on the functions and processes carried out at each preceding layer to prepare the data for delivery across the network to a destination node. Each layer relies upon lower layers to perform more elementary functions while the details carried out at these lower layers are masked from the higher layers. Each layer is separated by a boundary, known as an interface, which establishes a communication path for interaction between layers. The OSI model layered architecture is shown in Figure 2.1.

Figure 2.1 OSI Model

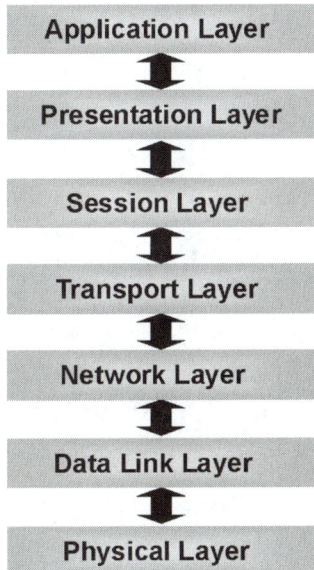

Layer Communication

Layers in the OSI model communicate through a vertical structure. However, virtual communication also occurs among layers horizontally, shown in Figure 2.2. In the OSI model, there are no direct horizontal links between layers, except at the Physical Layer. At the other layers however, virtual communication takes place where each layer on the source computer acts as if it communicates with an associated layer on the destination computer.

Figure 2.2 OSI Layer Communication

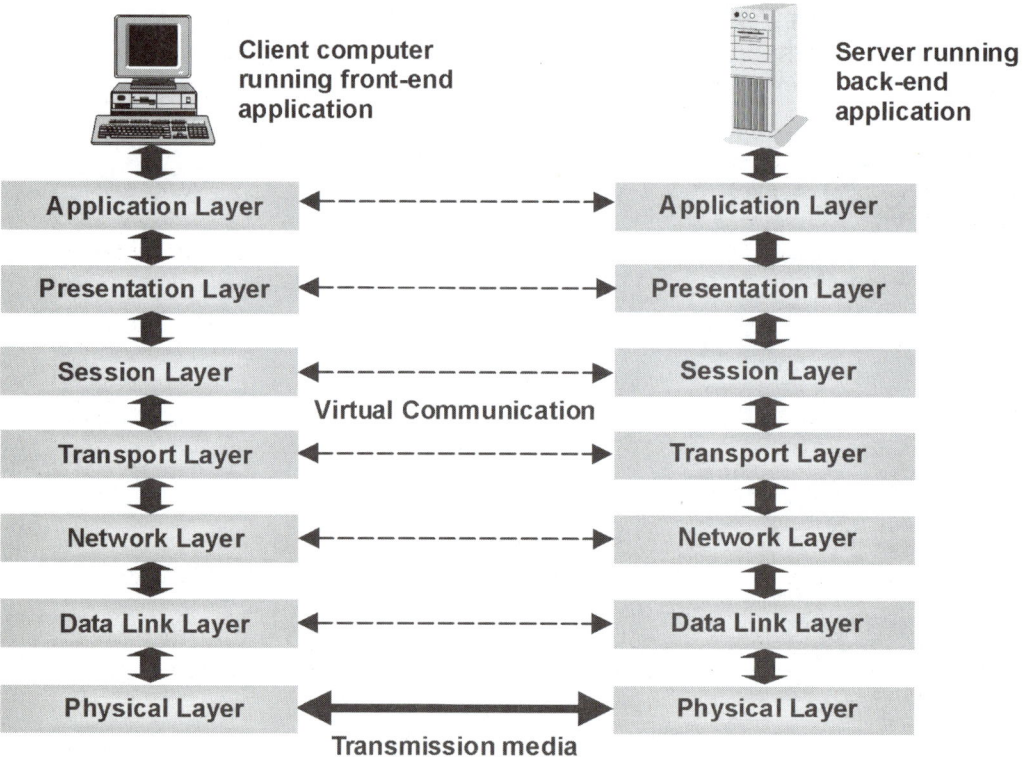

For example, as a data message emerging from the client's front-end application travels down the protocol stack, "encapsulation" occurs. During encapsulation, each OSI layer adds header information according to the syntax of the protocol residing at that layer. When the message arrives at the destination, it travels up the protocol stack and de-encapsulation occurs. At each successive OSI layer, the header (and trailer) information is removed and the data passed there by the corresponding source layer protocol is processed. In this manner, virtual communication occurs between OSI layer source and destination nodes.

Layer Tasks

Each OSI layer carries out a specific task in support of network communications, as follows:

Application Layer—Provides general network access, flow control, and error recovery functions. This layer represents the services that directly support user applications such as those involving file transfer, database access, or e-mail communications. The Application Layer does not contain end-user applications; rather, it supports them and acts as a window for user applications to access network services. Examples of Application Layer protocols include X.400 and X.500, which respectively offer interoperability and synchronization among different e-mail systems.

Presentation Layer—Provides an interface between user applications and the services at this layer needed by those applications. It handles such things as protocol conversion, data translation, data compression, encryption, and character set modification. Also operating at the Presentation Layer are redirectors, which redirect server I/O operations.

For example, when data is exchanged between communicating computers, Presentation Layer protocols determine the format. The sending computer Presentation Layer converts the data format (such as ASCII) passed from the Application Layer, to an intermediate format (such as EBCDIC). At the destination computer, the Presentation Layer translates the intermediate format into a form recognized and used at the destination Application Layer.

Session Layer—Establishes, controls, and terminates communication sessions between different computers by implementing an interactive dialogue between user applications on those computers. For example, the Session Layer performs name recognition and security functions to permit two applications to engage in network communications. It also places synchronization checkpoints in the data stream to enable efficient data retransmission in the event of communication failures.

Transport Layer—Provides end-to-end error recovery, flow control, and the functions necessary for reliable packet transmission and reception. It enables error-free packet delivery in sequence with no losses or duplications. Transport Layer protocols provide mechanisms for organizing multiple network packets in sequence to create a coherent message. The Transport Layer also breaks down messages into smaller packets for efficient network transmissions.

For example, the source computer Transport Layer can gather small packets together into one message and then add sequence information to guide the destination computer's reassembly of the data. At the destination, the Transport Layer unpacks the message, reassembles the original data from smaller packets, and sends an acknowledgment of receipt.

Transport Layer protocols are usually supplied with network operating systems and are closely linked with network layer protocols. For example, TCP/IP uses Transmission Control Protocol (TCP) to provide reliable services for the Internet Protocol (IP).

Network Layer—Establishes, controls, and terminates end-to-end network links. To enable computers in different networks to communicate, Network Layer protocols provide network addressing schemes and support internetwork routing of Network Layer data packets. To accomplish this, the Network Layer protocols do the following:

- Address messages

- Translate logical addresses and names into physical addresses

- Determine the route from source to destination computer based on network conditions and priorities

- Manage network traffic problems including packet switching, routing, and congestion

Network Layer protocols are part of a network operating system's protocol stack which sometimes includes multiple protocols for heterogeneous client/server computing environments.

Note: Packets are usually associated with Network Layer
protocols, and frames are associated with Data Link
Layer protocols.

Physical Layer—Establishes, controls, and terminates the physical connection
between communicating computers. Physical Layer protocols define the electrical,
mechanical, functional, and procedural elements of data transmission. For example,
the RS-232 specification defines a Physical Layer protocol for serial data
transmission.

The Physical Layer transmits the unstructured raw bit stream, including transmission
control signals generated at higher layers, over a physical link such as a network
cable or optical medium. To ensure reliable data transmission across the network
media, the following specifications are defined by the Physical Layer:

- The network adapter card's cable interface connections, including pinout
functions

- Data transmission techniques, including parallel to serial format conversions

- Data encoding, bit length, and bit synchronization, including translating bits into
appropriate electrical or optical pulses for transmission

Data Link Layer—Provides protocols that establish transmission reliability to the
upper OSI layers for the point-to-point connections made at the Physical Layer. The
Data Link Layer organizes the data (bit stream) into structured frames (logical
organized structures) to add address and error control information when sending
the frames to the Physical Layer. Information added to the front of the data is called
a header and information added to the back is called a "trailer."

Data Link Layer frames are built within the network adapter card according to a
structure that suits the network architecture and the adapter interfaces. The unique
hexadecimal address of the source network adapter card is embedded in the frame
along with the address of the destination network adapter.

The Data Link Layer then transfers error-free frames to the Physical Layer where a
raw bit stream is sent over the network media to the destination computer. At the

destination computer, the raw bits from the Physical Layer are repackaged into data frames by the network adapter card, which also acknowledges receiving the data. If damaged frames are detected at the destination, they are re-sent. A Data Link Layer frame looks similar to the illustration in Figure 2.3.

Figure 2.3 Data Link Layer Frame

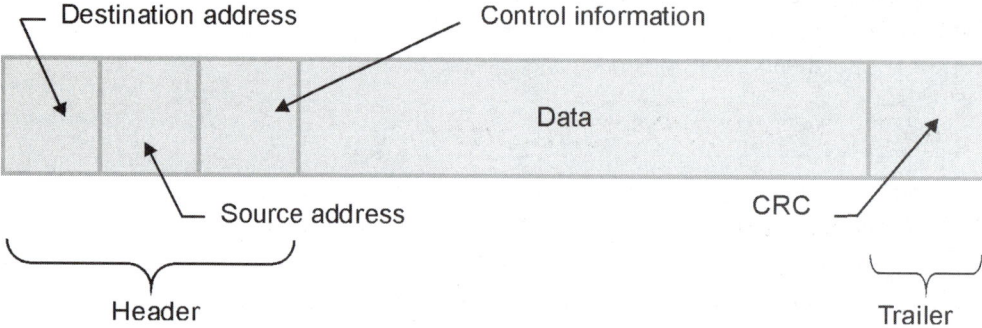

The header's control section consists of frame type, routing, and segmentation information. The trailer's cyclical redundancy check (CRC) section consists of error correction and verification information.

Project 802 Specifications

Project 802 was launched by the Institute of Electrical and Electronics Engineers (IEEE) in February 1980 to identify network standards for the physical aspects of a network such as the network adapter, coaxial and twisted-pair cabling, and WAN components. The 802 specifications revise the Data Link and Physical Layers of the OSI model and define the way network adapter cards access and transfer data over physical media. The categories established by IEEE for the 802 specifications are listed in Table 2.1.

Table 2.1 IEEE 802 Categories

IEEE Specification	Network Standards
802.1	Internetworking
802.2	Logical Link Control (LLC)
802.3	Ethernet LAN, CSMA/CD
802.4	Token Bus LAN
802.5	Token-ring LAN
802.6	Metropolitan Area Network (MAN)
802.7	Broadband Technical Advisory Group
802.8	Fiber Optic Technical Advisory Group
802.9	Integrated Voice/Data Networks
802.10	Network Security
802.11	Wireless Networks
802.12	Demand Priority Access, 100 BaseVG-AnyLAN

OSI Model Enhancements

IEEE 802 standards modified the Data Link Layer into two sublayers to further clarify its subfunctions, shown in Figure 2.4.

Figure 2.4 OSI Model Data Link Sublayers

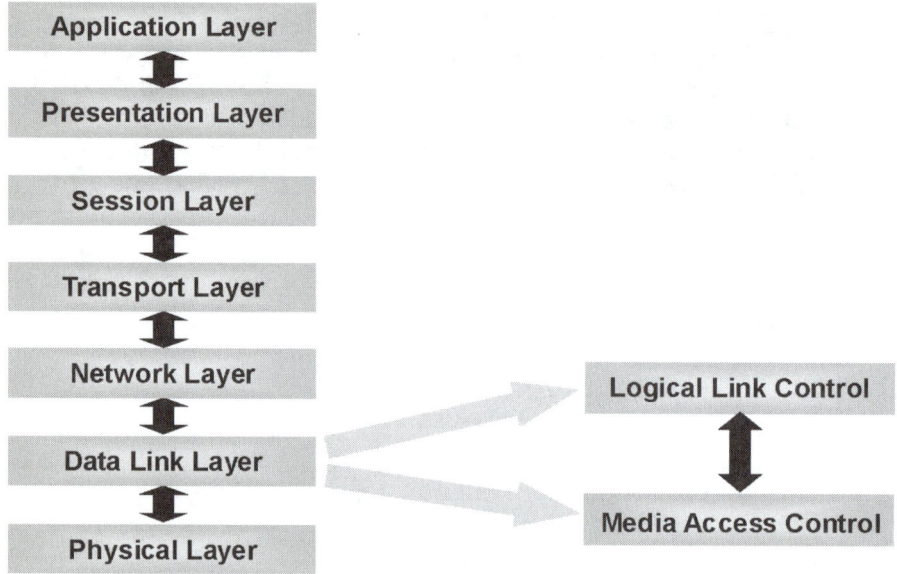

Logical Link Control (LLC) Sublayer

The LLC Sublayer is the upper layer of the Data Link Layer and interfaces the Network Layer. The LLC Layer is represented by a single protocol defined by IEEE 802.2 and serves two functions: a) provides transparency to the Network Layer above by masking the details of the LLC Sublayer to the Network Layer, and b) enables the MAC Sublayer protocols to vary depending on the network type in use. This allows NetWare to run equally well with Ethernet or token ring network adapter cards.

The specific function of the LLC Sublayer is to create and terminate communication links, control frame traffic, sequence frames, and acknowledge frames. It also defines logical interfaces known as service access points (SAPs) that destination computers use to transfer information from the LLC Sublayer to upper OSI layers.

Media Access Control (MAC) Sublayer

The MAC Sublayer interfaces to the Physical Layer and is represented by the protocols that define how computers access the network media. For example, different network types use different access methods and require specific IEEE protocols at the MAC Sublayer, as follows:

- Ethernet, IEEE 802.3

- Token bus, IEEE 802.4

- Token ring, IEEE 802.5

- Demand priority, IEEE 802.12

The specific function of the MAC Sublayer is to manage the media access method, recognize addresses embedded in Data Link Layer frames, delimit frames, and to ensure error-free data delivery by checking for frame errors.

 Note: The MAC Sublayer directly communicates with the network adapter card using a network adapter driver that is sometimes referred to as a MAC driver.

Packet Structure and Function

When communicating application data between source and destination computers, data must be broken down into small sections to prevent the network media from being inundated with large amounts of data. Breaking the data down into small sections also simplifies error recovery since only small sections of data need to be retransmitted when errors occur. Smaller sections of data are also more manageable and permit timely interactions between computers. For these reasons, network communications are faster with small pieces of data.

Small sections of data transmitted across a network are known as packets or frames, as shown in Figure 2.5. These basic network communication units allow faster data transmissions and quicker network access for network computers ready to send data.

Figure 2.5 Data Packets

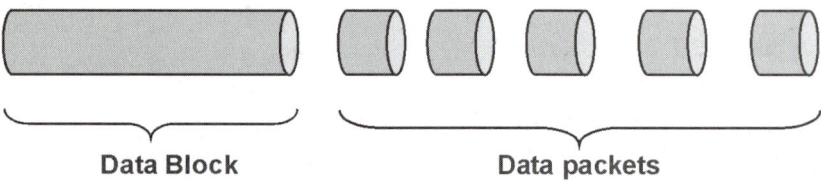

Data Block   Data packets

Packet Structure

Packets consist of data and other information that can include addresses, controls, commands, error correction codes, packet assembly instructions, routing instructions, or service requests.

Packet data can vary in size between 512 bytes and 4 KB per packet, depending on the network type in use. Packet size limits determine how many packets the network operating system generates from a file or message. To transfer large files, many packets must be generated and transmitted.

Header Information

Typical information embedded in a data packet header consists of the following:

- The sending computer's (source) address
- The receiving computer's (destination) address
- Clock synchronization information
- Control signals for packet transmission

The general structure of a data packet is shown in Figure 2.6.

Figure 2.6 Packet Structure

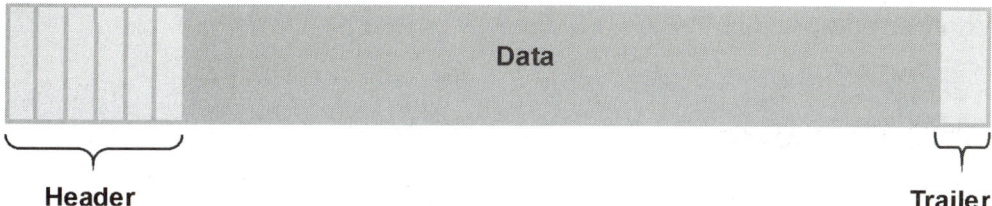

Header   **Trailer**

Trailer Information

The communication protocol usually determines the data packet trailer content. However, most trailers contain a common component called cyclical redundancy check (CRC). The CRC is predetermined from a mathematical calculation performed on the source packet. An equivalent calculation is performed on the packet at its destination and is compared to the original. A difference in these values indicates an error that causes the CRC to signal the source computer to transmit the packet again.

Packets and OSI Layering

Packet creation begins when data generated by a user application is introduced at the Application Layer of the OSI model. As the data descends through each layer, it adds information related to each layer. The added information is encapsulated and removed later at the corresponding layer of the destination computer, shown in Figure 2.7.

The data is actually broken into packets at the Transport Layer while the structure of each packet is defined by the transport protocol in use. Sequence information is also added at the Transport Layer to guide the receiving computer when reassembling the data from packet form. Network address information is added at the Network Layer while MAC addresses are added at the Data Link Layer. When the packet reaches the Physical Layer, it consists of the original data plus all the information encapsulated at each OSI layer.

Figure 2.7 Encapsulating Packet Information

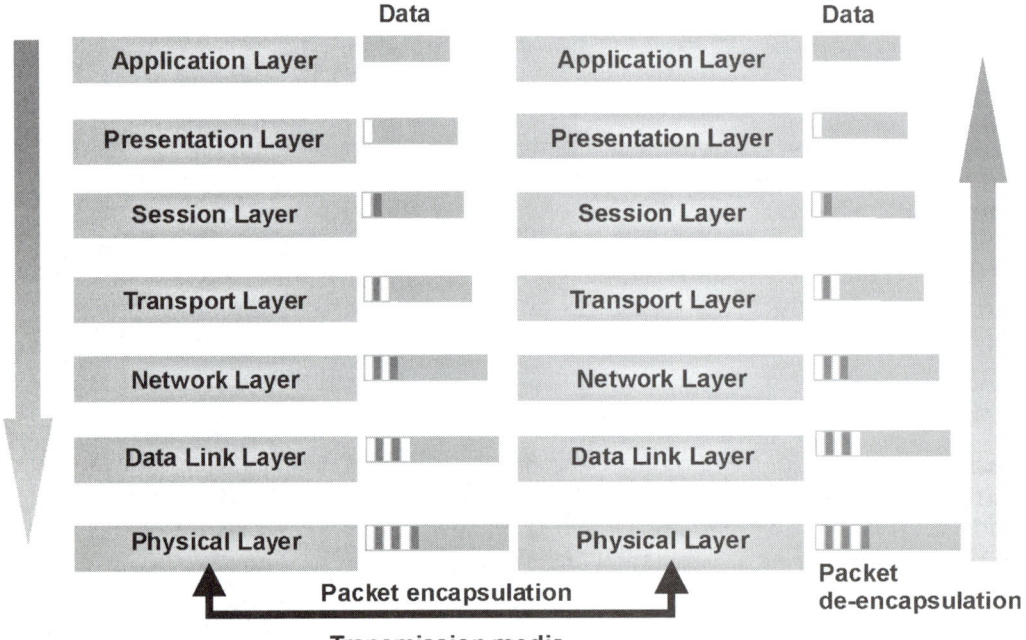

Packet Addressing

Data packets are addressed to only one computer on the network. As a result, only one computer responds to the packet by interrupting its CPU to process the incoming packets. However, each network computer listens to all packet traffic on its segment through its network adapter card.

In WANs or internetworks, packet-addressing information is used by routers to determine the best path to a destination. In this process, certain functions are utilized in directing packets to their destination. One function is packet forwarding, which enables packets to be sent to the next network component based on an address specified in the packet header. Another is packet filtering, which enables packets to be selected based on an address specified in the header.

Packet Processing Scenario

The following shows typical packet processing during network communications:

1.   A user application on the source computer has data to send.

2.   The source computer makes a connection with the destination computer.

3.   The data travels down the protocol stack with the source computer breaking large data blocks into data packets and encapsulating each packet with a source address, destination address, control information, and error recovery codes.

4.   The network adapter card prepares the data for transmission across the network media.

5.   The destination computer's network adapter receives and analyzes all packets (in a buffer) transmitted on its network segment. If the hexadecimal destination address in the packet header matches the network adapter card's unique hexadecimal address, the network adapter card interrupts the destination computer's CPU and pulls the data off the network media.

6.   The data travels up the protocol stack while the destination computer reassembles the packets into their original form, strips the encapsulated information, and stores the data in memory for use by the application. The destination computer passes the reassembled data to the receiving computer's application in a usable form.

Network Protocols

Network protocols are the rules, technical procedures, or discrete systematic steps followed by computers when communicating data across a network. Protocols implement the network communication functions defined by the layers of the OSI model. Each layer has specific steps that cannot be performed at any other.

At the source computer, the steps are performed from the top to the bottom of the OSI model. At the destination computer, the steps are performed from the bottom up. Since the steps of each protocol must be carried out consistently on all computers in a network, all the computers must use common protocols.

Three points to consider about protocols in a networked environment are the following:

- Multiple protocols provide basic network communications although each protocol has a different purpose, advantage, or restriction

- The OSI model layer at which a protocol works is related to its function

- Several protocols working together are known as a protocol stack or suite. The levels in a protocol stack map to the OSI model layers

Protocol Operation

The processes that control data transmission on a network must be broken down into orderly and systematic steps and carried out consistently for all network computers. Each step is controlled by its own set of rules and procedures, known as protocols. Both the source and destination computers must perform their protocol operations in the exact same manner to facilitate proper recognition of data packets when they arrive at their destination. For example, two different protocols may have certain processes in common, such as breaking data blocks into packets and adding sequence or timing information, but each protocol performs differently. Since the protocols differ in this example, the computers using them will not be able to communicate.

Protocol Routing

Some protocols are routable while others are not. Routable protocols allow a protocol to be passed from one LAN to another, enabling communication between the LANs. This is important when LANs are internetworked to form WANs.

Layered Architecture Protocol Stacks

A protocol stack is a grouping of protocols that function within the OSI model architecture. The protocols in a stack function at each layer to handle different components of the communication process, shown in Figure 2.8.

Figure 2.8 Layered Protocol Functions

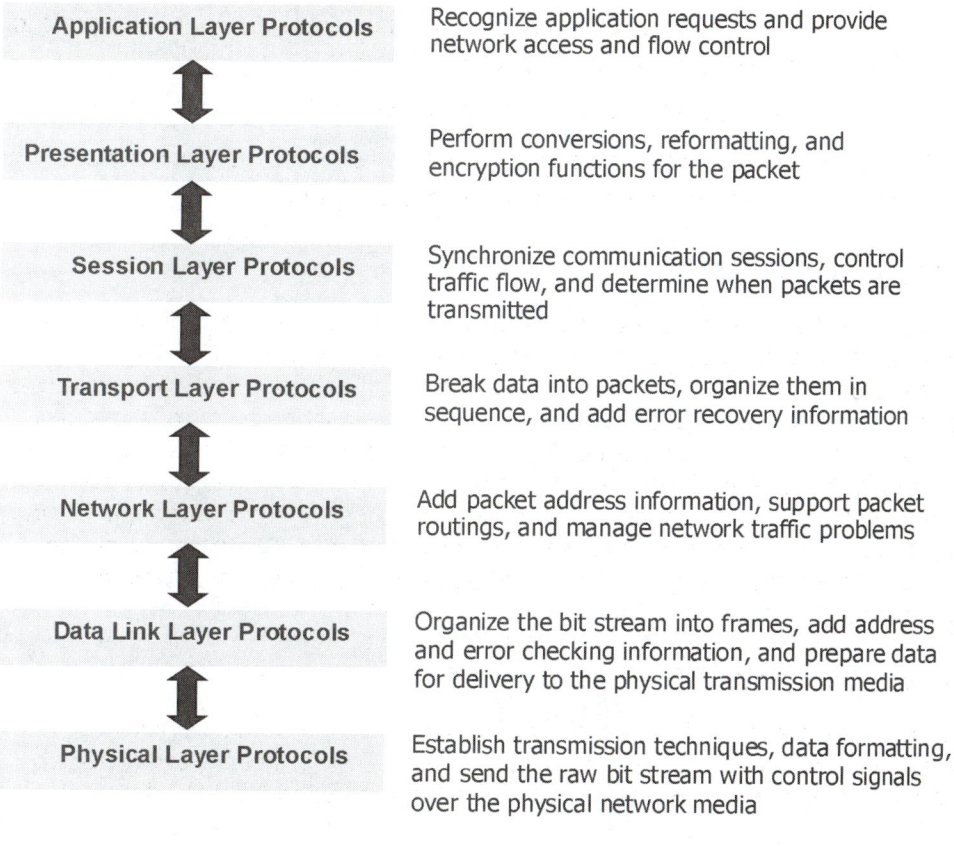

Application Layer Protocols	Recognize application requests and provide network access and flow control
Presentation Layer Protocols	Perform conversions, reformatting, and encryption functions for the packet
Session Layer Protocols	Synchronize communication sessions, control traffic flow, and determine when packets are transmitted
Transport Layer Protocols	Break data into packets, organize them in sequence, and add error recovery information
Network Layer Protocols	Add packet address information, support packet routings, and manage network traffic problems
Data Link Layer Protocols	Organize the bit stream into frames, add address and error checking information, and prepare data for delivery to the physical transmission media
Physical Layer Protocols	Establish transmission techniques, data formatting, and send the raw bit stream with control signals over the physical network media

Examples of common protocol stacks are TCP/IP and Internetwork Packet Exchange/Sequenced Packet Exchange (IPX/SPX).

Binding

Binding is a process where protocols are bound to their storage location on a network adapter card and become physically associated with it. A network adapter card can have multiple protocol stacks binded, for example, TCP/IP and IPX/SPX. Typically, binding occurs at the time of network operating system installation, but can be performed later.

The protocol stack binding order determines which stack the network operating system attempts to run first when making a network connection. For example, if TCP/IP is the first protocol stack bound to the network adapter card, it is used first upon a network connection. If the first network connection fails, the next protocol stack in the binding order attempts the connection. The most important protocol stacks designated as standard protocol models are as follows:

- TCP/IP

- IPX/SPX

- ISO/OSI

- SNA

- DECNet™

- AppleTalk®

Protocol Types

Network communication tasks can be broken down into four major types including those that support application, transport, network, and physical-level functions. The protocol types that implement these tasks are application, transport, network, or physical-level protocols. These protocol functions map to the OSI model as shown in Figure 2.9.

Figure 2.9 Mapping Protocol Types

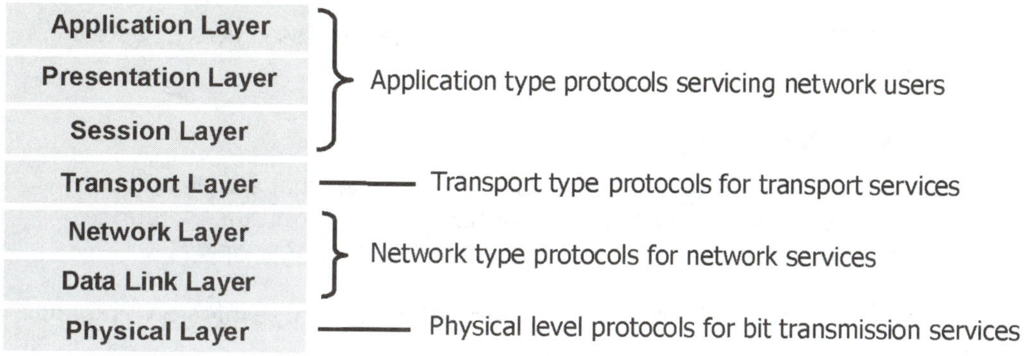

| Application Layer |
| Presentation Layer | } Application type protocols servicing network users
| Session Layer |

Transport Layer —————— Transport type protocols for transport services

| Network Layer |
| Data Link Layer | } Network type protocols for network services

Physical Layer —————— Physical level protocols for bit transmission services

Application Protocols

Protocols working at the OSI model's upper layer are application protocols, since they allow interaction and data exchange among applications. The following describes several application protocols:

AppleTalk—A remote file access protocol from Apple

Directory Access Protocol (DAP)—A DECNet file access protocol

File Transfer Access Management (FTAM)—An OSI file access protocol

File Transfer Protocol (FTP)—An Internet protocol

Simple Mail Transfer Protocol (SMTP)—An Internet e-mail protocol

Simple Network Management Protocol (SNMP)—Monitors networks and components

Telnet—An Internet protocol, establishes communication sessions with remote hosts

X.400—A Consulting Committee for International Telephone and Telegraph (CCITT) protocol for international e-mail transmissions

X.500—A CCITT protocol for file and directory services accessed across several systems

Transport Protocols

Protocols working at the Transport Layer of the OSI model that enable reliable packet delivery between computers are considered transport protocols. Some examples of protocols that exist in this category include the following:

Advanced Program To Program Communication (APPC)—IBM's SNA protocol used in peer-to-peer networks and on AS/400® mainframes

Transmission Control Protocol (TCP)—Provides guaranteed delivery of sequenced data

Sequenced Packet Exchange (SPX)—A protocol from Novell's IPX/SPX suite ensures message completeness and uses IPX to deliver packets

NWLink—Microsoft's implementation of IPX/SPX

NetBEUI—Microsoft's NetBIOS Extended User Interface network protocol is a connection-oriented protocol used for establishing communication sessions between subnet computers while providing rapid data transport service

Network Protocols

Protocols operating at the OSI model's lower layers are considered network protocols. The functions implemented by network protocols are called link services and include addressing information, routing information, error checking, and error recovery with associated requests for data retransmission. Network protocols also define certain network environment communication procedures such as those applying to Ethernet and token-ring networks. Some examples of protocols that exist in this category include the following:

IP—An Internet protocol from the TCP/IP suite that performs packet forwarding and routing

IPX—NetWare's internetwork packet exchange protocol used for packet forwarding and routing

Physical-level Protocols

Protocols operating at the OSI model's lowest layer. IEEE protocols that exist in this category perform the following tasks:

- 802.3 Ethernet protocols regulate bit transmission in a logical bus network using a 10-Mbps data rate and CSMA/CD (detects collisions between two simultaneously transmitting computers and halts transmissions until the cable is clear)

- 802.4 token-passing protocols regulate bit transmission in a bus network using a token that polls bus computers in numerical order to determine which computer can transmit

- 802.5 token-ring protocols regulate bit transmission in a logical ring network handling 4- to 16-Mbps data rates using a token traveling around the ring to determine which computer can transmit

---

 Note: Protocols operating at the physical level are sometimes called Media Access Control (MAC) protocols since they determine which computer can access the media and transmit on the network.

Commonly Used Network Protocols

This section describes some of the most commonly used protocols for network communication.

TCP/IP

The TCP/IP suite of protocols is an industry standard that supports heterogeneous environment communications. TCP/IP protocols are routable protocols suitable for enterprise networks requiring Internet access and support speeds equivalent to Novell's IPX/SPX. Since nearly all networks support TCP/IP, interoperability is one of its most prominent features. The TCP/IP suite is very large which can cause problems with MS-DOS clients, but most graphical user interface (GUI)-based operating systems are not affected by its size. Examples of protocols in the TCP/IP suite are the following:

- SMTP is used for e-mail

- FTP is used for file transfer

- SNMP is used for network management

- ARP is used for IP to MAC address resolutions

NetBEUI

NetBIOS Extended User Interface is a fast and efficient Transport Layer protocol with a small stack size. It is provided with most Microsoft network products and offers compatibility with all Microsoft-based networks. NetBEUI was developed from an IBM Session Layer application programming interface known as network basic input/output (I/O) system (NetBIOS) and provides tools that rapidly establish sessions with other network applications. NetBEUI is not a routable protocol, however, it is widely supported by many applications.

X.25

X.25 is a set of protocols incorporated in packet-switching networks and originally used to connect remote terminals to mainframe host systems. The X.25 protocol suite defines the interface between a synchronous packet mode host and the public data network (PDN) over a dedicated or leased line circuit.

 Note: An X.25 network consists of switches, circuits, and routes as they become available and provides best routing at any particular moment.

IPX/SPX

IPX/SPX is a fast protocol stack of small size used in Novell networks. It is an offshoot of the XNS protocol developed by Xerox and supports routing in a NetWare environment since IPX has its own routing software.

 Tip: Novell's IPX protocol is not routable on the Internet.

NWLink

NWLink is Microsoft's implementation of IPX/SPX. NWLink is a routable Transport protocol that typically provides for communication from Windows platforms to the NetWare environment. NWLink can also be used as a transport protocol to carry data between Windows-based networked computers.

APPC

Advanced Program-To-Program Communication is an IBM transport protocol developed for the SNA environment. APPC enables direct data exchange between applications on different computers.

AppleTalk

AppleTalk is a set of proprietary protocols for Macintosh computers designed to support network-based file and print sharing.

OSI Protocol Suite

The OSI protocol suite is a protocol stack that maps exactly to each layer of the OSI model. It includes routing and transport protocols, IEEE 802 protocols, and Session, Presentation, and Application Layer protocols that provide full network functionality including file access, printing, and terminal emulation.

DECNet

DECNet is Digital Equipment corporation's proprietary protocol stack that implements the Digital Network Architecture (DNA). It defines communications over Ethernet LANs, Fiber Distributed Data Interface Metropolitan Networks (FDDIMAN), and WANs using public or private transmission facilities. DECNet is a routable protocol that can also use OSI and TCP/IP protocols.

Devices Operating at OSI Layers

To support the operation of the various protocols residing at specific layers of the OSI model, network components such as those discussed in Lesson 1 must be utilized. For example, to implement the functions of the IP protocol in a network, a hardware routing device must be used to handle the network addresses that identify destination nodes. The following table shows common network components and the OSI layer at which they work.

Table 2.2 Network Components and OSI Layers

Network Component	OSI Layer Location
Bridge	Data Link (MAC Sublayer)
Brouter	Data Link, Network
Gateway	Transport, Session, Presentation and Application
Mulitplexer	Physical
Repeater	Physical
Router	Network
Switch	Data Link

Network Architecture Models

Network architecture defines the overall structure of a network and all the components, including hardware and software that make a network functional. The standard network architecture models described in this section consist of Ethernet, Token Ring, AppleTalk, ARCNet, and FDDI. In addition, some relatively new high-speed network architecture models are also discussed, including 100BaseVG-AnyLAN and 100BaseX Ethernet (Fast Ethernet). All network architecture models have certain elements in common which can be applied to the standard and high-speed models described in this section.

Common Network Architecture Elements

Although not all network architectures are standardized by organizations such as the IEEE, most network architectures consist of similar logical and physical elements. To define network architecture, the following network elements must be identified:

- Access method

- Logical topology

- Physical topology

- Transmission media

Network Access Method

Since multiple users in a network can simultaneously place data onto the network transmission media, it is necessary to implement controls that organize when and how users gain access to the network. A network's media sharing capability, known as its access methodology, consists of the following:

- Carrier Sense Multiple Access with Collision Detection (CSMA/CD)

- Carrier Sense Multiple Access with Collision Avoidance (CSMA/CA)

- Token passing

- Demand priority

Network Logical Topology

The logical topology element of network architecture is associated with the manner in which data passes from one workstation to another on the way to its destination. The two types of logical topology are the following:

Sequential—In a sequential topology, data passed from one workstation to another is examined by each workstation to determine the data's destination address. If the data is destined for another workstation, it is passed along to the next node in the same logical segment. Otherwise, the data is accepted by the receiving workstation.

Broadcast—In a broadcast topology, data simultaneously sent to all workstations is examined by each workstation to determine the data's destination address. If the data is destined for another workstation, it is not passed on. Otherwise, if the address matches, the data is accepted by the receiving workstation. With this method, there is no need to pass the data along to the next node.

Network Physical Topology

Physical topology is the way that clients and servers are physically laid out and connected in a network. It is also associated with the transmission media that connects network components. The configuration of a network's physical topology impacts network performance and reliability. Bus, ring, and star physical topologies are used most often in networks today.

Network Transmission Media

The transmission media element of network architecture is a critical network component since it is the foundation of network communication. It consists of the type of interconnecting cables and connectors used in a network. The network media determine the speed the data travels to its destination and the distance it transmits without signal loss.

Standard Network Architectures

This section describes the logical and physical elements of network architecture that combine to produce the standard network architectures currently used in LANs.

Ethernet Architecture

Ethernet is currently a very popular network architecture. The IEEE 802.3 Ethernet specification incorporates the identical functions defined in the Data Link and Physical Layers of the OSI model. Generally, Ethernet networks consist of the elements described in Table 2.3.

Table 2.3 Ethernet Network Specifications

Network Element	IEEE 802.3 Specification
Topology	Linear bus, sometimes star bus
Transmission type	Baseband
Access method	CSMA/CD
Data transfer speed	10 Mbps or 100 Mbps
Cable types	Thicknet, Thinnet, UTP

Ethernet frames have a different structure than packets used in other networks. An Ethernet frame is a package of information between 64 and 1,518 bytes long. When adding 18 bytes of control information to the frame, a single Ethernet frame contains between 46 and 1,500 bytes of data. The structure of an Ethernet frame is shown in Figure 2.10.

Figure 2.10 Ethernet Frame

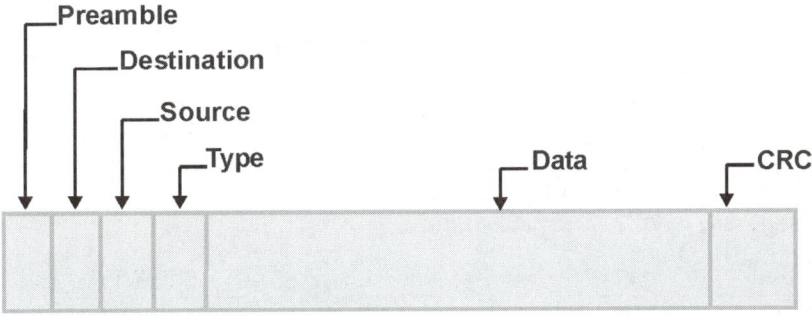

Note: The "preamble" of an Ethernet frame marks the frame's beginning point and the "type" identifies the Network Layer protocol, such as IP or IPX.

Several IEEE standards are defined for Ethernet networks, which combine different cable types, data transmission rates, and topology variations, as follows:

10BaseT—10 Mbps data rate, Baseband signal transmission type, and twisted pair wiring. Refer to Table 2.4 for additional 10BaseT specifications.

Table 2.4 10BaseT Specifications

Network Elements	IEEE 802.3 Specifications for 10BaseT Architecture
Topology	Star bus with multiport hub repeater
Access method	CSMA/CD
Data rate and transmission type	10 Mbps, baseband
Cabling	Category 3, 4, and 5 unshielded twisted pair (UTP) with RJ-45 connectors, or shielded twisted pair (STP)
Maximum cable segment length (network adapter card to hub)	100 meters
Minimum cable segment length (network adapter card to hub)	2.5 meters
Maximum number of computers in a 10BaseT LAN	1024, each with separate transmit and receive wires

 Note: Ninety percent of all new network installations use 10BaseT with Category 5 UTP to support future upgrade to 100-Mbps data transmission rates.

10Base2—10 Mbps data rate, baseband signal transmission type, and roughly twice the cabling distance of 10BaseT (185 meters), is described in Table 2.5.

Table 2.5 10Base2 Specifications

Network Elements	IEEE 802.3 Specifications for 10Base2 Architecture
Topology	Bus
Access method	CSMA/CD
Data rate and transmission type	10Mbps, baseband
Cabling	Thinnet coax
Maximum cable segment length	185 meters; up to 925 meters with repeaters
Minimum cable segment length	0.5 meters
Maximum computers per segment	30
Connectors	BNC barrels, terminators, and T-connectors. Network adapter cards connect directly with T-connectors only.

 Note: The Thinnet network "5-4-3" rule specifies that a Thinnet network can combine up to 5 cable segments, 4 repeaters, and 3 segments having up to 30 workstations attached per segment.

10Base5—10-Mbps data rate, baseband signal transmission, and roughly 5 times the cabling distance of 10BaseT (500 meters), is described in Table 2.6.

Table 2.6 10Base5 Specifications

Network Elements	IEEE 802.3 Specification for10Base5 Architecture
Topology	Bus (referred to as standard Ethernet)
Access method	CSMA/CD
Data rate and transmission type	10-Mbps, baseband
Cabling	Thicknet coax
Maximum cable segment (backbone) length	500 meters; up to 2,500 meters with repeaters
Max computer to transceiver segment length	50 meters (usually Thinnet used)
Maximum nodes (computers, repeaters, devices) per backbone segment	100
Connectors	All BNC components. Network adapter cards connect directly with T-connectors only

 Note: The Thicknet network "5-4-3" rule specifies that a
Thicknet network can combine up to five backbone
segments, four repeaters, and three segments populated
with computers.

10BaseFL—10-Mbps data rate, baseband signal transmission, and fiber-optic media, is used for Ethernet networks with fiber optic cable connecting computers and repeaters over distances up to 2,000 meters.

Two high-speed Ethernet standards are emerging to meet increased demands for speed with applications such as computer-aided design (CAD), computer aided manufacturing (CAM), video, and imaging. The standards are known as 100BaseVG-AnyLAN and 100BaseX Ethernet (Fast Ethernet). These new standards are five to ten times the speed of standard Ethernet, respectively, and support 10BaseT cabling compatibility.

100BaseVG-AnyLAN—100-Mbps data rate, baseband signal transmission, and voice grade compatibility, is described in Table 2.7. 100BaseVG-AnyLAN combines features of Ethernet and token ring networks such as transmission of Ethernet frames and token ring packets.

Table 2.7 100BaseVG-AnyLAN Specifications

Network Elements	IEEE 802.12 Specifications for 100BaseVG-AnyLAN Architecture
Topology	Cascaded star with central parent hub. Expanded using child hubs attached to the parent hub to add more computers
Access method	Demand priority, using low and high priority levels
Data rates and transmission type	100-Mbps, baseband
Data format	802.3 Ethernet frames and 802.5 token ring packets combined
Cabling	10BaseT-compatible using Categories 3, 4, and 5 UTP or fiber-optic cable
Maximum cable segment length (100BaseVG hub to node)	250 meters

Note: 100BaseVG-AnyLAN uses address filtering at the hub for
                       privacy

100BaseX Ethernet (Fast Ethernet)—100-Mbps data rate, baseband signal
transmission, and extension of existing Ethernet standards, is described in Table
2.8.

Table 2.8 100BaseX Ethernet Specifications

Network Elements	IEEE 802.3 Specifications for 100BaseX Ethernet Architecture
Topology	Star-wired bus similar to 10BaseT configuration
Access method	CSMA/CD
Data rates and transmission type	100 Mbps, baseband
Data format	Standard 802.3 Ethernet frames
Cabling UTP	Category 5 data-grade cable
Incorporates three additional media specifications:	
100BaseT4	Twisted-pair with four telephone-grade pairs.
100BaseTX	Twisted-pair with two data-grade pairs.
100BaseFX	Fiber optic link with two fiber-optic strands.

 Note: Ethernet networks use communication protocols such as
                       TCP/IP. Ethernet also works with Windows 95, Windows
                       NT Workstation, Windows NT Server, Windows
                       Workgroups, LAN Manager, Novell NetWare, AppleShare,
                       and IBM LAN Server.

Token-Ring Architecture

Token-ring architecture was introduced in 1984 by IBM as a networking solution for PCs, mid-range computers, and mainframes to communicate in the SNA environment. However, the basic design of token-ring architecture is currently used in the client/server environment. The use of token passing distinguishes the token ring architecture more than any other feature. The token-ring architecture uses unique frames and is defined by the IEEE 802.5 standard (Table 2.9).

Table 2.9 Token-Ring Specifications

Network Elements	IEEE 802.5 Specifications for Token Ring Architecture
Topology	Star wired ring
Access method	Token passing
Data rates and transmission type	4 and 16 Mbps, baseband
Cabling	STP and UTP (IBM Types 1, 2, and 3)
Impedance	STP: 150 ohms; UTP: 100-120 ohms
Maximum UTP cable segment length (MSAU to computer)	45 meters
Maximum STP cable segment length (MSAU to computer)	100 meters
Maximum Type 1 cable segment length (MSAU to computer)	101 meters; up to 730 meters with a pair of repeaters for Type 1 and 2
Maximum Type 3 cable segment length (MSAU to computer)	46 meters; up to 365 meters with a pair of repeaters
Minimum cable segment UTP or STP	2.5 meters
Maximum cable segment MSAU to MSAU	152 meters
Maximum computers per ring with STP	260
Maximum computers per ring with UTP	72
Maximum hubs per ring	33

The following connector types and components should be used with token-ring media:

- Media interface connector (MIC) is used with Type 1 and 2 twisted-pair cable

- RJ-45 is used with Type 3 twisted pair cable

- RJ-11 is used with Type 3 twisted pair cable

- Media filters for Type 3 twisted-pair cable connecting token-ring network adapter cards to RJ-11/45 wall jacks for line noise reduction

The structure of a token ring frame is shown in Figure 2.11.

Figure 2.11 Token-Ring Frame

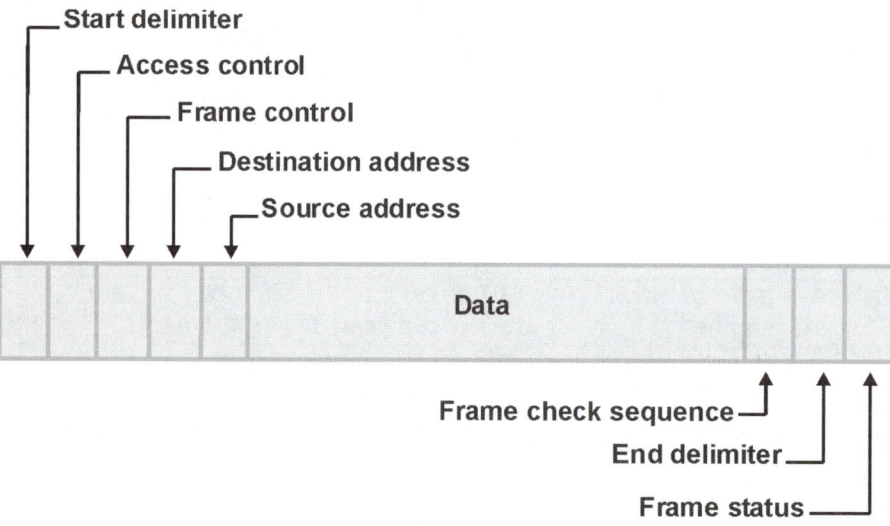

The frame components are defined as follows:

- Start delimiter indicates frame start point

- Access control indicates frame priority and whether the frame is a free token or contains data

- Frame control contains MAC information for all computers or end station information for one computer

- Frame check sequence consists of CRC error-checking code

- End delimiter indicates the frame end point

- Frame status indicates whether frame was recognized, copied, or if the destination address was available

The hub in token ring architecture contains the internal wiring that forms the ring. Token ring hubs are also known as either a Multistation Access Unit (MAU or MSAU) or a Smart Multistation Access Unit (SMAU).

An IBM MSAU has 10 connection ports with eight computers attached. A token-ring network is not limited to one ring and can have up to 33 hubs per ring. For expansion, MSAUs are connected in series using their "ring in" and "ring out" ports.

MSAUs have the fault tolerance ability to sense when a ring computer's network adapter card is faulty and disconnect from that computer to prevent the entire network from going down. Otherwise, if a computer fails, the entire network goes down.

Network adapter cards for token-ring architecture must have either 4- or 16-Mbps transmission rates. Network adapters supporting 16 Mbps are backward compatible to 4 Mbps. However, 4-Mbps network adapters do not support 16-Mbps token ring networks. In addition, 16-Mbps adapters accommodate a longer frame length, allowing fewer transmissions for the same amount of data.

AppleTalk Architecture

AppleTalk is simple proprietary network architecture for small groups using Macintosh computers with built-in networking functions. It consists of the following components:

AppleTalk—AppleTalk is the Apple network architecture included with all Macintosh operating systems. It is a group of protocols mapping to the OSI model. When a device attached to LocalTalk network comes online, the following occurs:

- The device assigns itself an address from a range of allowable addresses

- The device broadcasts address to see if other devices are using it

- The device stores addresses for future use, if no other devices are using it

LocalTalk—Refers to a single physical network in an AppleTalk network that supports up to 32 devices. It uses CSMA/CA as the access method in a bus or tree topology with STP cable, but can also use UTP or fiber-optic cable.

AppleShare—Allows file and print sharing on an AppleTalk network.

 Note: Single LocalTalk networks can be connected by using Zones to reduce traffic, expand the network, or join other networks together such as Token Ring and AppleTalk. Zones facilitate easy file server access across several LANs.

EtherTalk—EtherTalk allows AppleTalk network protocols to run on Ethernet coaxial cable. For example, EtherTalk NB cards allow a Macintosh II computer to connect to an 802.3 Ethernet network.

TokenTalk—Allows a Macintosh II computer to connect to an 802.5 token-ring network.

ARCNet Architecture

The Attached Resource Computer Network (ARCNet) was developed in 1977 by Datapoint and is used for workgroup-sized LANs. ARCNet architecture loosely maps to the IEEE 802.4 specification for token-passing bus or star bus networks using broadband cable.

ARCNet architecture uses the token-passing access method in a star bus topology with a 2.5-Mbps data rate. ARCNet + supports data rates up to 20 Mbps. With ARCNet architecture, the token passes in numerical order from one computer to another regardless of the location of computers on the network, as shown in Figure 2.12.

Figure 2.12 ARCNet Architecture

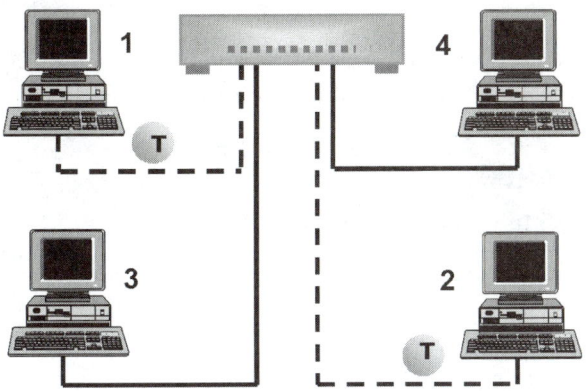

Note: In an ARCNet network, computers are connected to either an active, passive, or smart hub. Passive hubs distribute the signal, active hubs restore the signal, and smart hubs have additional features such as reconfiguration detection and operator control of port connections.

An ARCNet packet contains a destination address, source address, and 508 bytes of data (up to 4096 bytes in ARCNet +), as shown in Figure 2.13.

Figure 2.13 ARCNet Packet

Destination address ⌐   ⌐ **Source address**

508 bytes of data

Table 2.10 summarizes the IEEE 802.4 specifications that apply to ARCNet architecture.

Table 2.10 ARCNet Specifications

Network Elements	IEEE 802.4 Specifications for ARCNet Architecture
Topology	Star bus for workgroup size networks
Access method	Token passing in numerical order
Data transmission rate	2.5 Mbps; up to 20 Mbps with ARCNet +
Standard cable	93-ohm, RG-62 A/U coaxial or 75-ohm, RG-59 coaxial
Maximum coaxial cable length with star topology	610 meters with BNC connectors and active hubs
Maximum coaxial cable length with bus topology	305 meters
Maximum UTP cable length	244 meters with RJ-11 or RJ-45 connectors on star and bus topologies

 Note: ARCNet supports STP and fiber-optic cable.

Fiber Distributed Data Interface (FDDI) Architecture

FDDI architecture describes networks using a 100 Mbps data rate in a token-passing ring network with fiber-optic cable media. FDDI architecture is used when high-end computers need greater bandwidth than 10-Mbps Ethernet or 16 Mbps token-ring networks can provide. You can use the FDDI architecture in the following network scenarios:

■   Metropolitan Area Networks (MANs) connecting networks in the same city with a high-speed fiber-optic connection up to 100 kilometers (62 miles) apart

- Networks in high-end environments requiring connection of components, such as large mainframes and minicomputers
- Backbone networks to which low-capacity LANs connect
- LANs requiring high data rates to support high bandwidth applications like video, CAD, and CAM

The IEEE 802.8 specifications described in Table 2.11 are defined as standard FDDI architecture.

Table 2.11 FDDI Specifications

Network Elements	IEEE 802.8 Specifications for FDDI Architecture
Topology	Token-passing ring network or star ring network
Access method	Shared network access
Data transmission rate	100 Mbps
Network media	Fiber-optic cable
Maximum ring distance	100 Km
Maximum number of ring computers	500

FDDI architecture utilizes a standard token-passing scheme but differs from the definitions rendered in the IEEE 802.5 specification for token-ring LANs. In an FDDI network, a computer in possession of the token transmits as many frames as possible within a predetermined time interval and then releases the token. At that time, there can be many frames circulating around the ring, in contrast to a token-ring network, that only allows one frame to pass at a time. For this reason, FDDI offers higher throughput than a standard token-ring network.

FDDI networks use a dual-ring configuration, as shown in Figure 2.14, and a shared network access method where more than one computer can transmit simultaneously. FDDI can also use a star ring topology with point-to-point hub links.

Traffic in an FDDI network flows in two streams in opposite directions around two counter-rotating paths known as the primary and secondary rings. Traffic normally flows on the primary ring while the secondary ring is for redundancy. If the primary ring fails, the FDDI network automatically configures the secondary ring for data flow in the opposite direction. In this configuration, each ring cannot exceed the 100-kilometer distance specification and a maximum of 500 computers per ring can be supported. In addition, a fiber-optic repeater should be used at least every 2 kilometers.

Figure 2.14 FDDI Dual-Ring Configuration

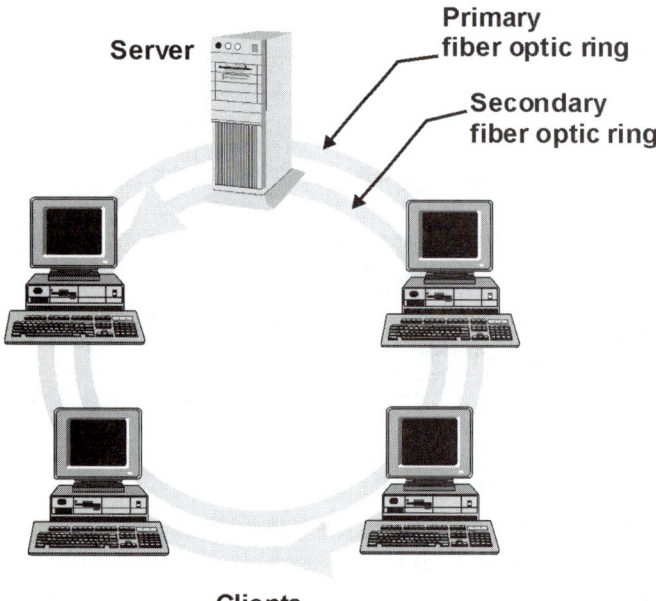

In an FDDI network, computers connect to one or both rings. Those that connect to one ring are known as Class B workstations while those that connect to both rings are Class A workstations. If there is a ring failure, Class A workstations can help configure the network while Class B workstations cannot.

All ring computers monitor the token-passing process. If a ring computer detects an error, it sends beaconing signals out on the network to notify other computers that token passing was interrupted by an error.

Vocabulary

Review the following terms in preparation for the certification exam.

Term	Description
access methods	Characterize the way computers transmit data onto the network media and standards specifications identifying network media configurations and their supporting data transmission rates
Application Layer	Provides general network access, flow control, and error recovery functions for services that directly support user applications such as file transfer, database access, or e-mail communications
communication protocols	Rules defining how data is formatted, transmitted, and recognized, as defined by the OSI model
Data Link Layer	Organizes data (bit stream) into structured frames (logical organized structures) to add address and error control information when sending frames to the Physical Layer
LLC Sublayer	Creates and terminates communication links, control frame traffic, sequence frames, and acknowledge frames
network architecture	The way communication protocols, logical and physical topologies, access methods, and standards combine to create a fully functional network

Network Layer	Provides network addressing schemes and supports internetwork routing of Network Layer data packets
OSI model	A network structure that uses layering techniques to facilitate open data communications
Physical Layer	Controls and terminates the physical connection between communicating computers
Presentation Layer	An interface between user applications and services that handles protocol conversion, data translation, data compression, encryption, and character set modification
protocols	The rules, technical procedures, or discrete systematic steps followed by computers when communicating data across a network
Session Layer	Controls and terminates communication sessions between computers by implementing an interactive dialogue between user applications on those computers
Transport Layer	Provides end-to-end error recovery, flow control, and necessary functions that guarantee reliable packet transmission and reception

In Brief

If you want to...	Then do this...
Regulate bit transmission in a bus network using a token that polls bus computers in numerical order to determine which computer can transmit	Use an 802.4 token passing standard
Use a connector that is compatible with Type 1 and 2 twisted-pair cable	Use a MIC
Allow AppleTalk network protocols to run on Ethernet coaxial cable	Use EtherTalk
Use a token passing access method in a star bus topology with a 2.5 Mbps data rate	Use an ARCNet architecture
Design a network that uses a data rate of 100 Mbps in a token-passing ring network with fiber-optic cable media	Use the FDDI architecture

Lesson 2 Activities

Complete the following activities to prepare for the certification exam.

1.   List the seven layers of the OSI model.

2.   Describe the purpose of the 802 specifications.

3.   List the information that is contained in a data packet header.

4.   List the transport protocols that work at the Transport Layer of the OSI model to enable reliable packet delivery between computers.

5.   Define the term broadcast and describe its function in a network topology.

6.   Describe a Thinnet network's "5-4-3" rule.

7.   List and describe the contents of an ARCNet packet.

8.   Describe the purpose and function of routable protocols.

9.   Define the term "protocol stack" and give examples of the most common protocol stacks in use today.

10.   Describe 10BaseT, and list which type of signal transmission type it uses.

Answers to Lesson 2 Activities

1. The seven layers of the OSI model are the Application Layer, Presentation Layer, Session Layer, Transport Layer, Network Layer, Data Link Layer, and the Physical Layer.

2. The 802 specifications revise the Data Link and Physical Layers of the OSI model and define the way network adapter cards access and transfer data over physical media.

3. A data packet header contains the sending computer's (source) address, the receiving computer's (destination) address, clock synchronization information, and control signals for packet transmission

4. The following transport protocols working at the Transport Layer of the OSI model enable reliable packet delivery between computers:

    ▪ APPC, Advanced Program To Program Communication is IBM's SNA protocol used in peer-to-peer networks and on AS/400® mainframes

    ▪ TCP, Transmission Control Protocol provides guaranteed delivery of sequenced data

    ▪ SPX, the Sequenced Packet Exchange protocol from Novell's IPX/SPX suite ensures message completeness and uses IPX to deliver packets

    ▪ NWLink, Microsoft's implementation of IPX/SPX

    ▪ NetBEUI, Microsoft's NetBIOS Extended User Interface network protocol is a connection-oriented protocol used for establishing communication sessions between subnet computers while providing rapid data transport service

5. In a broadcast topology, data simultaneously sent to all workstations is examined by each workstation to determine if the data is destined for it. If the data is destined for another workstation, it is ignored. Otherwise, the data is accepted by the receiving workstation. With this method, there is no need to pass the data along to the next node.

6.   The Thinnet network "5-4-3" rule specifies that a Thinnet network can combine up to five cable segments, four repeaters, and three populated segments.

7.   An ARCNet packet contains a destination address, source address, and 508 bytes of data.

8.   Routable protocols refer to the ability of a protocol to be passed from one LAN to another, enabling communication between the LANs.

9.   A protocol stack is a grouping of protocols that function within the OSI model layered architecture. Examples of common protocol stacks are Transmission Control Protocol/Internet Protocol (TCP/IP) and Internetwork Packet Exchange/Sequenced Packet Exchange (IPX/SPX).

10.   10BaseT has a 10 Mbps data rate, is a baseband signal transmission type, and uses Twisted pair wiring.

Lesson 2 Quiz

These questions test your knowledge of features, vocabulary, procedures, and syntax.

1. When a user application generates data, which layer of the OSI model is responsible for introducing the creation of packets?
    A.   Application Layer
    B.   Data Link Layer
    C.   Presentation Layer
    D.   Session Layer

2. If the LLC Sublayer is the upper layer of the Data Link Layer, which OSI model layer does the LLC Sublayer interface with?
    A.   Physical Layer
    B.   Data Link Layer
    C.   Network Layer
    D.   Presentation Layer

3. In WANs or internetworks, what type of information is used by routers to determine the best path to a destination?
    A.   Data link
    B.   Packet addressing
    C.   Session
    D.   Protocol

4. Which of the following protocols is considered an application protocol when it works at the OSI model's upper layer, allowing interaction and data exchange between applications?
    A.   DAP
    B.   AppleTalk
    C.   FTAM
    D.   FTP

5. Which of the following protocols is incorporated into packet switching networks and connects remote terminals to mainframe host systems?
    A.   X.25
    B.   TCP
    C.   IPX/SPX
    D.   100BaseT

6.   How many streams can the traffic in an FDDI network flow around two
      counter-rotating paths known as the primary and secondary rings?
      A.   One
      B.   Three
      C.   Four
      D.   Two

7.   Which of the following IEEE 802 specifications is associated with an Ethernet
      network?
      A.   802.4
      B.   802.3
      C.   802.5
      D.   802.12

8.   10Base5 has roughly how many times more cabling distance than 10BaseT?
      A.   5
      B.   2
      C.   10
      D.   25

9.   Which layer of the OSI model breaks data into packets, organizes them in
      sequence, and adds error recovery information?
      A.   Transport Layer
      B.   Data Link Layer
      C.   Network Layer
      D.   Presentation Layer

10.   100BaseVG-AnyLAN combines which of the following network features?
      A.   TCP/IP and IPX/SPX
      B.   Ethernet and token-ring networks
      C.   AppleTalk and IPX
      D.   Bus and Star

Answers to Lesson 2 Quiz

1. Answer A is correct. Packet creation begins when data generated by a user application is introduced at the Application Layer of the OSI model.

    Therefore, answers B, C, and D are incorrect.

2. Answer C is correct. The LLC Sublayer is the upper layer of the Data Link Layer and interfaces the Network Layer.

    Therefore, answers A, B, and D are incorrect.

3. Answer B is correct. In WANs or internetworks, packet-addressing information is used by routers to determine the best path to a destination.

    Therefore, answers A, C, and D are incorrect.

4. Answer A, B, C, and D are correct. AppleTalk is a remote file access protocol from Apple, DAP, a Directory Access Protocol is a DECNet file access protocol, FTAM, File Transfer Access Management, is an OSI file access protocol, and FTP, a File Transfer Protocol, is used over the Internet.

5. Answer A is correct. X.25 is a set of protocols incorporated in packet switching networks and originally used to connect remote terminals to mainframe host systems.

    Therefore, answers B, C, and D are incorrect.

6. Answer D is correct. Traffic in an FDDI network flows in two streams in opposite directions around two counter-rotating paths known as the primary and secondary rings.

    Therefore, answers A, B, and C are incorrect.

7. Answer B is correct. Each network type uses different access methods and

requires specific IEEE protocols working at the MAC Sublayer. Therefore, an Ethernet network uses the IEEE 802.3 protocol.
Answer A is incorrect because a Token bus uses IEEE 802.4

Answer C is incorrect because a token ring uses IEEE 802.5.

Answer D is incorrect because Demand Priority uses IEEE 802.12.

8.   Answer A is correct. 10Base5 has roughly 5 times the cabling distance of 10BaseT (500 meters).

Therefore, answers B, C, and D are incorrect.

9.   Answer A is correct. The Transport Layer breaks data into packets, organizes them in sequence, and adds error recovery information.

Answer B is incorrect because the Data Link Layer organizes the bit stream into frames, add address and error checking information, and prepares data for delivery to the physical transmission media

Answer C is incorrect because the Network Layer adds packet address information, supports packet routings, and manages network traffic problems.

Answer D is incorrect because the Presentation Layer performs conversions, reformatting, and encryption functions for the packet.

10.   Answer B is correct. 100BaseVG-AnyLAN has a data rate of 100 Mbps, uses baseband signal transmission, and voice grade compatibility, and combines features of Ethernet and token-ring networks.

Therefore answers A, C, and D are incorrect.

Lesson 3: Network Operating Systems

A network operating system (NOS) is computer operating system software that includes networking as part of its function. It coordinates all the activities of a network and provides services from a central server to network clients. For example, the network operating system recognizes and responds to client requests for network access, file and print sharing, application use, data storage, and other resources. This lesson describes the features and functions of typical network operating systems.

After completing this lesson, you should have a better understanding of the following topics:

- Stand-Alone Operating Systems

- Network Operating System Functionality

- Network Operating System Installation

- Network Printing

- E-Mail Standards

- Interoperability

- Fault Tolerance

- Novell NetWare

Stand-Alone Operating Systems

Before elaborating on network operating systems, it is important to discuss the background against which they evolved. This begins with stand-alone operating systems, which allow computers not attached to a LAN to operate independently.

A stand-alone operating system interfaces and coordinates between resident application programs, such as a word processor and the computer's hardware. The operating system manages the allocation of resources to applications competing for CPU time, memory access, disk drive storage/retrieval, and other peripherals.

A stand-alone operating system allows the user to easily execute application programs without needing to understand computer hardware on which the application runs. This functionality also extends to programmers who write applications. For example, the operating system enables the programmer to include standard operating system commands or calls in their code. This avoids having to write unique low-level commands covering all possible computer hardware configurations on which their applications run.

 Note: Frequently, applications are specifically written for certain operating systems. This allows vendors to promote their applications, taking full advantage of an operating system's features.

The operating system is sometimes referred to as the "kernel." Applications requesting computer resources interface with the operating system through system calls. The operating system or kernel then interfaces to the various hardware components and their controllers using small programs called device drivers. The operating system can allocate resources to multiple applications and processes because of its ability to handle interrupts. For example, applications generating interrupts to request CPU time or disk access are granted those resources based on the interrupt's priority level.

Network Operating System Functionality

In earlier systems, when a stand-alone computer joined a network, a network operating system application was added to the stand-alone operating system to enable networking. The results were that the networked computer contained both a stand-alone operating system and a network operating system. For example, Microsoft's LAN Manager application was used to add network functionality to MS-DOS, UNIX, and OS/2 operating systems.

With advanced network operating systems such as Windows NT Server, the stand-alone and networking functions are combined into one operating system with provisions for running both network and stand-alone computer functions. The combined network operating system resides on all networked computers and is the basis of all computer activity. This network configuration utilizes the client/server model where a central server processes client requests and client computers share the processing burden. This environment requires compatible, interacting networking software on both the client and server sides, although client computers can still run as stand-alone devices when the networking portion of the operating system is not used.

Client Network Software

The software running on a client computer that enables a user to log on to a server-based LAN for communication and information exchange with other LAN clients is known as client network software. For successful networking, client network software must be compatible with network software running on both the server and other LAN clients. This compatibility is ensured if the same network operating system is installed on the server and all LAN clients. Client network operating systems offer three major categories of functionality, as follows:

- Operating system capabilities
- Peer-to-peer networking capabilities
- Software for communicating with a variety of server network operating systems

Note: LAN clients are sometimes referred to as service
                    requestors and servers are known as service providers.

Network Server Software

Since servers have the job of handling multiple-client service requests, the tasks
associated with a server are far more complex than those of a client computer that
makes a single request for a service. For this reason, the server version of the
network operating system is more complicated, expensive, and larger than the client
version of the same system. Some of the tasks a network operating system can
handle include the ability to support hundreds of users, provide security and
resource access, coordinate network computer and peripheral functions, and enable
distributed application scenarios.

Server network operating systems offer several major categories of functionality, as
follows:

- Operating system capabilities

- Software for communicating with a variety of client operating systems

- Software for providing various network services to clients

- Software for interacting with other network operating systems

Multitasking

A network operating system must have multitasking capability, which means it must
be able to perform multiple simultaneous tasks. By definition, multitasking implies
that a network operating system can run as many tasks as there are processors.

With multitasking, if the number of tasks is greater than the processors available,
the computer performs time slicing and alternately allocates processor time between
tasks until they are complete. Managing the time each processor dedicates to
performing tasks allows the computer to work on several tasks at once. Multitasking
can be classified in two different ways, as follows:

Preemptive

With preemptive multitasking (also known as cooperative multitasking), the operating system takes control of the processor without involving the current task. Preemptive multitasking is most suitable for client/server environments having constant interaction between the client's stand-alone operating system and the server's network operating system. With preemptive multitasking, the CPU can easily be shifted from a local task to a network task.

Non-preemptive

With non-preemptive multitasking, the processor is never taken away from a task. As a result, no other tasks can run until the non-preemptive task gives up the processor.

Client Requests and Redirectors

On a stand-alone computer, when a service request performs a local task, the command is sent on the computer's local bus to the CPU. However, when a service request is initiated by a networked computer to retrieve resources across a network path, it must be directed away from the local computer bus and out to the network. The part of a network operating system that handles this process is called a redirector. In a network, the redirector intercepts service requests made by a client computer and determines whether to forward the request to a network server or let it run on the local computer's bus.

A redirector, which is also referred to as the shell or requestor, can route requests to either computers or peripherals, such as a network printer. Redirectors also register mapped network drive designations so they can provide access to those resources when a service request is made. This allows the network client to be indifferent to the location of data, peripherals, and the complexities of making network connections.

 Note: Redirectors can translate requests to different operating systems. For example, Gateway Services for Netware (GSNW) installed on a Windows NT Server can redirect Windows NT Workstation client requests to resources on a NetWare server.

Resource Sharing

Network operating systems allow clients to share the server's data and peripherals. The network operating system also determines to what degree resources are shared by controlling the level of client access, such as read, write, delete, full access, and so on.

Directory Services

Network operating systems provide directory services to standardize the naming of files in a network. Microsoft Windows NT Server operating system provides a service called Windows NT Directory Services (NTDS) for this purpose. The directories are designed in an upside-down tree formation where the main directory, or root directory, is on top and each subdirectory branches out to other subdirectories. NTDS also allows the network administrator to manage file access and security and the location of network computers.

Shared Network Applications

Network operating systems allow applications to be shared with network clients. This is accomplished by placing an application in a server subdirectory and then sharing out the subdirectory. This is advantageous since it does not require the administrator to purchase separate site licenses for each client using the application, and ensures that all clients are using the same application version.

User Management

Network operating systems allow the administrator to determine who can or cannot use the network. The administrator can use various management programs built into the operating system to do the following:

- Create users on the network

- Grant or take away user rights and permissions

- Remove users on the network

Network Operating System Installation

Network operating systems contain an installation program that automates the installation environment and setup process. This program prompts the installer to provide input that customizes the installation based on the following variables:

- Computing environment

- Network size

- Domain responsibilities and functions, such as an application server, primary domain controller, or backup domain controller

- Server or domain name, to provide unique identification in the network environment

- File system type, such as file allocation table (FAT) or Windows NT File System (NTFS)

- Disk partitioning and identification of network operating system installation location

Server Configurations

During network operating system installation, the server is configured to perform its primary function in the network. For example, the Windows NT Server operating system, depending on its function, can be installed as one of several different types of servers, as follows:

- Primary domain controller (PDC), which maintains a master copy of domain information, including security policies and tracking users.

- Backup domain controller (BDC), which replicates the PDC's domain database, security policy, and logon authentication.

- Application server, such as Microsoft Exchange Server or SQL Server installed with Windows NT Server. As an application server, the Windows NT Server operating system does not act as a domain controller, but is a considered a "member" server.

Installation Configuration

During network operating system installation, several network elements are set up, including the network adapter configuration, TCP/IP configuration, TCP/IP utility installation, and other network services, such as name resolution service.

Setup prompts the installer to choose a network adapter type, a driver, and the protocols to be binded to it, such as TCP/IP, NetBEUI, DLC, and IPX/SPX. In addition, the TCP/IP installation on the server must be configured with an IP address, subnet mask, and default gateway.

 Note: During network operating system installation, it is necessary to verify that all computer hardware in use is compatible with the operating system. Most operating system vendors have a Web-based hardware compatibility list (HCL) for this purpose.

Dynamic Host Configuration Protocol Service (DHCP)

Client computers must also use the TCP/IP configuration. You can accomplish this by either manually configuring TCP/IP on each client computer or setting up the DHCP service. When DHCP is set up on the server, clients can receive their TCP/IP configuration dynamically, including the IP address, default gateway, and subnet mask address. This greatly simplifies installation tasks since manually configuring TCP/IP on many client computers can be complex and time consuming.

Once installed and configured, TCP/IP provides the following capabilities:

- A standard, routable, enterprise network protocol for Windows NT Server

- A single architecture for heterogeneous environments

- Internet access

TCP/IP Utilities

The TCP/IP protocol suite includes utilities, which can be used for diagnostics and informational purposes. These include the following:

- Address Resolution Protocol (ARP) displays a cache of local IP to MAC address resolutions

- IPCONFIG displays and verifies the current IP configuration of the local computer

- NETSTAT displays TCP/IP protocol statistics, current connections and all other protocol statistics

- NBSTAT displays NetBIOS name resolution connections that use NetBIOS over TCP/IP

- Packet Internet Groper (PING) tests connections with echo signals

- ROUTE displays or modifies the local routing table

- TRACERT checks the route to a remote system

Name Resolution Service

Identifying computers on a network can be done with a numerical address method, but computer names are more easily remembered. To support the use of computer names, a method of resolving computer names to IP addresses is required. To serve this function, network operating systems have a name resolution service. During setup, the server registers its name and IP address that uniquely distinguishes it on the network with the name resolution service. At startup, all network computers enable resource resolutions in a routed TCP/IP environment.

Drivers

A driver is a small program installed during setup that tells the network operating system how to work with a particular device so it performs its function. Some hardware device categories that require drivers are as follows:

Input devices—Devices such as a keyboard or mouse.

SCSI disk controller—A Small Computer System Interface (SCSI) controller is a parallel interface that connects multiple devices, such as printers and hard disks, in a daisy chain fashion. SCSI controllers require a SCSI host adapter card with drivers in order to function.

IDE disk controllers—An Integrated Device Electronics (IDE) disk controller requires a driver but since the control electronics are onboard, a separate adapter card is not required.

Disk storage media—Hard disk drives, floppy drives, and tape drives all require drivers to function.

Multimedia devices—Devices such as cameras, recorders, and microphones.

Printers—To enable network printing, the proper printer drivers must be installed.

Modems—To enable remote communication across networks such as the Internet, a modem and modem driver must be installed and configured.

Network adapter cards—Network drivers provide a communication interface between a network adapter card and the network redirector running in the computer. The redirector is part of the networking software that accepts input/output requests for remote files and redirects them over the network to another computer.

Network Adapter Card Drivers and the OSI Model

Network adapter card drivers reside in the media access control (MAC) sublayer of the OSI model's Data Link Layer. The network adapter driver ensures direct communication between the computer and the network adapter card, which in turn, links the computer to the rest of the network, as shown in Figure 3.1. Network adapter card drivers are often supplied to network operating system vendors so they appear during operating system installation. For example, Windows NT Server supplies over 100 drivers from various network adapter card manufacturers to select from during installation.

Figure 3.1 Network Adapter Interfaces

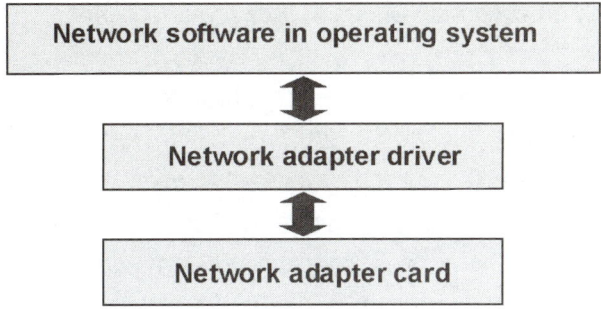

Network Printing

Network operating systems have provisions to support network printing. When you want to print to a shared network printer, the request is sent to a print server that feeds data to a shared printer. The redirector is engaged in this process because it must direct each print job away from the computer's local printer port and out to the network print server. The print server's software takes the print job off the cable using its Simultaneous Peripheral Operation On Line (SPOOL) feature and then sends it to the printer queue.

 Note: The spooler is a memory buffer in the print server's RAM where print job data is stored until ready to print. By using RAM, the print job can unload faster than if hard disk storage is used. If the spooler overflows with print jobs, only then is the print server's hard disk used to store the data.

Printer Configuration

Printers can be installed and configured at the time of network operating system installation. Connecting a printer to a network print server does not make the printer available to network clients. To enable client access to the printer, it must have a unique network identification name that lets the network operating system know it is available to network computers. To set up a shared printer configuration, the following specifications are necessary:

- Printer drivers must be installed on each network computer requiring print access.

- A share name must be created to allow network computers to recognize and access it. For example, the share name could be a NetBIOS name.

- The output destination must be identified so the redirector knows where to send the print job.

- Default output parameters are set so the network operating system knows how to handle and format the print job.

Printer Connection

After the printer is shared, the network operating system is used to connect to it. This requires the client requesting print services to supply two pieces of information—the print server name and the printer name. The printer name used is the one configured during the sharing process.

Network clients must be given the privilege to connect with a shared network printer to process their print jobs. The network administrator can grant this privilege along with others, including upgrading print job priorities and deleting print jobs.

E-Mail Standards

Since the International Standards Organization (ISO) standardized e-mail to be handled at the OSI model's Application Layer, different network operating systems can communicate electronic messages regardless of their differences. The common e-mail protocols or services that facilitate these standards are as follows:

X.400—Designed to be hardware and software independent. Features include different message priority levels, time/date stamping, delivery receipts, and multiple recipients. X.400 protocols were developed to provide the following:

- Routing information including message identifiers, display rules, destination address descriptions, and delivery/return receipt instructions

- Authorized user identification, recipient notification, and message subject indication
- Modification of routing and delivery parameters including passwords, message size, and testing to see if messages can be delivered

X.500—A set of directory services to help users in large distributed networks to locate and send messages to users on other remote networks. X.500 supplies a global directory of e-mail users with a hierarchical directory structure that employs agents to search for users and resources. X.500 uses the following to locate resources:

- Name services, to locate a particular network name

- Address books, to identify a particular network address

- Directory services, containing centrally managed network names and addresses for searching on an inter-network basis

SMTP—Part of the TCP/IP protocol stack designed for message transfer between remote network computers on the Internet. Typically, SMTP is used to send e-mail and POP3 is used for retrieving e-mail. SMTP makes it possible to review message contents, forward messages, send messages to groups, and print messages. SMTP also supports an address book feature.

When computers are communicating with SMTP, the following handshaking control signals are generated:

- Connection verification and other transmission parameters

- Message transmission

- Sender identification

Novell Message Handler Service (MHS)—A standard similar to X.400 in that one computer on the network, the MHS server, translates messages between computers using different e-mail systems.

Gateways—When communicating between different e-mail systems, a gateway can be used to translate incoming messages into a format understandable to the receiving system. For example, both Microsoft Exchange and Microsoft Mail contain gateways for this purpose.

GroupWare services—Provide communication and work coordination using network communications technology to allow real-time document management from a centralized location. Allows groups to have simultaneous access to information and each other. GroupWare is used for the following:

- Routing and sharing information

- Project document development and coordination

- Tracking projects and managing group processes

- Group discussion facilitation

- Tracking customer inquiries and managing customer relations

 Note: Some popular groupware products include the Microsoft Exchange/Outlook platform, Lotus Notes, and Novell GroupWise.

Interoperability

Many networks are comprised of multi-vendor products that mix different operating systems and redirectors. For example, a network might consist of a server running the Windows NT Server operating system and various clients running Windows 95, Novell NetWare, and Apple Macintosh systems. For proper network operation, the server operating system, client operating system, and redirectors must all be compatible. To facilitate compatible network operations, one of two different solutions can be employed—one on the client side and the other on the server.

Client Side Solution

When multiple operating systems are involved, the redirector is the key to interoperability. To facilitate communication between two different operating systems, a redirector that can understand the protocols of a certain operating system must be implemented. For example, you can place a redirector on a particular client computer operating system to access a server using another operating system, as shown in Figure 3.2.

Figure 3.2 Implementing a Redirector for Interoperability

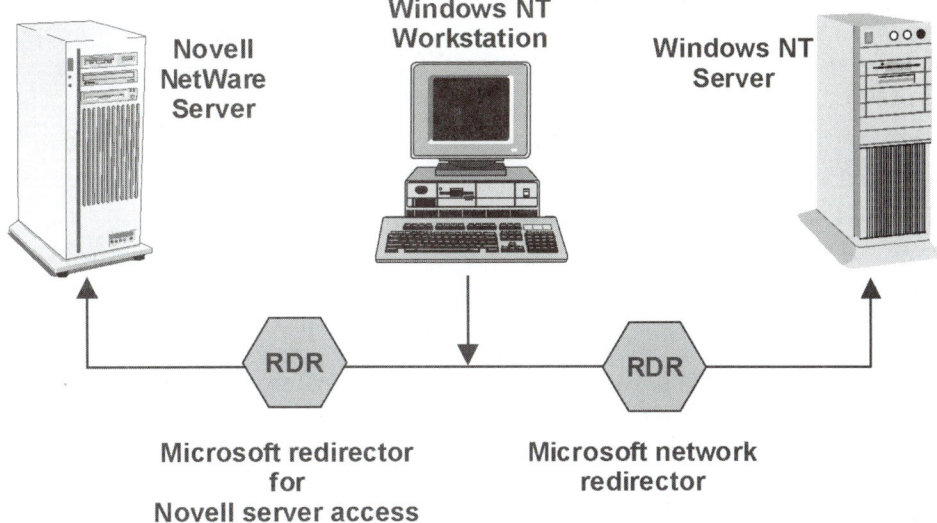

In Figure 3.2, a Microsoft redirector for accessing a Novell server is installed on the Windows NT Workstation. Since this redirector only handles packets sent in a language or protocol it understands, it directs client requests to the correct destination on the Novell Server. Client requests destined for the Windows NT Server are appropriately sent by the Microsoft network redirector.

When implementing a client side solution, you should first determine whether redirectors are available, that run on your network computers and can communicate with another network operating system.

Server Side Solution

In a multi-vendor operating system environment, a server side solution can be implemented to allow communication with dissimilar client operating systems. For example, when introducing a Macintosh computer into a PC environment, you can install Microsoft Services for Macintosh on the Windows NT Server. With this service installed, it makes the server appear to be running the operating system of the client. Macintosh clients can then use their familiar interfaces to access resources on the Windows NT Server. The service converts files so that Windows and Macintosh clients can use their own interfaces to share the same files.

Major Network Operating System Vendors

The three major network operating system vendors are Microsoft, Novell, and Apple. To facilitate interoperability between each operating system, each vendor provides utilities that do the following:

- Allow its client operating system to communicate with servers from the other two vendors
- Allow its servers to recognize clients from the other two vendors

Microsoft Redirector Drivers

Drivers for redirectors are built into Windows NT, Windows 98, and Windows for Workgroups operating systems. During operating system installation, setup loads the required drivers and edits the startup files so the redirector functions when a client computer starts up.

Microsoft and Novell Interoperability

For Microsoft Windows NT Server operating system and Novell NetWare interoperability, the following services are required.

NWLink and Client Services for NetWare (CSNW)—Needed to connect a Windows NT Workstation client to a Novell NetWare server.

 Note: NWLink is Microsoft's implementation of Novell's IPS/SPX protocol. CSNW is Microsoft's implementation of a NetWare requester (the Novell equivalent term for a redirector). Together these provide connectivity between Windows NT and Novell NetWare operating systems.

NWLink and Gateway Services for NetWare (GSNW)—Required to connect a Windows NT Server to a Novell NetWare server network.

---

the exception of partition locations. A duplexed mirror set has very high data reliability since the entire input/output subsystem is duplicated.

Figure 3.4 Disk Duplexing Configurations with RAID 1

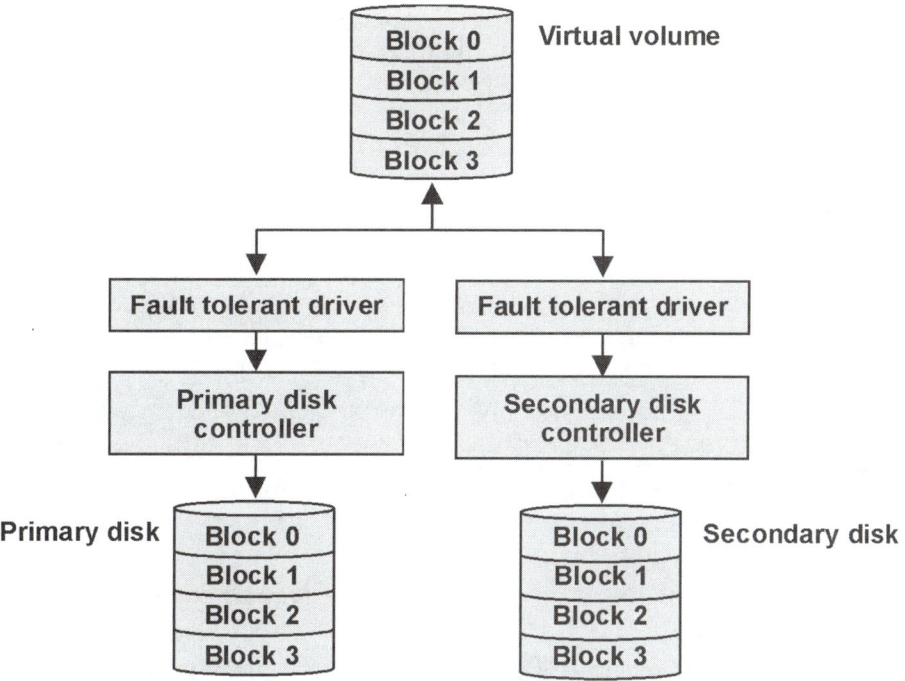

RAID 5—Disk Striping With Parity

Disk striping with parity is currently the most popular approach to fault tolerance. It writes and distributes parity information across all disks in the array while supporting between 3 and 32 disk drives. A stripe set with parity dedicates the equivalent of a strip of disk space for the parity information. The data and parity information are arranged on the volume so they are always on different disks, as shown in Table 3.2.

Table 3.2 Disk Striping with Parity Configuration

	Disk 1	Disk 2	Disk 3	Disk 4
Stripe 1	Parity 1	(Strip) 1	2	3
Stripe 2	4	Parity 2	5	6
Stripe 3	7	8	Parity 3	9
Stripe 4	10	11	12	Parity 4

The first strip on disk 1 is the parity strip for the three data strips included in stripe 1. The second strip on disk 2 is the parity strip for the three data strips on stripe 2, and so on. The parity strip is the exclusive OR (XOR) of all the data values for the data strips in the stripe. Parity is used to reconstruct the data of a failed physical disk from the existing information stored in the parity strips.

Disk Administration

Most network operating systems provide a utility that allows for administration of the hard disk system. A utility such as Microsoft's Disk Administrator can be used to implement fault tolerance and disk partitioning with the following configuration:

- Stripe sets with parity, accumulate multiple disk areas into one large partition, distribute data storage across all drives simultaneously, and add fault tolerant parity information

- Mirror sets, create a duplicate of one partition and place it on a separate physical disk

- Volume sets, accumulate multiple disk areas into one large partition and fill disk areas in sequence

- Stripe sets, accumulate multiple disk areas into one large partition and distribute data storage across all drives simultaneously

Novell NetWare

Novell NetWare is a group of network operating system products that allow file and system resource sharing for all devices on a network. Novell NetWare can be implemented on many platforms, including PCs, Macintosh, AS/400s, and UNIX.

The following section describes some of the protocols and services used on the Novell NetWare operating system.

Open Data-Link Interface (ODI)

ODI allows using multiple protocols on one network adapter card and conversely, lets you set up one protocol to use on multiple network adapter cards. With ODI, Novell NetWare drivers can be written without concern for the protocols running over them. ODI consists of the following three parts:

Multiple Protocol Interface (MPI)

MPI provides connectivity to the OSI network layer where data is addressed and the data route is determined.

Link Support Layer (LSL)

LSL coordinates numbers assigned to the Multiple-Link Interface Drivers that transfer data packets to the network. LSL also manages protocol stack assignments, which are the layers of software that define the protocols for interaction among network computers.

Multiple-Link Interface Driver (MLID)

MLID passes data to and from the network.

Internet Package Exchange/Sequenced Packet Exchange (IPX/SPX)

IPX is a connection-oriented network protocol originally created by Novell to transfer data over compatible networks. SPX is the connection-oriented transport protocol that guarantees delivery of data packets and works together with IPX. When data is properly received, the destination device sends confirmation. If data is not properly received, SPX resends the data. If the data transfer fails a second time, an error is logged.

NetWare Core Protocol (NCP)

Controls client and server operations by defining interactions between them. The file server operating system uses NCP to respond to requests. The NCP packet contains request type, sequencing, and connection information as well as task and function information.

 Note: NCP is the shell used by workstations in the NetWare environment.

Service Advertising Protocol (SAP)

SAP enables servers to broadcast their available services across the network. For example, file and print server information is broadcast on the network every 60 seconds to notify network devices of file and print server availability. In addition, SAP broadcasts are sent to the network to synchronize software routers. The responses received are used to update the router's Server Information Table (SIT), which keeps track of network device information availability.

Error Protocol

In the event that a node is unreachable or an error occurs in packet delivery, Error Protocol returns an error packet to the requesting host. This packet describes the error condition and status of the target host.

ECHO

EHCO is a protocol that checks the validity of a path to a destination. ECHO verifies that a path is functional and that the destination is available before sending a message.

Network Basic Input/Output System (NetBIOS)

NetBIOS is an Application Programming Interface (API) that allows communication between the network adapter card and applications. In NetWare, NetBIOS is implemented on top of IPX, much like SPX.

Novell Directory Services (NDS)

Novell's specific standards for network file design and management. NDS offers a catalog-type format so that information in directories is easily available for sorting and searches. NDS lets you access the network with your logon ID at any workstation on the network.

UNIX®

UNIX is a network operating system that supports multiple users and multitasking. The UNIX network operating system can be installed on more different types of computers than any other operating system. In the early 1970s, AT&T developed UNIX to run on their network system, which included TCP/IP protocols. Since that time, Novell acquired UNIX. The following are several types of UNIX systems:

Advanced Interactive Executive (AIX)—A UNIX system produced by IBM to work on their RISC-based computers.

A/UX—A UNIX system enhanced by Apple Computer to run on Macintosh computers.

Linux—A UNIX system developed by Linus Torvalds of Sweden and distributed with Linux-compatible products.

Vocabulary

Review the following terms in preparation for the certification exam.

Term	Description
ARP	Address resolution protocol that displays a cache of local IP to MAC address resolutions
BDC	Backup domain controller that replicates the PDC's domain database, security policy, and logon authentication
disk mirroring	Duplicates identical data from one physical disk to another to prevent data loss
fault tolerance	Data that is replicated to a physically separate and redundant source such as a different partition or hard disk to guard against data loss
groupware	Allows groups to have simultaneous access to information and each other
HCL	Hardware compatibility list maintained by Microsoft that lists approved operating system
IPX/SPX	IPX is a connection-oriented network protocol created by Novell to transfer data over compatible networks, and SPX is the connection-oriented transport protocol that guarantees data packet delivery of data
NCP	NetWare Core Protocol that controls client and server operations by defining interactions between them
NETSTAT	As a TCP/IP utility, NETSTAT displays TCP/IP protocol (and other protocol) statistics and current connections

network operating system	Coordinates all activities of a network and provides services from a central server to network clients
PDC	Primary domain controller that maintains a master copy of domain information, including security policies and tracking users
PING	Packet Internet Groper verifies configurations and tests connections with echo signals
preemptive multitasking	A function whereby the operating system takes control of the processor without cooperation from the current task
redirector	Routes requests to either computers or peripherals such as a network printer
SAP	Service Advertising Protocol that enables servers to broadcast their available services across the network
SCSI controller	Small Computer System Interface controller is a parallel interface that connects multiple devices, such as printers and hard disks, in a daisy chain fashion
SMTP	Part of the TCP/IP protocol stack designed for message transfer between remote network computers on the Internet
stand-alone operating system	Interfaces and communicates between resident application programs such as a word processor and the computer's hardware

In Brief

If you want to...	Then do this...
Register mapped network drive designations so they can provide access to those resources when a service request is made	Use a redirector
Display the current IP configuration of a local computer	Use the IPCONFIG utility with Windows NT systems, and on Windows 95 or 98 systems, use WINIPCFG.
Maintain a master copy of domain information, including security policies and tracking users	Use a PDC
Provide a communication interface between a network adapter card and the network redirector running in the computer	Install a network adapter card driver
Allow clients to share the server's data and peripherals	Implement a network operating system so users can exchange data across the network
Connect a Windows NT Workstation client to a Novell NetWare server	Use NWLink and Client Services for NetWare
Allow servers to broadcast their available services across the network	Use the Service Advertising Protocol (SAP)
Reconstruct the data from a failed physical disk to the existing information stored in the parity strips	Use fault tolerance with parity
Provide fault tolerance for disk and controller failure and reduce channel traffic	Implement duplexing

Lesson 3 Activities

Complete the following activities to better prepare you for the certification exam.

1.   List some of the popular groupware products available on the market today.

2.   List the three major network operating system vendors.

3.   Explain the purpose and function of client network software.

4.   List the three major categories of functionality for client network operating systems.

5.   Describe the purpose and function of directory services.

6.   List the three most common e-mail protocols or services in existence today.

7.   List the three parts that define the ODI, and then explain the purpose of each.

8.   Explain the purpose and function of stripe sets with parity.

9.   Describe the purpose and functionality of an error protocol.

10. Explain what occurs when DHCP is set up on a server.

Answers to Lesson 3 Activities

1.   Some popular groupware products include the Microsoft Exchange/Outlook platform, Lotus Notes, and Novell GroupWise.

2.   The three major network operating system vendors are Microsoft, Novell, and Apple.

3.   The software running on a client computer that enables a user to log on to a server-based LAN for communication and information exchange with other LAN clients is known as client network software.

4.   Client network operating systems offer the following three major categories of functionality: operating system capabilities, peer-to-peer networking capabilities, and software for communicating with a variety of server network operating systems.

5.   Network operating systems provide directory services to standardize the naming of files in a network.

6.   The most popular e-mail protocols or services are SMTP, X.400, and X.500.

7.   ODI consists of the following three parts:

     ■   Multiple Protocol Interface (MPI) provides connectivity to the OSI network layer where data is addressed and the data route is determined.

     ■   Link Support Layer (LSL), coordinates numbers assigned to the Multiple-Link Interface Drivers that transfer data packets to the network. LSL also manages protocol stack assignments, which are the layers of software that define the protocols for interaction among network computers.

     ■   Multiple-Link Interface Driver (MLID) passes data to and from the network.

8.   Stripe sets with parity accumulate multiple disk areas into one large

partition, distribute data storage across all drives simultaneously, and add fault tolerant parity information

9. In the event that a node is unreachable or an error occurs in packet delivery, Error Protocol returns an error packet to the requesting host. This packet describes the error condition and status of the target host.

10. When DHCP is set up on the server, clients can receive their TCP/IP configuration dynamically, including the IP address, default gateway, and subnet mask address.

Lesson 3 Quiz

These questions test your knowledge of features, vocabulary, procedures, and syntax.

1.   What is the primary goal of a stand-alone operating system?

      A.   To use MS/DOS
      B.   To access data stored on a UNIX server
      C.   To allow the user access to either Netscape or Explorer browsers
      D.   To allow the user to execute application programs without understanding the details of the computer hardware on which the application runs

2.   What are the two methods for classifying multitasking?

      A.   Preemptive
      B.   Non-preemptive
      C.   Servers and clients
      D.   Remote dial-up and stand-alone network operating systems

3.   What is another name for a redirector?

      A.   Token ring
      B.   Requestor
      C.   Router
      D.   Shell

4.   Which of the following network operating system components must be compatible with each other when you are trying to build a dependable network operation?

      A.   Server operating system
      B.   Client operating system
      C.   Redirectors
      D.   Hubs

5.   Which RAID number consists of stripe sets composed of strips of equal
     size on each disk in a volume, with 2 to 32 disks in the configuration?

     A.   0
     B.   5
     C.   1
     D.   7

6.   Which of the following best describes the functionality of NBSTAT?

     A.   Displays a cache of local IP to MAC address resolutions
     B.   Displays the current IP configuration of the local computer
     C.   Displays TCP/IP protocol statistics and current connections
     D.   Displays NetBEUI protocol statistics and connections that use
          NetBIOS over TCP/IP
     E.   Verifies configurations and tests connections

7.   Which e-mail protocol is part of the TCP/IP protocol stack designed for
     message transfer between remote network computers on the Internet?

     A.   X.400
     B.   SMTP
     C.   X.500
     D.   AOL.com

8.   When communicating between different e-mail systems, what can you use
     to translate incoming messages into a format understandable to the
     receiving system, such as Microsoft Exchange and Microsoft Mail?

     A.   Gateway
     B.   Router
     C.   Bridge
     D.   Hub

9.   Which of the following allows communication between the network
     adapter card and applications?

     A.   NetBIOS
     B.   NetBEUI

C.   IPX/SPX
D.   TCP/IP

10.   Which of the following devices requires an installed driver before the
      device can operate on a network? (Choose all that apply.)

      A.   Input device
      B.   SCSI disk controller
      C.   IDE disk controller
      D.   Disk storage media

Answers to Lesson 3 Quiz

1.   Answer D is correct. The primary goal of a stand-alone operating system is to allow the user to easily execute application programs without needing to understand the details of the computer hardware on which the application runs.

     Therefore, answers A, B, and C are incorrect, because they are bogus.

2.   Answers A and B are correct. The two classifications of multitasking are preemptive and non-preemptive. With preemptive multitasking, the operating system takes control of the processor without cooperation from the current task. With non-preemptive multitasking, the processor is never taken from a task and therefore no other tasks can run until the non-preemptive task gives up the processor.

     Therefore, answers C and D are incorrect.

3.   Answers B and D are correct. A redirector is also referred to as the shell or requestor.

     Therefore, answers A and C are incorrect.

4.   Answers A, B, and C correct. For proper network operation, the server operating system, client operating system, and redirectors must all be compatible.

     Therefore, answer D is incorrect.

5.   Answer A is correct. RAID 0 consists of stripe sets composed of strips of equal size on each disk in a volume, with 2 to 32 disks in the configuration.

     Therefore, answers B, C, and D are incorrect.

6.   Answer D is correct. NBSTAT, displays NetBEUI protocol statistics and connections that use NetBIOS over TCP/IP.

Answer A is incorrect. Address Resolution Protocol (ARP) displays a cache of local IP to MAC address resolutions.

Answer B is incorrect because IPCONFIG displays the current IP configuration of the local computer.

Answer C is incorrect. NETSTAT displays TCP/IP protocol statistics and current connections.

Answer E is incorrect. Packet Internet Groper (PING) verifies configurations and tests connections.

7.   Answer B is correct. SMTP is part of the TCP/IP protocol stack designed for message transfer between remote network computers on the Internet.

Therefore, answers A, B, and D are incorrect.

8.   Answer A is correct. When communicating between different e-mail systems, you can use a gateway to translate incoming messages into a format understandable to the receiving system, such as Microsoft Exchange and Microsoft Mail.

Therefore, answers B, C, and D are incorrect.

9.   Answer A is correct. Network Basic Input/Output System (NetBIOS) is an Application Programming Interface (API) that allows communication between the network adapter card and applications.

Therefore, answers B, C, and D are incorrect.

10.   Answers A, B, C, and D are correct. Each device listed requires an installed driver before the device will work: input devices, SCSI disk controller, IDE disk controller, and disk storage media.

Lesson 4: TCP/IP Fundamentals

TCP/IP stands for Transmission Control Protocol/Internet Protocol and is a set of instructions that governs communications between all computers using the Internet. TCP/IP dictates how small portions of data which carry the addresses of its sender and receiver, called packets, are sent across multiple networks. TCP/IP is the main default standard for worldwide network communication because it is a high quality, vendor-independent platform on the popular and successful Internet, the largest computer network in the world. TCP/IP is the protocol of choice for corporate intranetworks because corporations want to work and share information in the private, controlled environment and then release information to the Internet using the standard TCP/IP protocols.

After completing this lesson, you should have a better understanding of the following topics:

- TCP/IP Model and Network

- TCP/IP Terminology and Functions

- TCP/IP Protocols

- Windows Internet Name Service (WINS) Server

- TCP/IP Addressing

- TCP/IP Configuration

TCP/IP Model and Network

The set of TCP/IP protocols is also known as the TCP/IP stack, protocol stack, or the TCP/IP protocol suite. TCP/IP is the set of protocols that work together on different levels, or layers, to enable communication among networks. TCP/IP, the protocol suite on the Internet, is the main communication language. The layers of protocols provide specific services or functions for information exchange over a communications network, and pass information from one layer to the next.

The TCP/IP protocols map to a four-layer concept model. Each layer of the model corresponds to one or more layers of the seven-layer Open Systems Interconnection (OSI) model. The following model shows how the TCP/IP suite of protocols maps to conceptual model layers (Figure 4.1).

Figure 4.1 TCP/IP Protocol Architecture

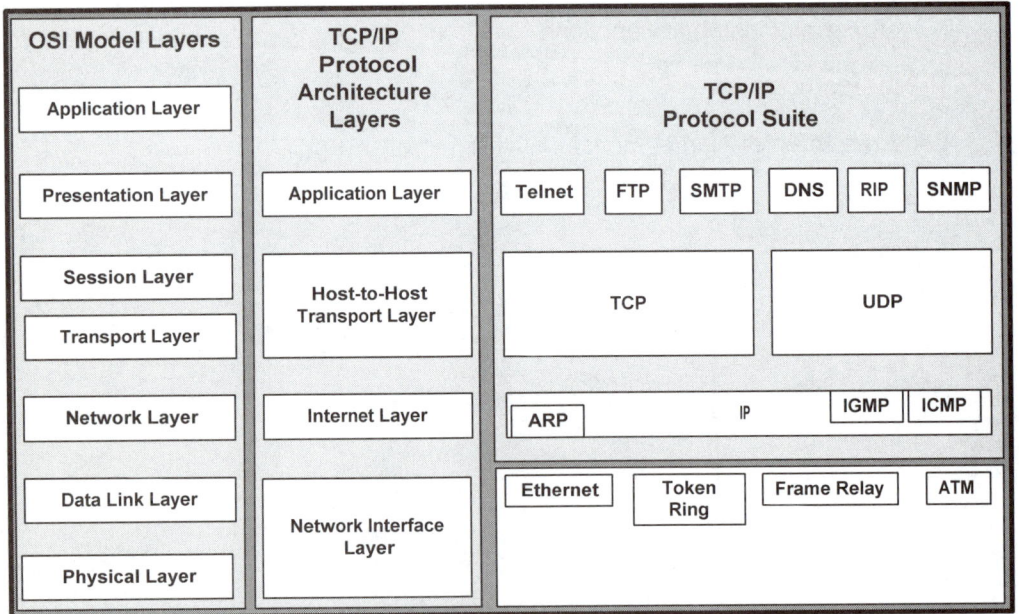

TCP/IP LAN and WAN Network Overviews

TCP/IP is the standard protocol language that allows networks of most product brand types and platforms to communicate. A network is a group of computers and associated devices that are connected by communications facilities, such as TCP/IP, and involves permanent connections, such as cables, or temporary connections made through telephone or other communication links. A Wide Area Network (WAN) is a network spread over a wide geographical area, and relies on communication capabilities to link the various network segments.

A Local Area Network (LAN) is a group of computers and other devices dispersed over a relatively limited area and connected by a communications link that enables any device to interact with any other on the network. LANs commonly include PCs and shared resources such as laser printers and large hard disks. Devices on a LAN are known as nodes, and cables through which messages are transmitted connect the nodes. TCP/IP uses the term hosts to indicate computers on the network.

An intranet is a private network based on the Internet protocols such as TCP/IP, but designed for information management within a company or organization. Its uses include such services as document distribution, software distribution, access to databases, and training.

An intranet looks like a World Wide Web (WWW) site and is based on the same technologies, but is internal to the organization and is not connected to the Internet. Some intranet LANs and WANs also offer access to the World Wide Web Internet, but such connections are directed through a server that protects the intranetwork from the external connections.

Communication with TCP/IP in an intranetwork through LANs and WANS and the Internet is achieved with IP addresses and cross reference or name resolution functions in a network system. You can change this number to mask or hide information from the rest of the Internet, or you can change this information to create more addresses for future subnetworks.

TCP/IP Terminology and Functions

TCP/IP is an industry standard set of protocols that provides communication among varied computer environments. Network protocols are rules that enable two networks or hosts on a network to communicate.

IP is an Internet Protocol that functions within the network layer. The IP is a protocol, but also functions like a network IP model layer. The IP gives addresses to hosts, and handles the routing of data as a single network interface. The IP delivers datagrams or packets through hosts or networks to destination addresses. IP addresses are unique numbered identifiers that specify the original location of data or the destination of data across a communications link.

With TCP/IP and the protocol functions that allow computer address assignment and name resolution service, you can connect your computer and transmit to networks, servers, and computers located anywhere in the world.

The following services and components facilitate the process of mapping computer names and address for communication:

- IP default gateway

- Dynamic Host Configuration Protocol (DHCP)

- Domain Name System (DNS)

- Windows Internet Name Service (WINS)/Server in the Microsoft environment (the NetBIOS name server in other environments)

- Host files

IP Default Gateway

The IP default gateway is the router that recognizes network IDs and forwards addresses or ID packets to other networks.

TCP/IP establishes communication among your hosts and hosts on different networks and sends messages back and forth. The IP determines destinations local or remote, and searches the routing table for a path to the remote host or network. If the address is not found on the local network, the IP routes the packet to the default gateway address.

Dynamic Host Configuration Protocol (DHCP)

DHCP assigns configuration information to devices in a network system, including default gateways, subnet masks, and IP addresses. Each IP address for a computer on the Internet is a unique address. DHCP prevents address conflicts and helps conserve address usage.

Following are the functions of DHCP:

- Automates the assignment and reassignment of IP addresses and related information
- Provides or leases a set period of time for computers to economize IP address usage in the TCP/IP network
- Provides centralized address allocation

Domain Name System (DNS)

DNS is a network of file servers that provides a service on the Internet to translate domain names into IP addresses. DNS is not updated dynamically, but can query dynamically updated databases such as WINS in the Microsoft environment.

DNS is a distributed database management system that maps to the application layer with the TCP and UDP protocols. The DNS looks up the domain name and cross-references or resolves it, to the IP address.

Following are the functions of DNS:

- Correlates alphabetical names to assigned IP addresses
- Manages a network's domain names and IP addresses

The following three components comprise DNS:

- Resolvers
- Name servers
- Domain name space

DNS client computers are called resolvers, and server computers are called name servers. DNS servers on the Internet convert domain names to IP addresses. For

example, www.microsoft.com is a domain name. DNS maps each alphabetical
domain name to its assigned numerical IP address.

The root level domains in Figure 4.2 show the hosts on an inverted tree structure in
a hierarchical group of names, with the root-level domain at the top, countries in
the middle (or top) level, and the local or second-level domain.

Figure 4.2 Root-Level Domains

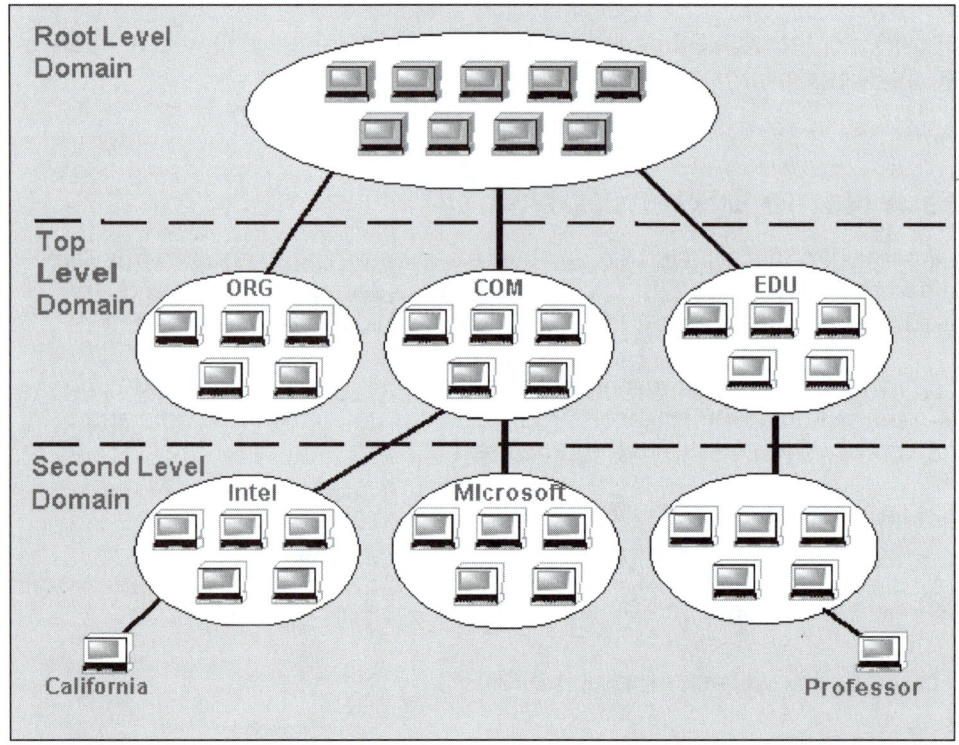

 Tip: You should configure a Windows NT® client to have at least one
server specified under DNS Service Search Order in the DNS
properties dialog box to use for host name to IP address resolution.

Think of the domain name structure as a hierarchical group of names in a tree structure. Domains define the various levels of authority in the domain name scheme of the hierarchical structure. Following are the three levels of domains:

■ Root-level domain—a .(dot) expresses the root level

■ Top-level domain—a two- or three-letter extension, like net, au, com, or edu

■ Second-level domain—a name like Microsoft, IBIDpublishing, or cityu

If you want to establish a Web site on the Internet, you can apply to InterNIC to register your chosen domain name. InterNIC is the organization that assigns and registers Internet Protocol (IP) addresses and correlates them to domain names, which are easy-to-remember address names.

InterNIC is a service operated as a cooperative project between the National Science Foundation, AT&T, and Network Solutions, Inc. (NSI). Since 1993, Network Solutions, Inc. has been the sole provider of domain name registration services for the .com, .net, and .org top-level domains because of a cooperative agreement with the United States Government. In addition, InterNIC provides database information services, such as WHOIS, that allows you to query the records of the currently registered domain names.

For more information, see www.internic.net and www.networksolutions.com.

Since October 1998, the agreement between the U.S. Government and NSI was amended to reflect the commitment to develop a protocol and its associated software to support a system known as the Shared Registration System. Registrars other than NSI will provide registration services for .com, .net, and .org level domains. A non-profit organization, Internet Corporation for Assigned Names and Numbers (ICANN) is responsible for the registration accreditation process for other registration provider companies.

For more information, see www.icann.org.

Table 4.1 describes typical domain types and country designations.

Table 4.1 Common Domain Types

Domain Reference	Description
com	Commercial organization
edu	Education institution
gov	Government organizations
mil	Military organizations
net	Network service providers
org	Organizations (usually non-profit)
au	Australia
fr	France
uk	United Kingdom
us	United States

Windows Internet Name Service (WINS)

WINS resolves names and IP addresses in Microsoft Windows NT, and allows you to register domain names and IP addresses and NetBIOS names into its database. You can use WINS to reduce NetBIOS broadcast traffic problems or storms.

 Note: Broadcast storms are communications sent to more than one recipient that cause multiple hosts to respond at the same time. A broadcast storm may occur as TCP/IP routers are mixed with routers that support newer or different protocols.

The Network Basic Input Output System (NetBIOS) is a protocol that allows applications to communicate over a local area network. It provides a standard application-programming interface (API), and is a set of routines that an application program uses to manage the session.

In WINS, the computer name, or NetBIOS name, should be available so that the computer name and IP are unique to the network in the WINS environment. A NetBIOS name is a unique 16-byte address used to identify a NetBIOS resource on the network. For example, a Windows NT server service registers a unique NetBIOS name based on the computer name when your computer starts up. That exact name is the 15-character computer name plus a 16th character, 20 hexadecimal. Other network services also use the computer name to build NetBIOS names and use a unique 16th character to identify a specific service.

NetBIOS can run on any protocol software that supports the NetBIOS interface. However, UNIX® and Novell® do not support WINS.

The table below describes some valid NetBIOS names you might see in a WINS database (Table 4.2):

Table 4.2 Valid NetBIOS Names in a WINS Database

Registered Name	Description
\\computer_name[00h]	A name registered for the Workstation service on the WINS client
\\computer_name[03h]	A name registered for the Messenger service on the WINS client
\\computer_name[20h]	A name registered for the Server service on the WINS client
\\username[03h]	A name of a person logged on to the computer. This name is registered by the Messenger service so that the person can receive net send commands.
\\domain_name[1Bh]	The domain name registered by the Windows NT Server. Use this name to remote browse. As the WINS server is queried for this name, it returns the IP address of the computer that registered this name.

WINS implements NetBIOS Name Server, and eliminates the need for broadcasts to resolve computer names to IP addresses and provides a dynamic database that maintains mappings of computer name to IP addresses.

In summary, you can use WINS to:

■   Provide name resolution of NetBIOS computer names.

■   Communicate with a WINS server to send resource notification, release its NetBIOS name, or locate a resource.

Host Files

Before the TCP/IP address resolution protocol (ARP) resolves IP addresses to hardware addresses, host names first need to be resolved to IP addresses. You can use two types of host files to resolve names to IP addresses within an organization: HOSTS and LMHOSTS. Both hosts files are static text files on your computer, and are manually kept up to date much like DNS. WINS in the Microsoft environment, however, is dynamically updated, and is able to more quickly browse addresses across network routers and manage name resolutions.

Remote networks use the HOST file. It is a text file table with lines of IP addresses, host names, aliases, and comments. A valid entry in this file is a string of up to 256 characters, with comments placed to the right of a pound sign (#) in both the UNIX and Windows NT environments.

Warning: After you update a HOSTS file, make sure you do not save it with a file extension. The HOSTS file must have a blank file extension, and it must be saved in the correct place for your TCP/IP installation.

The LMHOSTS file on small local Windows NT networks that use NetBIOS over TCP/IP provides the name resolution for local computer names. The LMHOSTS works in much the same way as HOSTS file. The LMHOSTS file relies on broadcasts for name resolution, which can be a problem on larger networks. If you have WINS servers on your network, it is not necessary to use LMHOSTS, since WINS doesn't use broadcasts for name resolution.

Tip: The NetBIOS name and the host name of the computer should be the same.

In summary, the industry standard suite of TCP/IP protocols provides communications in a varied environment. The IP default gateway points to a server to resolve addresses destined for other networks. The IP default gateway, DHCP, DNS, WINS, and host files work together to manage, resolve, and serve addresses.

TCP/IP Protocols

The Transmission Control Protocol/Internet Protocol (TCP/IP) is a communication standard that allows you to connect to the Internet using computers with different operating systems. TCP/IP protocols include methods for terminal emulation, file transfer, packet transportation, application session communications, and e-mail.

Network communication sessions can be formal or informal. A connection-oriented communication session is a more formal communication session among network computers, whereas connectionless communication is more informal. The formal connection-oriented TCP protocol, for example, first establishes the connection among devices and gains agreement on the parameters and protocols for the communication session. Then the computers engage in dialog and transfer data. At the end of the TCP connection-oriented session, the computers release communication in a systemic manner. The informal, connectionless UDP protocol for example, allows a computer to transmit data out to a receiver or receivers, but doesn't allow any two-way establishment of communications with receivers.

Transmission Control Protocol (TCP)

TCP is a connection-oriented communication protocol that adds reliable byte-stream delivery services to the less reliable IP datagram delivery service in a communications session among network computers. A byte stream is data that is transferred continuously from beginning to end in a steady flow. You can send or receive large volumes of data from your applications with TCP. It connects computers with full-duplex transmission. Full-duplex transmission is simultaneous communication in both directions among computer senders and receivers. TCP allows applications to send and receive data streams or control information back and forth at the same time.

In addition, the TCP protocol assigns a sequence number to each data segment. The application checks successive sequence numbers to ensure that segments you receive and process are in order. The receiver then sends an acknowledgment to the sender. TCP is a reliable application because it tracks and checks byte-stream delivery service and allows you to verify your data transmission activity.

Tip: TCP is more reliable than the UDP for packet delivery because it verifies your datagrams' correct order and arrival at the correct destination, while UDP does not verify any information.

User Datagram Protocol (UDP)

UDP is a connectionless protocol. Unlike the TCP, UDP does not send data as a stream of bytes, and it does not require that you establish a connection with another program to exchange information. However, UDP transmits data in discrete units called datagrams similar to IP datagrams. The UDP is much faster than TCP because it does not contain connection and checking applications and can deliver a minimum number of packets in a quick manner. However, UDP is unreliable because you cannot verify the arrival of datagrams at the correct destination.

Warning: UDP is fast but does not sequence packets and therefore could send or receive your data out of order.

Post Office Protocol Version 3 (POP3)

The Post Office Protocol version 3, (POP3) is a mail server protocol that retrieves e-mail from an e-mail server and delivers the e-mail to your local client computer. It does not send mail, and assumes SMTP will provide mail-sending service. POP3 is useful for computers without a permanent network connection, like mobile laptops or home computers, which require a server to hold mail until it is retrieved. POP3 is not compatible with older versions of POP.

The Simple Mail Transfer Protocol (SMTP)

SMTP uses TCP/IP protocols to send mail over the Internet. SMTP delivers e-mail to an e-mail server and transfers mail among mail servers.

 Note: You can use POP3 to retrieve e-mail from an e-mail server and deliver it to your local computer. You can use SMTP to send or relay e-mail to an e-mail server.

Simple Network Management Protocol (SNMP)

SNMP is the standard and primary protocol for managing your network. It collects and manages information and statistics about TCP/IP network devices and forwards it to a network management console. For example, SNMP warns you when a host is running out of hard disk space.

SNMP is an optional part of the TCP/IP protocol suite. SNMP provides the ability to monitor and communicate status information among:

- Computers running Windows NT
- LAN manager services
- Routers or gateways
- Minicomputers or mainframe computers
- Terminal servers
- Network hubs

There are third-party SNMP management tools available from IBM®, Hewlett-Packard®, and Sun®.

File Transfer Protocol (FTP)

FTP transfers files among computer systems. FTP is both a protocol and a utility application. The FTP utility on your computer uses the FTP protocol to send or receive files from one computer to another, regardless of the type of computer system. You can browse lists of files on other systems, and transfer them to your computer with FTP.

Hypertext Transport Protocol (HTTP)

HTTP is not part of the TCP/IP stack, but it uses TCP to make a connection between your HTTP client workstation browser and an Internet Web server. HTTP transfers the contents of Web pages in hypertext markup language (HTML), such as plain text, formatted text, binaries, graphics, sound, animation, or Java-based applications among Web browsers. For example, HTTP transfers small programs written in the Java programming language to add and run objects and multi-media activity to Web pages such as background music, real-time video displays, animations, calculators, and interactive games.

There are HTTP clients (Web browsers) for many uncommon systems, and there are HTTP clients for all common UNIX and desktop operating systems. The Internet has surged in popularity because HTTP clients transferring data across TCP/IP networks via the Web browsers are common. To locate a Web document, use its Uniform Resource Locator (URL). A URL consists of a protocol identifier and a location. The HTTP URL for the Microsoft home Web page is http://www.microsoft.com. If you type this URL in your Web browser, HTTP makes a connection-oriented TCP/IP connection, and transfers data for each object on the page.

HTTP defines the method for transferring hypertext data across TCP/IP networks. This hypertext data is organized into documents with the HyperText Markup Language (HTML). Any HTML document can reference any other HTML document, which makes navigation through related documents easy.

The Internet Protocol (IP)

IP contains the information that allows you to route packets over the network. IP is a connectionless-oriented protocol that sends and receives information on the network layer. It is connectionless because packets are not sent in streams, but are sent independent of one another.

IP packets or datagrams have a header section and data section. The header section contains the IP addresses of the sender and receiver. Routers use the information in the header section to forward and deliver these packets to their destination. Packet delivery is not guaranteed by this service.

 Tip: TCP numerically sequences IP packets or datagrams before sending them over the network, and makes sure they are put in order at the destination.

In summary, Transmission Control Protocol/Internet Protocol (TCP/IP) is a set of rules that standardizes your connection to other network systems that includes the services of terminal emulation, file transfer, packet transportation, session communications, and e-mail.

Windows Internet Name Service Server (WINS)

WINS is a server software system that determines the IP address associated with a particular network computer in the Microsoft environment. WINS supports other platform protocols such Novell's Netware. WINS supports network client and server computers running Windows and can provide name resolution for other computers. A WINS-based server checks to see if a requesting computer name is unique on the network, and will respond with a positive or negative WINS name registration response message.

Before two NetBIOS-based hosts can communicate, the destination NetBIOS name must be resolved to an IP address. You can use the Windows Internet Name Service (WINS) protocol as your resolution service for Windows NT, and resolve NetBIOS names to an IP address. WINS servers handle name registrations and queries, and maintain databases that map the names of WINS clients to their IP addresses. A WINS client requests an IP address and a WINS server retrieves the IP address from its database and routes the resolved name to the clients.

Hierarchies

NetBIOS names are registered and resolved on the network by NetBIOS-to-IP address name resolution methods. The methods computers use to resolve their names are called the node types. The following methods can resolve names in a hierarchical sequence to provide flexible NetBIOS-to-IP name resolution:

- **b-node**—also called the broadcast method, uses broadcasts to resolve names
- **p-node**—also called the name query method, uses point-to-point communications with a name server to resolve names
- **m-node**—first uses the b-node, and upon failure then uses p-node
- **h-node**—a hybrid method, first uses p-node and upon failure, then uses b-node to resolve names

You can configure a hierarchy of name-resolution mechanisms in Windows NT. First, make sure that the default NetBIOS node is set to the h-node. The default node is the b-node. In the h-node, WINS searches for IP addresses by NetBIOS name, and uses a distributed database that is automatically updated with the names of computers and the IP address assigned to each one.

WINS allows you to resolve NetBIOS names to IP addresses by searching the sources in this order:

- Local NetBIOS name cache
- WINS server database
- Broadcast
- LMHOSTS file
- HOSTS file
- DNS system

Once the NetBIOS name is resolved to the IP address destination, the computer can communicate within the network, or to other remote networks.

In summary, TCP/IP and its set of protocols and utilities, allows you worldwide accesses, and the WINS NetBIOS over TCP/IP name resolution hierarchical methods allow local and worldwide access.

TCP/IP Addressing

TCP/IP uses the IP address as a unique identifier to determine the original location of your data or to find the destination of data to be transmitted across a communications link to you. The IP address identifies a system's location on the network with a unique number identifier.

The TCP/IP protocol address resolution protocol (ARP) allows you to find a physical device number or media access control number from a device on the network. In addition, TCP/IP allows the addition of more IP addresses. You can subdivide and overlay or mask your IP address by an addressing process called subnet masking. Moreover, application-to-application communications request the source and destination ports of the programs.

Address Resolution Protocol (ARP)

ARP correlates an IP address with the Media Access Control (MAC) address, a unique hardware address on the network interface card of a device. Use ARP to map an IP address to a physical address on broadcast networks such as Ethernet®, Token Ring®, or ARCnet®. ARP creates an entry in its address-resolution cache every time the node broadcasts an ARP request and receives a response. The entry maps the IP address to the physical address.

IP Addresses and Classes

The IP address has two parts—a network ID number and a host ID number. The network ID identifies all the hosts or nodes on the same physical network. Hosts on the same network require an identical network ID, and the ID must be unique to the Internetwork.

In addition, the host ID identifies a workstation, server, router, or other TCP/IP host within a network. The host ID must be unique to the network ID. Each TCP/IP host is identified by a logical IP address.

You can reference IP addresses in two formats—binary and dotted decimal notation. Each IP address is 32 bits and composed of four 8-bit fields, called octets. Each octet translates into a decimal equivalent. The decimal equivalent is more reader-friendly.

Tip: An IP address contains a series of numbers divided into four groups
         by periods or dots, such as 123.121.4.5.

For example, a binary format and its comparable dotted decimal format are shown
below:

Binary Format: 1000001100 01101011 00000011 00011000
Dotted Decimal Format: 131.107.3.24

For more information on how to convert a binary address to the dotted decimal
format, see Valid IP and Subnet Masking at the end of this lesson.

Address Classes

The following three address classifications define network ID bits and the host ID
bits.

- Class A accommodates large networks

- Class B accommodates medium networks

- Class C accommodates small networks

The 32-bit addressing scheme supports a total of 3,720,314,628 hosts (Table 4.3).

Table 4.3 Network Numbers and Sizes

Class	First Number	Maximum Number of Hosts
A	1-126	16,387,064
B	128-191	64,516
C	192-223	254

Class A addresses are assigned to networks with a large number of nodes or hosts.
This allows a capacity of 126 networks and approximately 17 million hosts per
network.

Class B addresses are assigned to medium to large networks. This allows a capacity of approximately 16,384 networks and approximately 5,000 hosts per network.

Class C addresses are used for small Local Area Networks (LANs). This allows for approximately 2 million networks and 254 hosts on each network.

Warning: You cannot use 0 or 255 in the IP address octets because they are reserved. The network ID cannot be 127. This ID is reserved for loopback and diagnostic functions.

Subnet Mask Numbers

The subnet mask shows the number of bits that represent the network ID. The portion that holds the network ID is set to all binary 1s, and the remainder (the host ID portion of the number) is set to 0s. This changes the way your software interprets the binary values of the address. The interpretation of the address is called the subnet mask. All devices on a subnet must support subnet masking.

The scheme of the subnet mask indicates whether the destination is local or remote. In addition, subnet masks allow devices on the net to withhold access of individual workstation IP address numbers to intruders from outside the local network, while it allows data to reach the server. The server then takes the data and transfers it to the local IP address. For more information, see Lesson 10.

Warning: The 127 designation has a special use for loopback tests. It is reserved for the local host computer to test for connections. If you send a message to 127.0.0.1, then the message will come back to you at 127.0.0.0, unless there is something wrong in the network connection.

Default Subnet Masks

The default subnet mask allows splitting the network into multiple subnets. The default subnet mask number can be identified by looking at the leftmost number of the address which is covered or masked by using the dotted decimal number 255, and the rightmost number is viewed with a 0.

You can determine whether or not your network is a subnetwork, or whether it is local or remote. You can find the different classes of networks in Table 4.4 below.

Table 4.4 Network Classes

Class	Decimal Range	Default Subnet Mask	Maximum Hosts Per Network
A	1-126	255.0.0.0	16,777,214
B	128-191	255.255.0.0	65,534
C	192-223	255.255.255.0	254

For example, Table 4.4 above shows IP address 13.1.12.34.9 as a Class A address. The first number 13 is less than 127 and falls into the decimal range between 1 and 127. This mask indicates that the ID with the first three decimal numbers 255.255.255 is local to your network. The same is true if you have the last two decimal numbers assigned, such as x.x.0255.0255.

The default subnet mask column in Table 4.4 indicates that the 255.255.00 subnet mask number is remote to your network. A number with the same first two numbers is located on your local network.

 Note: Class D and E networks are reserved for special
                      purposes.

For more information, see www.microsoft.com or www.globetrotting.com.
In addition to IP addresses and subnet mask addresses, TCP uses port numbers to enable IP packets to be sent to a particular process on a computer connected to the

Internet. A port is an interface through which data is transferred between a computer and other devices, a network, or another computer. A port appears to the processing computer as an address in memory that sends and receives data. Some port numbers, or well-known port numbers are permanently assigned to TCP protocols.

Hypertext Transfer Protocol (HTTP) Port Numbers

HTTP is the protocol used for the Web browser attached to your Web server service. HTTP uses Port 80 of the TCP protocol to make the Internet connection.

File Transfer Protocol (FTP) Port Numbers

The FTP protocol allows you to upload and download files stored on the network. The service monitors Port 21 for incoming messages and requests. Your client software opens a TCP session on Port 21 with an FTP server.
Some common FTP commands include:

- Get
- Put
- Binary
- ASCII

 Note: UNIX hosts also use the FTP protocol.

Simple Mail Transfer Protocol (SMTP) Port Numbers

With SMTP, you can send e-mail messages to another server that differs from your type of server. Different types of e-mail servers can send messages to each other on a TCP/IP network. SMTP service uses Port 25.

In addition, TCP/IP allows you to place many more IP addresses on your network IP address by subdividing your IP address. You can submask your IP addresses to make additional subnetwork addresses. Moreover, application-to-application communications request the source and destination ports of the programs.

In summary, TCP/IP uses the IP address as a unique identifier to determine the original location of sent data or the destination of data to be transmitted across a communications link. The IP address identifies a system's location on the network with a unique number identifier. The TCP/IP protocol Address Resolution Protocol (ARP) allows you to find a physical device number or media access control number from a device on your network.

TCP/IP Configuration

To connect to the Internet with the TCP/IP set of protocols, or stack of protocols, you need to install the TCP/IP protocol with the correct network adapter drivers. Then you can configure the protocols.

To configure TCP/IP you should understand the following topics:

- IP proxy
- Workstation configuration parameters

IP Proxy

The IP proxy, or network address translation, is a network service that queries the Internet from your workstation and translates the address between local area networks (LANs) and the Internet into a proxy address. The proxy replaces its own address in the response with the address of the station for which the response is destined. Then it sends the packet onto the internal network.

The IP proxy allows you to manage the Web connection on a port-per-port basis, and allows secure communication over the Internet. IP proxy does not allow workstation IP addresses to be sent to the Internet.

TCP/IP Workstation Configuration Parameters

To set up TCP/IP on a Windows 95 workstation, follow these steps:

1. From the Start menu, select Settings.

2. Then choose Control panel and then select Network.

3. Choose the TCP/IP protocol and then select Properties.

You can configure TCP/IP with a few optional parameters to add extra TCP/IP functions to a workstation. The configuration parameters for the Host IP address and subnet mask are, DNS information, Default gateway, WINS (optional), DHCP (optional), Host name, and the Internet domain name.

 Tip: To install a computer on an IP network, you need a NIC, an IP address, and a subnet mask.

IP Address and Subnet Mask Workstation Configuration Parameters

To configure TCP/IP, you must set parameters for the IP address and a subnet mask on the TCP/IP stack of the workstation. The IP address on your workstation must be unique for identifying the network. The subnet mask indicates the type of IP address on your workstation.

Use the parameters in TCP/IP protocol configuration specific to your operating system. Reboot the system, and you should be able to use TCP/IP communications. If your network is connected to the Internet, use the block of TCP/IP addresses, a range of dotted decimal numbers, and a subnet mask given to you from the Internet Service Provider (ISP). If your network is not connected to the Internet and you don't plan to connect, use your chosen address scheme. However, the class of address and subnet mask must be the same.

 Tip: To configure your TCP/IP settings, choose Start and select Settings, Control Panel, and Network. Then choose TCP/IP protocol, and select Properties.

DNS Workstation Configuration Parameters

To configure a workstation for DNS, use at least three of the following parameters:

- IP address of the DNS server

- Local host name

- Internet host domain name

The location for configuring these parameters depends on your operating system.

Default Gateway Workstation Configuration Parameters

You can configure the default gateway to send IP packets to other networks. Fill in the default blank parameter with the IP address of the router interface (connection) that connects your workstation's network segment to the Internet.

Again, the location to configure this parameter depends upon which operating system you use.

 Tip: For Windows, set the default gateway information on the DNS property page for TCP/IP Protocols.

IP Proxy Workstation Configuration Parameters

Most IP proxies do not require workstation configuration. Each type of IP proxy server has a unique method of configuration. However, the server is configured with two Network Interface Cards (NICs); one internal NIC and one external NIC to the network for:

- Internet connection IP address

- Internal network address

WINS Workstation Configuration Parameters

A WINS server handles name registrations and queries, and maintains a database that maps the NetBIOS computer names of WINS clients or workstations to their IP addresses.

Name registrations and inquiries are important in matching or resolving NetBIOS names to an IP address. Client workstations without a specified WINS server, try to resolve names with a HOSTS file, broadcasts, an LMHOSTS file, or a DNS server. A WINS client is a PC running a Microsoft enterprise TCP/IP network client software that uses WINS for NetBIOS name resolution. The WINS server makes sure that no other computer uses the same name at the same time. If the WINS server sees there is another computer on the network with the same, it tells the workstation to not use the name.

When a WINS workstation requests an IP address, the WINS server retrieves the IP address from its database and routes it to the workstation. Windows-based networking clients or workstations can use WINS directly, but non-WINS computers that use broadcasts can access WINS through proxies. Proxies are WINS-enabled computers that listen to name-query broadcast messages, forward the request to the WINS server, and then respond for names that are not on the local subnet.

You can configure the workstation to query the WINS server. Enter the primary WINS server IP address. It indicates to the Windows computer the method to resolve NetBIOS computer names over TCP/IP.

To configure WINS on a workstation:

- Enable WINS

- Specify a primary WINS server IP address

To configure WINS server service, you need your IP address for both the workstation and the server. To install WINS server service on a computer running Windows NT server, and to configure it to automatically resolve NetBIOS names to IP addresses for WINS clients or workstations, do the following steps:

1. Log on as Administrator.
2. From the Control Panel, Select Network Program.
3. Choose Services on the Properties page, and select Add.

4.   Choose Windows Internet Name Service, and enter your full path, such as
      \\instructor\ntsrv, and then select continue.
5.   From the Protocols Properties page, Choose TCP/IP Protocol, and Select the
      WINS Address Properties page.
6.   Type your IP address into the Primary WINS Server box and select OK.
7.   From the Network Properties page, select Close.
8.   From the Network Settings Change Properties Page, choose Yes to restart the
      Windows NT server.

The WINS Server service installation performs all WINS management and
configuration for Windows NT with WINS Manager. WINS manager is found in the
Administrative Tools group on computers running Windows NT Server where the
WINS Server service has been installed. WINS Manager gives access to detailed
information about the WINS servers and displays the WINS database.

DHCP Workstation Configuration Parameters

The DHCP server automatically assigns an IP address, a subnet mask, and any other
configured parameters on the DHCP server during installation. DHCP automatic
configuration allows the client workstation to receive a valid IP address rather than
a manually assigned address, and eliminates some difficulties in tracing network
problems.

In most Windows TCP/IP stacks, you can choose an option to automatically obtain
an IP address.

Warning: If you do not opt to automatically obtain an IP
                             address, configure the workstation with a static IP
                             address and subnet mask.

To configure a DHCP server the DHCP server service must be installed on a
computer running a Windows NT Server, and the DHCP server must be configured
with a static (unchanging) IP address, subnet mask, and optionally, a default
gateway. The DHCP scope—a range of IP addresses that the DHCP server can
assign or lease to DHCP clients must also be assigned.

To configure the DHCP server to automatically assign the WINS server address and net BIOS node, enter the name of the DHCP client workstation (Figure 4.3) and proceed with the following steps:

Figure 4.3 DHCP Default Scope Property Page

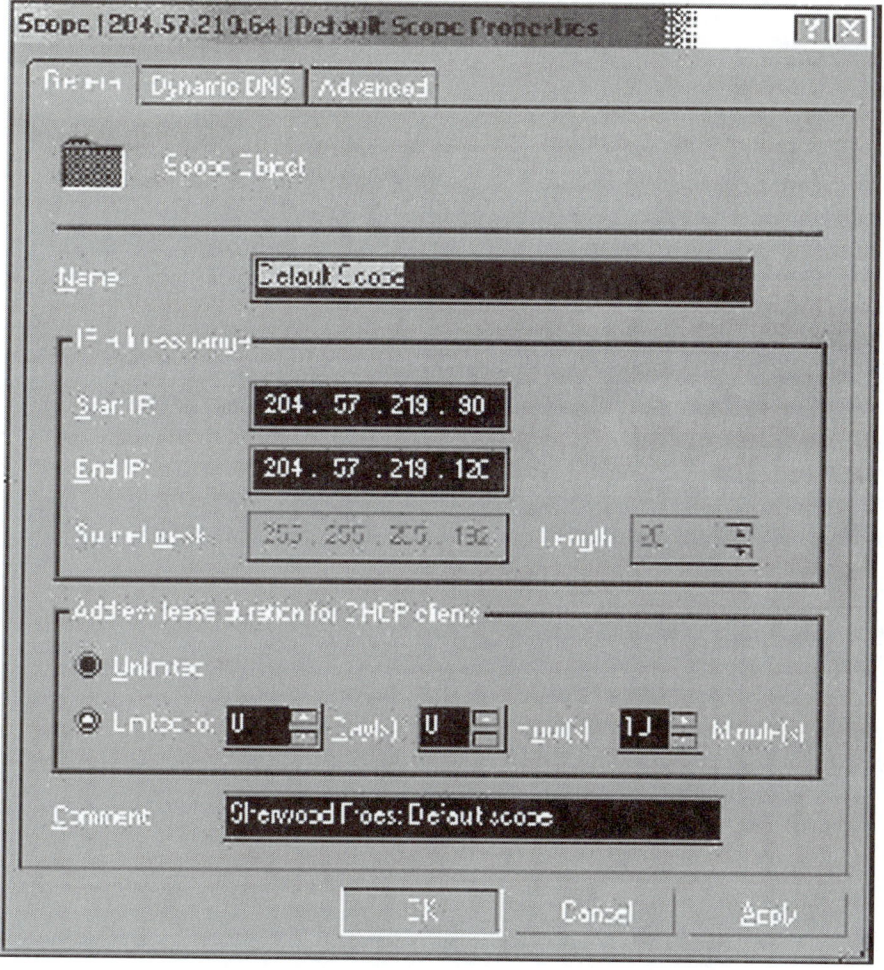

1.   Log on as Administrator.

2.   From the Administrative Tools (Common) menu, choose DHCP Manager.

3.   Choose Local Machine, and select the IP address for the local scope.

4.   Choose the IP Address for the local scope from the Option Configuration Properties page.

5.   From DHCP options, choose Scope, then select Unused Options, select 044 WINS/NBNS Servers, and select Add

6.   From the DHCP Manager properties page, here you must add the option 046 WINS/NBT Node type for WINS to function properly, and then choose OK

7.   From the DHCP Manager properties page with the 044 WINS/NBNS Servers option showing under the Active Options, Choose Value.

8.   From the DHCP Scope: Options Properties page, Choose Edit Array.

9.   From the IP Address Array Editor Properties page, type in your IP address in the New IP Address box, and choose Add, and Choose OK to return to the DHCP Options: Scope Properties page (Figure 4.4).

10.   From the DHCP Options, Scope Properties page, Choose Unused Options, Select 046 WINS/NBT Node Type, and choose Add.

11.   At the Byte box, type 0x8 and choose OK.

12.   From the DHCP Manager, you should see the Option Configuration. The active scope options appear as 003 Router, 044 WINS/NBNS Servers, and 046 WINS/NBT Node Type.

13.   Exit the DHCP Manager.

Figure 4.4 Setting the DHCP Scope

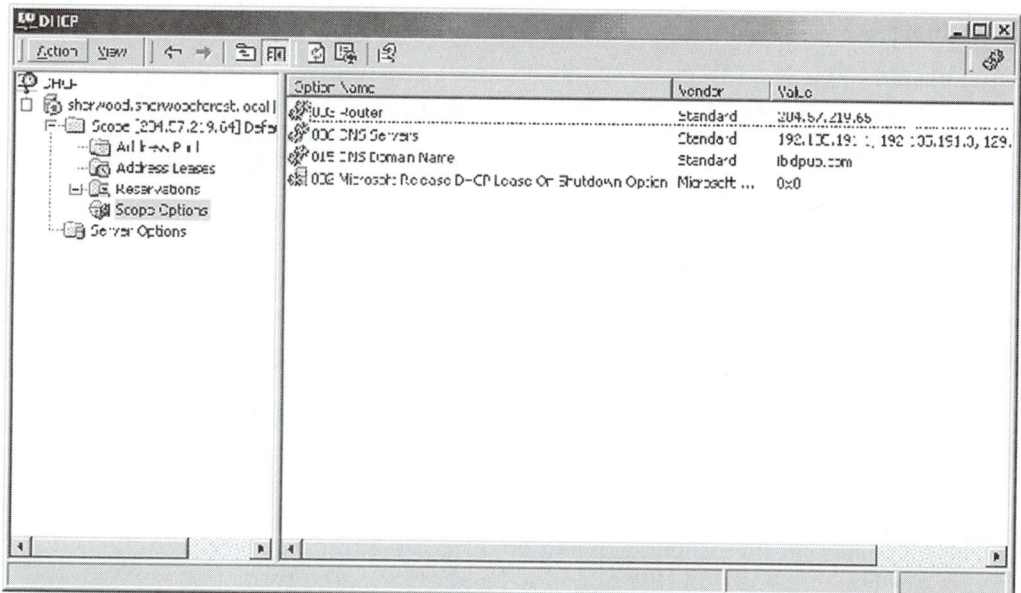

Host Name Workstation Configuration Parameters

To configure the host name of the workstation, use the DNS property page of the TCP/IP properties dialog box, and type in your host name.

There is no default for this parameter. The host name is the DNS name of a workstation. Configure the host name so other workstations on the network can identify a workstation by the DNS name in addition to the IP address.

Internet Domain Name Workstation Configuration Parameters

To configure the host name of the workstation, use the DNS property page associated with the TCP/IP settings, and type in your domain name. There is no default for this parameter, and the field remains blank.

Configure the Internet domain name of the network on which the workstation resides. Obtain the name from your Internet Service Provider (ISP) if you don't have your own domain name. This parameter is optional. The workstation recognizes its Internet domain and can use it in TCP/IP communication.

In summary, the TCP/IP configuration allows you to set your protocols to communicate through an IP proxy to the Internet. The IP proxy changes your IP address and does not allow your address to be distributed anywhere to the Internet. Through configuration management and the use of automatic service functions such as WINS, DHCP, and DNS to assign addresses, you can use TCP/IP protocols to keep your Internet accesses correct and protected.

A Small Office Workstation Scenario

In Figure 4.5, you see a conceptual view of a small business intranetwork network. The diagram shows the connection of two local area networks and the Internet. In this conceptual network, a workstation on one Local Area intranetwork is able to communicate with a workstation on another Local Area Network with an Internet Protocol (IP) address and subnet masking.

To allow the workstations to communicate with servers using TCP/IP, a network administrator has set up, configured, and enabled the intranet server with the appropriate Internet Protocol (IP) addresses and other name resolution services—DNS, WINS, and DHCP.

In the network administrator's plans for the future expansion of this intranet, such as connecting this intranet to the external world wide Internet, the TCP/IP server is configured with a valid Internet address supplied by InterNIC.

The network administrator configures the intranet server databases with a domain name and maps it to the valid (InterNIC registered) IP address in the DNS, WINS, and DHCP databases.

Figure 4.5 TCP/IP in A Small Office Intranetwork

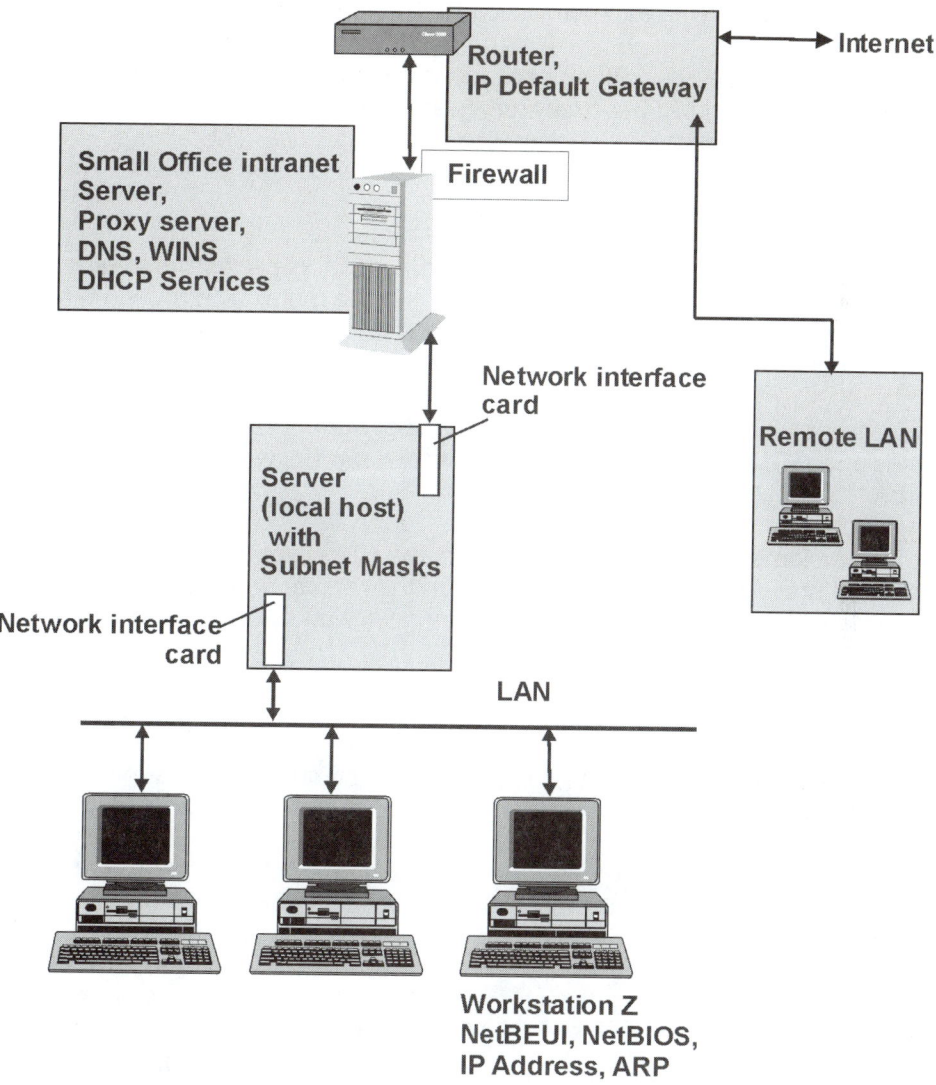

To make further provisions for future LAN expansion, and make LAN communications more efficient, the network administrator divides each LAN into appropriate size subnets.

A default gateway address is specified for each LAN.

To exchange files in this scenario, the workstations rely on the File Transfer Protocol (FTP) client applications and FTP services on a server on the opposite network. This server has LAN service applications enabled like the FTP service. This allows workstations to save and retrieve files from this file server using their client FTP applications.

When workstation Z sends a packet with an IP address that is not on the local area network, the packet is sent to the Default Gateway on the router. The routing table determines that the destination IP address is located on the other network on a file server. The router forwards the packet to that destination IP address.

When the packet arrives at the destination IP address, the desired file server, the router sends a reply packet back to the original workstation Z, with connection confirmation.

Now that the session is established, you can start the FTP client program from the workstation and browse the list of existing files on the remote server, download files you want to your workstation, or upload files from your workstation to the remote server.

Valid IP and Subnet Setup

You can add new networks into an intranet, no matter the size, or to the Internet itself, without interruption to the network service. The additional networks that are part of a larger network are subnets. To design a network to expand with subnets at another time, you can set up IP address numbers in advance. You can add another subnet with almost instant Internet availability because your subnets will conform to a valid format. Only the TCP/IP protocol suite allows for this kind of flexibility.

In a small office with three workstations as shown in the illustration in Figure 4.5, you can set up your IP address with valid subnets. This small office could operate without TCP/IP if it never connects to the outside Internet or doesn't expand— NetBEUI could be to transmit only to each other. However, the network in this scenario will expand later, and may become geographically placed. The benefits of subnetting include:

- Reduced network traffic

- Fast network performance

- Easy management of the network

- Efficient routing paths

To set up a valid IP address and subnet addresses in your TCP/IP internal network software for a small Class C network, you must use the 32-bit binary addresses given to you by InterNIC or an Internet Service Provider (ISP).

Valid IP Address

Here is your binary address:
11001010000011111010101000000001

The 32-bit binary number address is difficult to remember, and can be converted into a more friendly decimal number. First, break the binary number into four 8-bit sets, called octets:
11001010 00001111 10101010 00000001

Use your calculator to convert each octet to its decimal equivalent. Put the first binary octet into its byte position under the decimal equivalent in the following chart, from left to right. Add the total decimal equivalents of only the 1 bit to get the decimal number of the first octet in the dotted quad as shown in Table 4.5:

Table 4.5 Binary Octets to Decimal Octet Conversion

Byte Position Value	128	64	32	16	8	4	2	1	Decimal Octet Sum
Binary Number	1	1	0	0	1	0	1	0	
Equiv- alent	128	64	0	0	8	0	2	0	202

The conversion result is the first binary number in the dotted decimal notation IP address: 202. The binary octet 11001010 is converted to the decimal octet 202.

The binary number 11001010 00001111 10101010 00000001 octets convert to a dotted decimal notation as follows:
202.15.170.1

The decimal form of the IP address is known as dotted decimal or dotted quad notation because the four sets of numbers are connected by dots.

In addition, you can identify this dotted quad notation by two parts: the network portion and the host portion. The first three quads of the address identify the network portion of the address, as shown with the underline here: 202.15.170.1. The last quad identifies the host portion of the address for a small Class C network, as shown here: 202.15.170.1.

The network portion of the ID must be masked to hide information from the external networks. Subnet masking allows the workstations' address to be hidden from any device outside of the intranet.

Subnet the Network and Mask the IP Address

The ANDing process determines the first subnet number from the IP address. The first three quads in a Class C address are masked with 255. Subnet the last 8-bit quad with the ANDing process.

202.15.170.1

Moreover, you can assign or setup your network address to accommodate subnets. The TCP/IP name for the range of addresses is a subnet. You must first hide, or mask the network identifications with the standard value 255, and then convert the host portion of the IP address to a subnet address: The value for each 8 bit can range from 0 to 255.

Small Class C network subnets get their addresses from the host portion of your IP address, and you convert your Class C host portion of the address.

Each subnet will have address numbers that are available only for the subnetwork number, the router number, and the broadcast number. The remaining numbers can be assigned to workstations on the subnetwork. Subnet your network according to the following table (Table 4.6).

Table 4.6 Subnetting Class C Networks

Number of Desired Subnets	Subnet Mask	Network Number	Router Address	Broadcast Address	Remaining Number of Hosts
1	255.255.255.0	x.y.z.0	x.y.z.1	x.y.z.255	253
2	255.255.255.128 255.255.255.	x.y.z.0 x.y.z.128	x.y.z.1 x.y.z.129	x.y.z.127 x.y.z.255	125 125
4	255.255.255.192 255.255.255. 255.255.255. 255.255.255.	x.y.z.0 x.y.z.64 x.y.z.128 x.y.z.192	x.y.z.1 x.y.z.65 x.y.z.129 x.y.z.193	x.y.z.63 x.y.z.127 x.y.z.191 x.y.z.255	61 61 61 61
8	255.255.255.224 255.255.255. 255.255.255. 255.255.255. 255.255.255. 255.255.255. 255.255.255. 255.255.255.	x.y.z.0 x.y.z.32 x.y.z.64 x.y.z.96 x.y.z.128 x.y.z.160 x.y.z.192 x.y.z.224	x.y.z.1 x.y.z.33 x.y.z.65 x.y.z.97 x.y.z.129 x.y.z.161 x.y.z.193 x.y.z.225	x.y.z.31 x.y.z.63 x.y.z.95 x.y.z.127 x.y.z.159 x.y.z.191 x.y.z.223 x.y.z.255	29 29 29 29 29 29 29 29

Determine how many subnets you want. Here we want to subnet the network with four subnets. Use the table 4.6 to determine your beginning subnet mask network number. For four subnets, our first subnet IP address number will be 202.15.170.0. The router number is 202.15.170.1, and the broadcast address is 202.15.170.63. This subnet can accommodate up to 61 hosts.

Now, the system administrator can set the TCP/IP, the gateway, and routers, and name resolving protocols and utilities with all the applicable subnet numbers.

Vocabulary

Review the following terms in preparation for the certification exam.

Term	Description
address class	Address classes A, B, and C define which IP address bits are used for the network ID and which are used for the host ID
ARP	Address Resolution Protocol correlates an IP address with the Media Access Control (MAC) address
byte stream	Continuous flow of data from beginning to end
connectionless communications	Communications or data transfers that don't require direct connection among senders and receivers
connection-oriented communications	Communication protocols that require and maintain a two-way or sending and receiving packets until a systematic release of the session is communicated
datagram	Any IP packet or unit of information along with relevant delivery information such as the destination address, that is transferred across a network
default subnet mask	An IP network address number that is subdivided, and identifies the default class networks as Class A, Class B, or Class C
DHCP	A system that assigns TCP/IP internet protocol configuration information to devices in a network system
DNS	A network of file servers that translates domain names into IP addresses

domain	The different levels of authority in the domain name scheme of the hierarchical structure for Internet access
FTP	The standard protocol and utility application for file transfer between systems
full-duplex transmission	A data transmission that communicates simultaneously in both directions among computer senders and receivers
host files	Static files that contain information to resolve names to IP addresses
host name	An assigned text identifier (called an alias) that's used to designate a specific TCP/IP host in a logical way
HTTP	Hypertext Transfer Protocol connects the Internet Web server and the Web browser on the workstation
IP	A network layer protocol that moves and delivers datagrams or packets through hosts or networks
IP address	The Internet Protocol address is a 32-bit number which identifies the host and network address and the host node address on the network
IP default gateway	The router on the network that forwards packets to other gateways and the specified destination address
IP proxy	A network service that queries the Internet for your workstation
Internet	The worldwide collection of networks and gateways that use the TCP/IP suite of protocols to communicate among each other

intranet	A private network based on the Internet protocols such as TCP/IP but designed for information management within a company or organization. It looks like a World Wide Web site, and could be connected to the Internet only through a proxy server outside the organization network
LAN	A Local Area Network (LAN) is a group of computers and other devices dispersed over a limited area and connected by a communications link that enables any device to interact with any other on the network. Cables through which messages are transmitted connect the devices
layer	A series of software functions for data communications
loopback	A method that sends signals to a device and receives it back to test system devices
MAC	A unique hardware address on the network interface card of a device
NetBIOS	Communicates between IBM clients and servers and can be implemented over TCP/IP and many other protocols
NetBIOS name	A NetBIOS name is the computer name that you gave it when you installed the operating system
network adapter driver	The software that interfaces with the hardware and operating system for the expansion card or device used to connect a computer to a local area network (LAN)
network interface card (NIC)	The expansion board plugged into computers and servers used to control the flow of information over the network
packet	A unit of information that can be transmitted over a network

POP3	Post Office Protocol version 3, is a mail server protocol that retrieves e-mail from an e-mail server and delivers the e-mail to your local client computer
port	An interface through which data is transferred between a computer and other devices, a network, or another computer, and appears to the processing computer as an address in memory to send and receive data
protocol	Uniform industry standards of rules and procedures established for accomplishing remote connectivity
protocol stack	The set of protocols that work together on different levels to enable communications on a network
protocol suite	A set of protocols designed by a vendor as complementary parts of a protocol stack
router	Directs packets to and from one network to another based on the destination ID
routing protocols	Protocols that decide which path data should take if the two computers are geographically separated
SMTP	Simple Mail Transfer Protocol that transfers mail between servers
SNMP	The Simple Network Management Protocol is the standard and primary protocol that manages your network
stack	Another name for a set of protocols
subnet	A network that is a part of a larger network
subnet mask	Allows you to split a single TCP/IP network address into multiple subnets

TCP	Transport layer protocol used to send messages across a network, and manage and deliver error free messages from the beginning of a connection to the end
TCP/IP	An industry standard protocol used that exists on most computers, and enables computers with differing operating systems to communicate
UDP	A connectionless protocol which provides simple but unreliable datagram services
WAN	Stands for Wide Area Network (WAN)—a network spread over a wide geographical area, and relies on communication capabilities to link the various network segments. A WAN can be one large network, or it can consist of a number of linked LANs.
WINS	The Windows Internet Name Server (WINS) resolves NetBIOS names and IP addresses, and reduces NetBIOS broadcast traffic problems
WWW	World Wide Web

In Brief

If you want to...	Then do this...
Use a set of protocols that standardizes the connection to the Internet and UNIX environments	Use TCP/IP protocols
Route packets to a remote network	Use the IP default gateway
Automate IP address assignments and economize IP address usage	Use the Dynamic Host Configuration Protocol (DHCP)
Assign alphabetical names to IP addresses	Use the domain name system (DNS) and the common domain types
Provide name resolution of NetBIOS computer names, and IP addresses	Use the Windows Internet Name Service (WINS)
Resolve names to IP addresses with a static file	Use the host files LMHOSTS and HOSTS
Send or receive large volumes of data with reliable packet or datagram delivery service	Use the Transmission Control Protocol (TCP)
Use the TCP/IP protocols for Internet connection	Install the TCP/IP and the correct network adapter, and then configure them
Send a fast transmission of a minimum number of packets with low reliability	Use UDP instead of TCP
Retrieve e-mail from mail server and deliver it to your local computer	Use POP3

Transfer e-mail between servers	Use Simple Mail Transfer Protocol (SMTP) because SMTP deals only with the delivery of the mail
Collect and manage information and statistics about devices on your TCP/IP network	Use the Simple Network Management Protocol (SNMP)
Browse lists of files that are available for you from other systems	Use the File Transfer Protocol (FTP) utility
Use a network service that queries the Internet for your workstation	Install the proxy server
Make a connection between a workstation browser and an Internet web server	Use the Hypertext Transport Protocol (HTTP)
Route packets in a connectionless protocol with no guarantee of delivery	Use the IP
Search for NetBIOS name databases in hierarchical order	Use WINS
Map an IP address to a physical address on broadcast networks like Ethernet, token ring or ARCnet	Use ARP
Split the network into multiple subnets	Use default subnet masks
Increase the number of addresses on your network	Use a custom subnet mask
Browse the web	Use The HTTP protocol for the web browser to the Web server service. The HTTP uses Port 80
Upload and download files from a remote server	Use FTP. FTP uses Port 21

Use the network protocol service that queries the Internet for your workstation and translates the address between local area networks (LANs) into a proxy address	Use the IP proxy
Configure the TCP/IP on your workstation	Set the parameters for the IP address and a subnet mask on the TCP/IP stack of the workstation, and use the parameters in TCP/IP protocol configuration specific to your operating system
To configure a workstation for DNS	Use three parameters: the IP address of the DNS server, the local host name, and the host Internet domain name
Configure the default gateway to send IP packets to other networks	Fill in the default blank parameter with the IP address of the router interface (connection) that connects your workstation's network segment to the Internet
To configure WINS on a workstation	Enable WINS, and specify a primary WINS server IP address
Configure the DHCP automatically	Accept the default setting and any other configured parameters on the DHCP server

Lesson 4 Activities

Complete the following activities to better prepare you for the certification exam.

1.   Explain the purpose and function of TCP/IP.

2.   Describe the IP gateway on a network and what it does.

3.   Explain the purposes of DHCP, WINS, and Host Files.

4.   Describe the difference between TCP and UDP.

5.   List two e-mail protocols and their functions.

6.   Explain IP.

7.   List the hierarchy of name resolution mechanisms in WINS for Windows NT.

8.   Explain Default Subnet Masks and how to identify the subnet mask number.

9.   List the port numbers for HTTP, FTP, and SMTP.

10. Memorize the Network Classes Table for subnet masking.

Answers to Lesson 4 Activities

1.   TCP/IP is an industry standard suite of protocols that provides communication
      among varied environments. It is the standard protocol used for interoperations
      among different types of computers. Network protocols embody rules that
      enable two nodes or hosts on a network to communicate. These rules are
      incorporated into software programs that perform communication links and
      other link services.

2.   IP default gateway is the router on your network that recognizes other network
      IDs and forwards addresses or ID packets destined for other networks in a
      secure manner.

3.   DHCP automates the assignment and reassignment of IP addresses and related
      information, provides or leases a set period of time to computers to economize IP
      address usage in the TCP/IP network, and provides centralized address
      allocation.

      WINS provides name resolution of NetBIOS computer names, communicates
      with a WINS server to send resource notification, and releases its NetBIOS
      name or locate a resource.

      Host files resolve names to IP addresses. For example, if a NetBIOS name in
      the LMHOSTS file is the same as the host name of the computer, the host
      name is resolved to the IP address.

4.   The User Datagram Protocol (UDP) is a connectionless protocol. UDP
      exchanges your data in discrete units called datagrams similar to IP datagrams.
      The UDP is much faster than TCP because it doesn't have connection and
      checking applications, but UDP can deliver a minimum number of packets in a
      quick manner. UDP is unreliable because you can't verify the arrival of the UDP
      at the correct destination.

      The Transmission Control Protocol (TCP) is a connection-oriented
      communication protocol. TCP adds reliable byte stream delivery services to the
      less reliable IP datagram delivery service. You can send or receive large
      volumes of data from your applications with the TCP. It makes the connection

between computers with full-duplex transmission. Your applications can send and receive data streams or control information back and forth at the same time.

5.   The Post Office Protocol version 3, (POP3) is a mail server protocol that retrieves e-mail from an e-mail server and delivers the e-mail to your local client computer.

The Simple Mail Transfer Protocol (SMTP) uses TCP/IP protocols to send mail over the Internet. SMTP delivers e-mail to an e-mail server, and it transfers mail between mail servers.

6.   The Internet Protocol (IP) contains the information to route packets over the network. IP is a connectionless-oriented protocol that sends and receives information on the Network layer. It is connectionless because packets are not sent in streams, but are sent independent of one another.

7.   You can use WINS to search NetBIOS name databases in this order: Local NetBIOS name cache, WINS server database, Broadcast, LMHOSTS file, HOSTS file, DNS system.

8.   The default subnet mask allows you to split the network into multiple subnets. Here is how to identify the default subnet mask number: The leftmost number of the address is covered or masked by using the dotted decimal number 255, and the right most number is viewed with a 0.

9.   The port numbers are:
      HTTP Port 80
      FTP Port 21
      SMTP Port 25

10.   Memorize the Default Subnet Mask addresses and classes:

Class	Decimal Range	Default Subnet Mask	Maximum Nodes
A	1-126	255.0.0.0	16,777,214
B	128-191	255.255.0.0	65,534
C	192-223	255.255.255.0	254

Lesson 4 Quiz

These questions test your knowledge of features, vocabulary, procedures, and syntax.

1. Which three components comprise the Domain Name System (DNS)?

    A.   Resolvers, name servers, and domain name space
    B.   Root level, top level, second level
    C.   Class A, B, and C
    D.   TCP, IP, UDP

2. Which TCP/IP protocols are connection-oriented protocols?

    A.   TCP
    B.   UDP
    C.   IP
    D.   WINs

3. Which protocol dynamically assigns IP addresses for a network?

    A.   DHCP
    B.   WINS
    C.   DNS
    D.   UNKNO

4. What is the name of the file where static IP addresses are stored in a Windows NT server?

    A.   IPDUMP
    B.   HOSTS
    C.   LMHOSTS
    D.   IPHOSTS

5.   Why does the UDP protocol deliver packets faster than TCP?

      A.   TCP sends large volumes of data
      B.   TCP sends acknowledgement to the sender
      C.   UDP checks sequence numbers
      D.   UDP does not check sequence numbers

6.   Which standard protocols are included in the TCP/IP stack?

      A.   FTP
      B.   HTTP
      C.   IP
      D.   TCP

7.   Which of these numbers in the IP address octets can you not use because they
      are reserved for subnet masks?

      A.   0
      B.   255
      C.   127
      D.   72

8.   What is the port number for the File Transfer Protocol (FTP)?

      A.   80
      B.   25
      C.   21
      D.   99

9.   Which network service queries the Internet for your workstation?

      A.   IP proxy
      B.   NIC
      C.   IP address
      D.   Subnet mask

10.   How many hosts are allowed on a Class A network?

    A.   254
    B.   16,387,064
    C.   64,516
    D.   16

Answers to Lesson 4 Quiz

1.   Answer A is correct. DNS is a distributed database management system, and
      maps to the application layer with the TCP and UDP protocols. The DNS looks
      up the name of a domain name and cross-references or resolves it to the IP
      address.

      Answer B is incorrect because root level, top level, and second level are domain
      name levels in a hierarchical structure.

      Answer C is incorrect because these are the subnet mask or IP address classes.

      Answer D is incorrect because these are examples of protocols.

2.   Answer A is correct. TCP is connection-oriented.

      Answer B is incorrect because UDP is connectionless.

      Answer C is incorrect because IP is connectionless.

      D. WINs is incorrect because it is not a TCP/IP protocol.

3.   Answer A is correct. DHCP dynamically assigns IP addresses.

      Answer B is incorrect because WINS is the Windows to IP name resolution
      standard.

      Answer C is incorrect because DNS looks up the name of a domain and
      cross-references or resolves it to the IP address.

      Answer D is incorrect because UNKNO is fictitious.

4.   Answer B is correct. HOSTS is a static IP file.

      Answer A is incorrect because it is a fictitious file name.

      Answer C is incorrect because LMHOSTS is the static database for the WINS information.

Answer D is incorrect because it is a fictitious file name.

5.   Answer A is correct. TCP sends large volumes of data. TCP sends large volumes of data because it supports application-to-application transfer, and UDP does not. UDP is faster.

Answer B is correct because TCP sends acknowledgment to the sender. UDP is faster because it does not sent acknowledgment.

Answer C is incorrect because UDP does not check sequence. It is faster because it does not check sequence numbers.

Answer D is correct because UDP does not check sequence numbers is correct. It does not establish communication with another program in order to exchange information.

6.   Answer A is correct. FTP is a standard protocol in the TCP/IP stack.

Answer B is incorrect because HTTP is not a part of the TCP/IP protocol stack, but it does use TCP to make a connection.

Answer C is correct because IP is the part of TCP/IP. It is a network layer protocol.

Answer D is correct because TCP is a standard protocol in the TCP/IP stack.

7.   Answer A, the number 0, is correct. The number 0 masks an IP number for the rightmost ID.

Answer B, the number 255, is correct because it masks an IP number for the leftmost network ID.

Answer C is incorrect because the number 127 is reserved for loopback and diagnostic functions.

Answer D, the number 72, is incorrect because it is fictitious.

8.   Answer A, Port 80 is incorrect. HTTP uses port 80 of the TCP protocol to make an Internet connection.

      Answer B, Port 25, is incorrect because SMTP uses Port 25 to send e-mail across different types of servers.

      Answer C, Port 21 is incorrect because the FTP protocol uses Port 21 for incoming messages and requests.

      Answer D, 99, is incorrect.

9.   Answer A, IP proxy is correct. IP proxy queries the Internet for your workstation.

      Answer B is incorrect because NIC is a Network interface card and has a static number address.

      Answer C is incorrect because protocols and application programs use IP addresses.

      Answer D is incorrect because subnet mask is a covered, subdivided IP address.

10.   Answer B is correct. The largest network, Class A, can have 16,387,064 hosts.

      Answer A is incorrect because the smallest network, Class C, can have 254 hosts.

      Answer C is incorrect because the medium network, Class B, can have 64,516 hosts.

      Answer D is incorrect because up to 254 hosts are allowed on a Class C network.

Lesson 5: TCP/IP Utilities Suite

TCP/IP is a suite of protocols designed for internetworks. TCP/IP utilities work with TCP/IP protocols to provide additional functionality for testing, validating, and troubleshooting TCP/IP connectivity.

After completing this lesson, you should have a better understanding of the following:

- ARP

- Telnet

- NBTSTAT

- TRACERT

- NETSTAT

- IPCONFIG

- FTP

- PING

Testing, Validating, and Troubleshooting IP Connectivity

Address Resolution Protocol (ARP)

ARP is a diagnostic utility you can use to view a cache of IP addresses to local hardware address resolutions that are on your local computer. You can use ARP to view the ARP cache table to find invalid entries and to purge the cache. The following screen shows the functions of the ARP utility (Figure 5.1).

Figure 5.1 ARP Screen

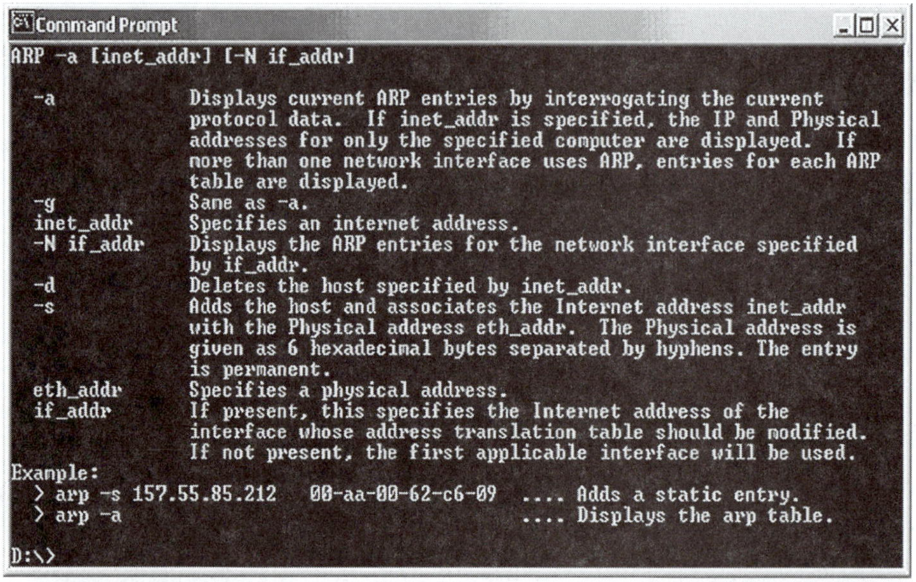

```
Command Prompt                                                    _ □ ×
ARP -a [inet_addr] [-N if_addr]

   -a              Displays current ARP entries by interrogating the current
                   protocol data.  If inet_addr is specified, the IP and Physical
                   addresses for only the specified computer are displayed.  If
                   more than one network interface uses ARP, entries for each ARP
                   table are displayed.
   -g              Same as -a.
   inet_addr       Specifies an internet address.
   -N if_addr      Displays the ARP entries for the network interface specified
                   by if_addr.
   -d              Deletes the host specified by inet_addr.
   -s              Adds the host and associates the Internet address inet_addr
                   with the Physical address eth_addr.  The Physical address is
                   given as 6 hexadecimal bytes separated by hyphens. The entry
                   is permanent.
   eth_addr        Specifies a physical address.
   if_addr         If present, this specifies the Internet address of the
                   interface whose address translation table should be modified.
                   If not present, the first applicable interface will be used.
Example:
   > arp -s 157.55.85.212   00-aa-00-62-c6-09  .... Adds a static entry.
   > arp -a                                    .... Displays the arp table.

D:\>
```

 Tip: Each Network Interface Card (NIC) has a unique ID address built into it by its manufacturer. These local hardware addresses are also known as Media Access Control (MAC) addresses.

Telnet

Telnet is a connection utility application that provides remote terminal emulation. Telnet allows you to access and run applications on a remote computer, network device, or private TCP/IP network. Use Telnet if the computer you want to access has an active Telnet service, and you have a valid account on that remote server.

NetBIOS Over TCP/IP Status (NBTSTAT)

NBTSTAT is a diagnostic utility (Figure 5.2) that examines the state of current NetBIOS over TCP/IP (NBT) connections.

Use NBTSTAT to view:

- NetBIOS names registered on a remote client or on your local computer

- All the client and server sessions your local computer processes

- All the attempts of your local computer to resolve an IP address to a host name

Figure 5.2 NBTSTAT Screen

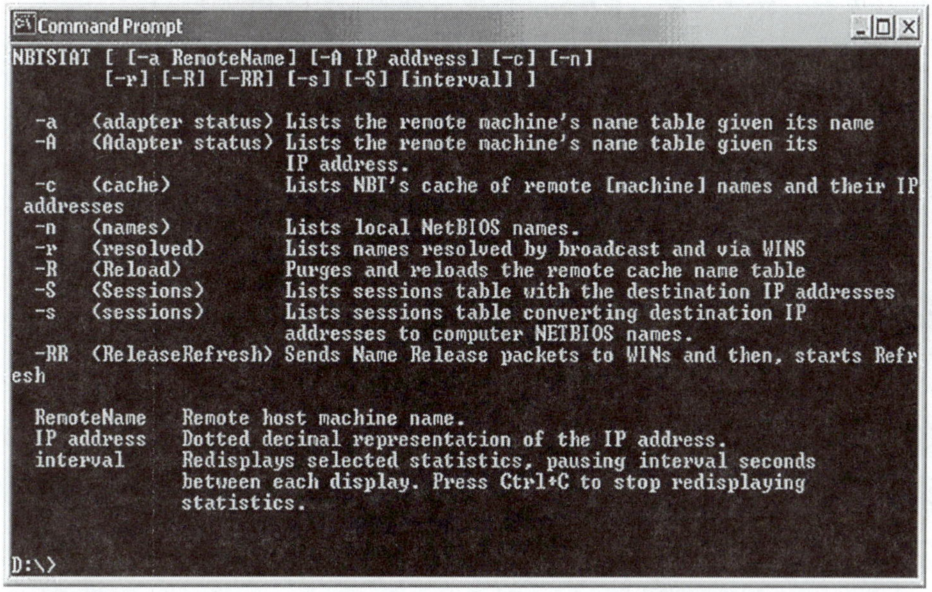

Trace Route (TRACERT)

TRACERT is a diagnostic utility that traces the path of a packet as it travels in real time through routers (hops) from the local host to a remote host.

Use TRACERT to determine:

- If the packet reached its destination host

- The last router the packet passed through (hop)

- The time it takes for a packet to travel between routers

- The devices along the path to the destination

For more information about TRACERT, see Lesson 10. The following screen shows TRACERT command options (Figure 5.3).

Figure 5.3 TRACERT Screen

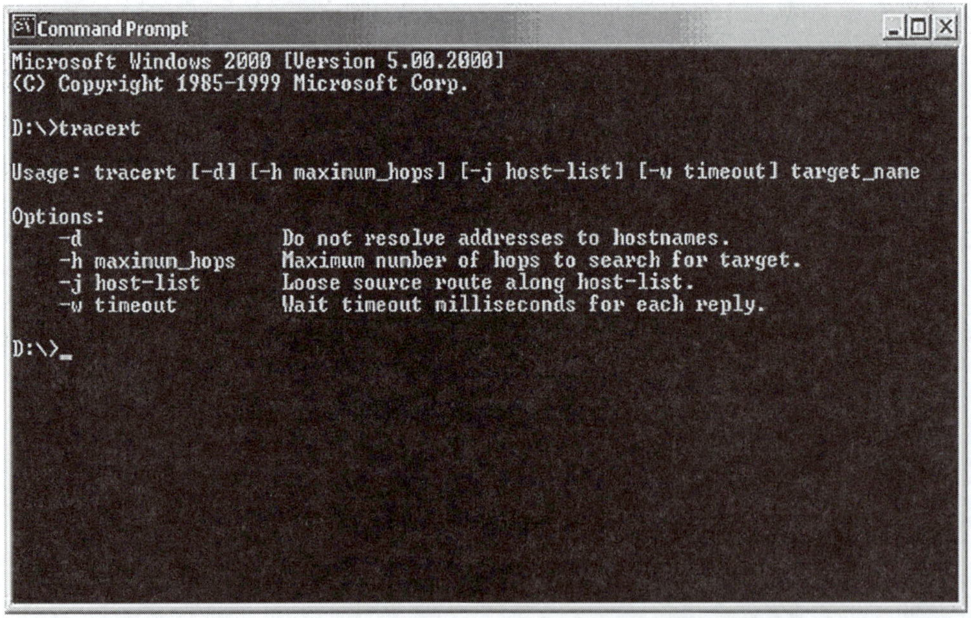

```
Command Prompt                                                    _ |□| x|
Microsoft Windows 2000 [Version 5.00.2000]
(C) Copyright 1985-1999 Microsoft Corp.

D:\>tracert

Usage: tracert [-d] [-h maxinum_hops] [-j host-list] [-w timeout] target_nane

Options:
    -d                      Do not resolve addresses to hostnames.
    -h maxinum_hops         Maximum number of hops to search for target.
    -j host-list            Loose source route along host-list.
    -w timeout              Wait timeout milliseconds for each reply.

D:\>_
```

Network Status (NETSTAT)

NETSTAT is a diagnostic utility that views all the TCP/IP protocol statistics and the current state of the TCP/IP connections with the NETSTAT command (Figure 5.4). NETSTAT displays all connections and listening ports.

The command to display all connections and listening ports:
```
NETSTAT -a
```

Figure 5.4 NETSTAT Screen

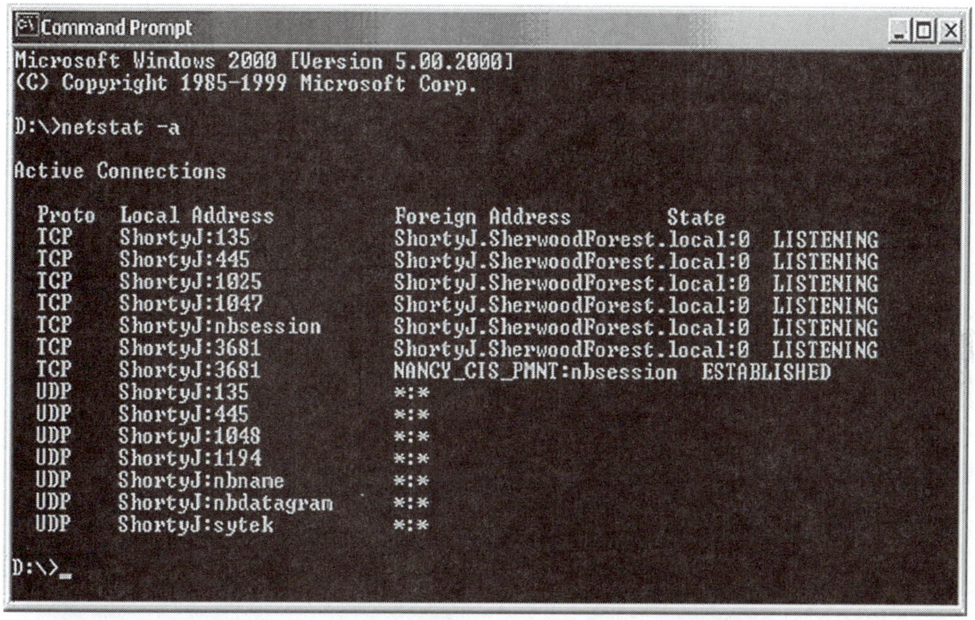

Internet Protocol Configuration (IPCONFIG)

IPCONFIG is a diagnostic utility that views the TCP/IP configuration of DHCP, DNS, and WINS server addresses.

You can display the configuration information with the command:

```
IPCONFIG /all
```

The default IPCONFIG configuration displays:

- The IP address

- Subnet mask

- Default gateway address

 Note: You can use the `IPCONFIG /all` command to display TCP/IP configuration information for DHCP, DNS, and WINS server addresses.

File Transfer Protocol (FTP)

FTP is a connection utility application on your client computer that allows logging onto a network and transferring files to or from an FTP enabled server and your client computer. You can configure the FTP service on the server:

- To require a username and password to logon

- To accept anonymous logon requests.

FTP is a:

- Protocol

- Client application

- Service on a server

An FTP server is easy to install and makes files on your server available to remote users. The available FTP files can be in any format, such as document files, multimedia files, or application files. The World Wide Web (WWW) replaces most functions of FTP. FTP requires you to log on to use the service. Once you are logged on, you can navigate the directories made available to the FTP service.

The Microsoft Internet Information Server includes Microsoft® Internet Explorer, which allows you to browse FTP servers. You can use a Uniform Resource Locator (URL) to connect to an FTP server. With this service, for example, you will reach the FTP site at the Microsoft Corporation if you type:

```
//ftp.microsoft.com/
```

 Note: If you configure an FTP service for browsers, set the FTP listing style to UNIX for maximum compatibility with browsers.

On older dedicated FTP clients, remote clients can copy files to the server and issue other FTP commands. Some FTP commands include those in the following Table 5.1.

Table 5.1 FTP Commands

Ftp: !	Ftp: glob	Ftp: put
Ftp: ?	Ftp: hash	Ftp: pwd
Ftp: append	Ftp: help	Ftp: quit
Ftp: ascii	Ftp: lcd	Ftp: quote
Ftp: bell	Ftp: literal	Ftp: recv
Ftp: binary	Ftp: ls	Ftp: remotehelp
Ftp: bye	Ftp: mdelete	Ftp: rename
Ftp: cd	Ftp: mdir	Ftp: rmdir
Ftp: close	Ftp: mget	Ftp: send
Ftp: debug	Ftp: mkdir	Ftp: status
Ftp: delete	Ftp: mls	Ftp: trace
Ftp: dir	Ftp: mput	Ftp: type
Ftp: disconnect	Ftp: open	Ftp: user
Ftp: get	Ftp: prompt	Ftp: verbose

FTP clients can FTP at the command prompt. The following screen figures show a sample FTP client to server session request to transfer a file, and the steps to retrieve a file from another computer during an online session through the Internet

The FTP command transfers files to and from a computer running an FTP server service, and is available only if the TCP/IP protocol is installed.

Figure 5.5 FTP Logon Screen

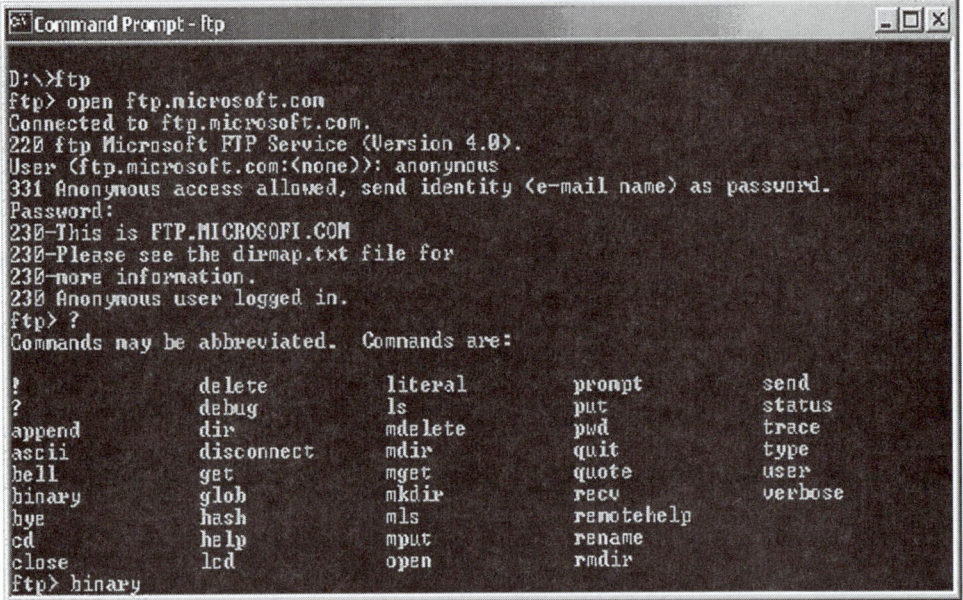

To open an FTP session and find a directory of files to transfer from the Microsoft FTP site, do the following steps at the DOS command prompt (Figure 5.5).

1. Create a temporary on your computer and call it FTPTEMP, and then change to the FTPTEMP directory.

2. Type

    `ftp` and press ENTER.

3. At the ftp> prompt, type

    `open ftp.microsoft.com`

4. At the User prompt, log on as:

    `Anonymous`

5. At the password prompt, type a fictitious e-mail address, such as:

    `anybody@nowhere.com`

6. An ftp> prompt will appear, with an acknowledgment that you are logged in to the site.

The screen figure and steps following, shows how you can get a list commands, and a directory of files to retrieve (Figure 5.6).

Figure 5.6 FTP Screen (Directory)

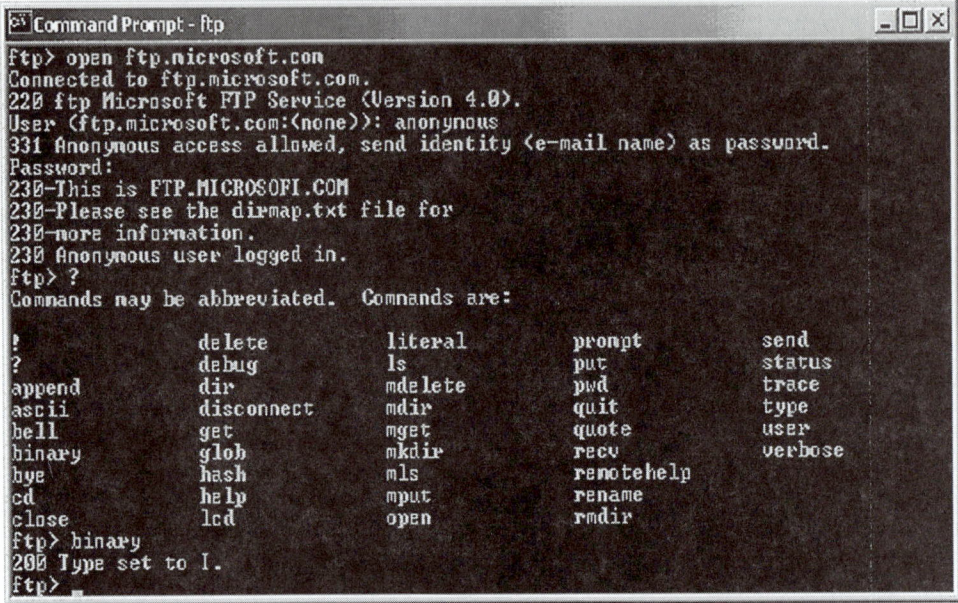

1.   To see a list of available commands and file types, at the ftp> prompt,
      type: ?
2.   Select a file binary file type. Type:
      `binary`
3.   Press ENTER
4.   To see a director of available files to transfer, at the ftp> prompt, type:
      `dir`
5.   Press ENTER

The following screen shows you the list of available files to retrieve, and the steps
following retrieve the file called DIRMAP.TXT (Figure 5.7).

Figure 5.7 FTP Get Screen

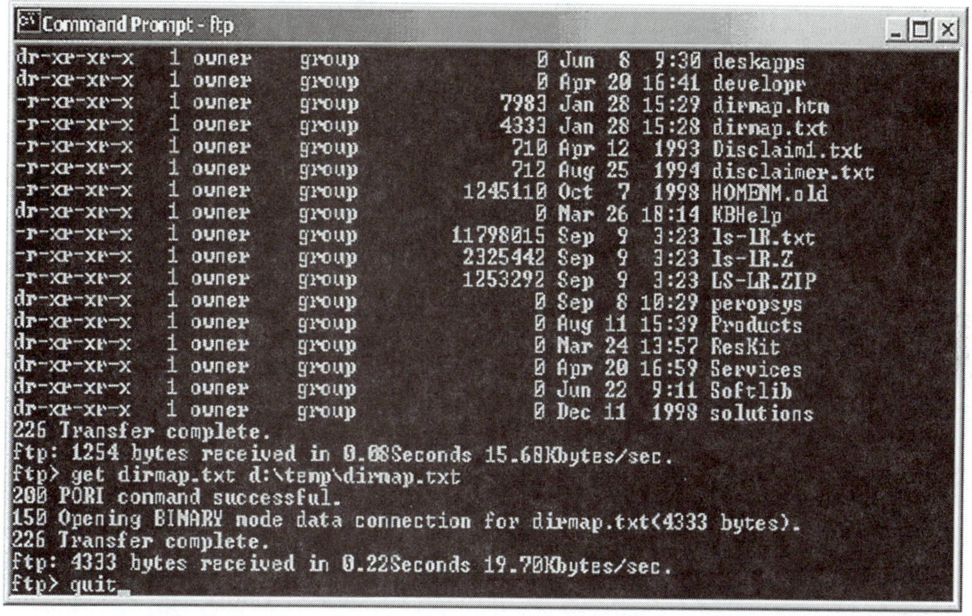

```
Command Prompt - ftp                                              _ □ x
dr-xr-xr-x   1 owner     group           0 Jun  8  9:30 deskapps
dr-xr-xr-x   1 owner     group           0 Apr 20 16:41 developr
-r-xr-xr-x   1 owner     group        7983 Jan 28 15:29 dirmap.htm
-r-xr-xr-x   1 owner     group        4333 Jan 28 15:28 dirmap.txt
-r-xr-xr-x   1 owner     group         710 Apr 12  1993 Disclaim1.txt
-r-xr-xr-x   1 owner     group         712 Aug 25  1994 disclaimer.txt
-r-xr-xr-x   1 owner     group     1245110 Oct  7  1998 HOMENM.old
dr-xr-xr-x   1 owner     group           0 Mar 26 18:14 KBHelp
-r-xr-xr-x   1 owner     group    11798015 Sep  9  3:23 ls-lR.txt
-r-xr-xr-x   1 owner     group     2325442 Sep  9  3:23 ls-lR.Z
-r-xr-xr-x   1 owner     group     1253292 Sep  9  3:23 LS-LR.ZIP
dr-xr-xr-x   1 owner     group           0 Sep  8 10:29 peropsys
dr-xr-xr-x   1 owner     group           0 Aug 11 15:39 Products
dr-xr-xr-x   1 owner     group           0 Mar 24 13:57 ResKit
dr-xr-xr-x   1 owner     group           0 Apr 20 16:59 Services
dr-xr-xr-x   1 owner     group           0 Jun 22  9:11 Softlib
dr-xr-xr-x   1 owner     group           0 Dec 11  1998 solutions
226 Transfer complete.
ftp: 1254 bytes received in 0.08Seconds 15.68Kbytes/sec.
ftp> get dirmap.txt d:\temp\dirmap.txt
200 PORT command successful.
150 Opening BINARY mode data connection for dirmap.txt(4333 bytes).
226 Transfer complete.
ftp: 4333 bytes received in 0.22Seconds 19.70Kbytes/sec.
ftp> quit
```

At the directory, you could choose any file to transfer to your computer's directory (Figure 5.7). Follow the steps below to retrieve a file called DIRMAP.TXT.

1. Retrieve the file and save it to the temporary directory. At the ftp> prompt, type:

    ```
    get dirmap.txt d:\temp\dirmap.txt
    ```

2. The FTP session will acknowledge that your file was received, and remain open for more activity.

3. To end the session, at the ftp> prompt type:
    ```
    quit
    ```

Packet InterNet Groper (PING)

Many people consider PING the most useful TCP/IP diagnostic utility. Use PING to send IP echo request packets to a destination host on a local or remote network to determine if a destination TCP/IP host is available, functional, or has a valid IP address. If PING is successful, a reply message displays.

 Tip: PING the IP address to eliminate a host name-resolution problem. PING the host name if you think the host name-resolution process is the problem that needs to be fixed.

The sequence in which you use the PING utility to diagnose a problem and determine the location of a failure is important.

Use PING in the following order to determine TCP/IP failures:

- Ping the loopback IP address on your local computer to determine if TCP/IP is on your computer and functional.

- Ping the IP address of the local host to make sure the IP address is valid.

- Ping a host on the local network.

- Ping the IP address of the default gateway router to determine if the router is functional.

- Ping the far side of the router.

- Ping the IP address of a host on a remote network to test if you can communicate across (hop) the router.

Vocabulary

Review the following terms in preparation for the certification exam.

Term	Description
ARP	The Address Resolution Protocol utility is a diagnostic utility that allows viewing the cache of IP addresses to local hardware address resolutions that your local computer makes.
FTP	File Transfer Protocol is a connection utility application on your client computer that allows you to log onto a network and transfer files to or from an FTP enabled server and your computer.
IPCONFIG	Internet Protocol Configuration is a diagnostic utility that allows viewing the TCP/IP configuration of DHCP, DNS, and WINS server addresses.
MAC	Media Access Control addresses are the unique ID addresses built into each Network Interface Card.
NBTSTAT	NetBIOS Over TCP/IP Status is a diagnostic utility that examines the state of current NetBIOS over TCP/IP (NBT) connections.
NETSTAT	Network Status (NETSTAT) is a diagnostic utility that displays all TCP/IP protocol statistics and the current state of TCP/IP connections with the command: NETSTAT -a

PING	PING is a TCP/IP diagnostic utility that sends IP echo request packets to a destination host on a local or remote network to determine if a destination TCP/IP host is available, functional, or has a valid IP address.
router	A device that contains tables (router tables) that can choose the routes along which it can send data in a network, depending on protocol and IP address.
Telnet	A connection utility application that allows running applications on a remote system. You can use Telnet if the computer you want to access has an active Telnet service, and you have a valid Telnet account on that server.
TRACERT	A diagnostic utility that traces the path of a packet as it travels in real time through routers (hops) from the local host to a remote host.

In Brief

If you want to...	Then do this...
Log onto a network and transfer files from an FTP enabled server to your computer	Use the File Transfer Protocol (FTP) utility
Determine if a local or remote destination TCP/IP host is available, functional, or has a valid IP address	Use the PING utility to send IP echo request packets to the destination host. If PING is successful, a reply message displays
Display the IP address cache to local hardware address resolutions on your local computer to allow you to find invalid entries or purge the cache	Use the ARP utility
Access and run applications on a remote system that requires remote terminal emulation	Use the Telnet connection utility to access the remote computer with an active Telnet service and your valid Telnet account
Trace the path a packet travels in real time as it passes through routers (hops) from the local host to a remote host	Use the TRACERT diagnostic utility
View the NetBIOS names registered on a remote client, or on your local computer	Use the NBTSTAT diagnostic utility `IPCONFIG /all`
View the TCP/IP configuration of DHCP, DNS, and WINS server addresses	Use the IPCONFIG diagnostic utility
View all the TCP/IP protocol statistics and the current state of the TCP/IP connections	Use the NETSTAT diagnostic utility with the command: NETSTAT `-a`

Lesson 5 Activities

Complete the following activities to better prepare you for the certification exam.

1. Describe how you can view the TCP/IP configuration of the DHCP, DNS, and WINS server addresses.

2. Explain the difference between a diagnostic and a connection utility.

3. Name the utility that allows you to access and run applications on a remote system.

4. Describe the default configuration of the IPCONFIG utility.

5. Explain how you can use TRACERT to determine a router failure.

6. Explain the difference between an FTP utility and an FTP service.

7. Name the utility you can use to view all the TCP/IP protocol statistics and the current state of the TCP/IP connections.

8. Describe how you can determine if a local or remote destination TCP/IP host is available, functional, or has a valid IP address.

9. Explain why you want to configure the IPCONFIG utility.

10. Describe the diagnostic utility you can use to examine the state of current NetBIOS over TCP/IP (NBT) connections.

Answers to Lesson 5 Activities

1.   Use the IPCONFIG diagnostic utility.

2.   A diagnostic utility can configure parameters of a workstation, network, or host computer. A connection utility is an application you use to access and run applications on a remote system or access and transfer files between a client and server.

3.   Telnet.

4.   The default IPCONFIG configuration displays the IP address, subnet mask, and default gateway address.

5.   The indication of a router failure with TRACERT is that a packet stops at a router and does not reach its target host IP address destination.

6.   An FTP connection utility is an application on your client computer that allows you to log on to a network and transfer files to or from an FTP-enabled server and your client computer. An FTP service is located on a server that you can configure two ways: To require a username and password to logon or to accept anonymous logon requests.

7.   The NETSTAT diagnostic utility with the NETSTAT is a command.

8.   Use the PING utility to send IP echo request packets to the destination host. If PING is successful, a reply message displays.

9.   You want to configure the Internet Protocol Configuration IPCONFIG utility to make it easy to view the TCP/IP configuration of DHCP, DNS, and WINS server addresses.

10.   NBTSTAT is the diagnostic utility you can use to examine the NetBIOS over TCP/IP connections.

Lesson 5 Quiz

These questions test your knowledge of features, vocabulary, procedures, and syntax.

1. Which utility should you use to transfer files to and from a host and remote server?

    A.   Telnet

    B.   IPCONFIG

    C.   PING

    D.   FTP

2. Which tool allows you to determine how much time it takes a packet to move from one router to another?

    A.   NETSTAT

    B.   TRACERT

    C.   PING

    D.   IPCONFIG

3. Which utility displays the IP address cache to local hardware address resolutions that are on your local computer and enables you to find invalid entries or purge the cache?

    A.   PING

    B.   IPCONFIG

    C.   ARP

    D.   NBTSTAT

4.   The local hardware addresses that are built into each Network Interface
      Card (NIC) by the manufacturer, and are known as these types of addresses.

      A.   Internet Protocol

      B.   Media Access Control (MAC)

      C.   Transmission Control Protocol

      D.   Host name

5.   Which items use ARP to view or purge?

      A.   IP addresses to local hardware address resolutions

      B.   IP addresses to host names

      C.   A default gateway address

      D.   A router table

6.   Which utility displays all the TCP/IP protocol statistics and the current state
      of the TCP/IP connections?

      A.   NBTSTAT diagnostic utility

      B.   ARP diagnostic utility

      C.   NETSTAT diagnostic utility

      D.   MAC diagnostic utility

7.   Which tool can examine the NetBIOS over TCP/IP connections?

      A.   PING connection utility

      B.   IPCONFIG diagnostic utility

      C.   NETSTAT diagnostic utility

      D.   NBTSTAT is the diagnostic utility

8.   Which command displays the TCP/IP configuration information for DHCP,
      DNS, and WINS server addresses?

      A.   NETSTAT –A

      B.   IPCONFIG /ALL

      C.   PING

      D.   FTP

9.   Which is the first step in the sequence to test the loopback IP address on
      your local computer to determine that TCP/IP is on your computer?

      A.   TRACERT

      B.   NETSTAT

      C.   IPCONFIG

      D.   PING

10.   Which utility tests the IP address of the default gateway router to determine
      if the router is functional?

      A.   IPCONFIG

      B.   PING

      C.   NETSTAT

      D.   ARP

Answers to Lesson 5 Quiz

1. Answer D is correct. FTP is the connection utility application that allows you to access an FTP server and transfer files to and from a host and remote server.

   Answer A is incorrect because Telnet allows you to access and run applications on a Telnet service enabled remote system.

   Answer B is incorrect because IPCONFIG is the Internet Protocol Configuration diagnostic utility you can configure to view the TCP/IP configuration of DHCP, DNS, and WINS server addresses.

   Answer C is incorrect because PING is a diagnostic utility you can use to send IP echo request packets to a destination host on a local or remote network to determine if a destination TCP/IP host is available, functional, or has a valid IP address.

2. Answer B is correct. TRACERT allows you to determine how much time it takes a packet to travel between routers if the packet can reach its destination; and the last router the packet can pass across (hops).

   Answer A is incorrect because NETSTAT can display all TCP/IP protocol statistics and the state of TCP/IP connections with the netstat –a command.

   Answer C is incorrect because PING does not allow you to determine how much time it takes a packet to travel between routers.

   Answer D is incorrect because IPCONFIG is the Internet Protocol Configuration diagnostic utility you can configure to view the TCP/IP configuration of DHCP, DNS, and WINS server addresses.

3. Answer C is correct. The ARP diagnostic utility displays the cache of IP address to local hardware address resolutions that are on your local computer and enables you to find invalid entries or purge the cache.

   Answer A is incorrect because you use PING to send IP echo request packets to a destination host on a local or remote network to determine if a destination TCP/IP host is available, functional, or has a valid IP address.

Answer B is incorrect because Internet Protocol Configuration is a diagnostic utility you can configure to view the TCP/IP configuration of DHCP, DNS, and WINS server addresses.

Answer D is incorrect because NetBIOS Over TCP/IP Status Utility (NBTSTAT) is a diagnostic utility you can use to examine the state of current NetBIOS over TCP/IP (NBT) connections.

4.   Answer B is correct. Media Access Control (MAC) addresses are the unique ID addresses built into each Network Interface Card.

Answer A is incorrect because Internet Protocol (IP) addresses and routes packets between hosts, makes best effort attempts to deliver packets, but does not guarantee packet delivery.

Answer C is incorrect because TCP is a reliable packet delivery protocol that assigns sequence numbers to packets and puts them in order at their destination.

Answer D is incorrect because a host name simplifies the way people reference TCP/IP hosts because they are alphabetic names that people find easier to remember than numerical IP addresses. Host names correlate to IP addresses and are kept together in HOSTS files, or databases such as DNS or NetBIOS name servers.

5.   Answer A is correct. ARP displays the cache of IP address to local hardware address resolutions that are on your local computer to allow you to find invalid entries or purge the cache.

Answer B is incorrect because server systems such as DHCP, DNS, and WINS resolve IP addresses to host names.

Answer C is incorrect because a default gateway specifies the IP address of a router.

Answer D is incorrect because routers XE "Routers" contain tables (router tables) that have the IP addresses of router interfaces to other networks. Routers forward packets to other router interfaces listed in their router tables. ARP cannot view or purge these addresses.

6.   Answer C is correct. NETSTAT –a command is the diagnostic utility you can use to examine the state of current NetBIOS over TCP/IP (NBT) connections with the netstat –a command.

Answer A is incorrect because NBTSTAT is the diagnostic utility you can use to examine the state of current NetBIOS over TCP/IP (NBT) connections.

Answer B is incorrect because the ARP utility is a diagnostic tool that displays the cache of IP address to local hardware address resolutions that are on your local computer to allow you to find invalid entries or purge the cache.

Answer D is incorrect because MAC is not a software utility. MACs are Media Access Control addresses and are the unique ID addresses built into each Network Interface Card.

7.   Answer D is correct. The diagnostic utility is correct because the NetBIOS Over TCP/IP Status Utility (NBTSTAT) is the diagnostic utility you can use to examine the state of current NetBIOS over TCP/IP (NBT) connections.

Answer A is incorrect because the PING is a diagnostic utility, not a connection utility.

Answer B is incorrect because the IPCONFIG utility is a diagnostic utility you can configure to view the TCP/IP configuration of DHCP, DNS, and WINS server addresses.

Answer C is incorrect because NETSTAT is a diagnostic utility that can display all TCP/IP protocol statistics and the current state of TCP/IP connections with the NETSTAT –a command.

8.   Answer B is correct. You can use the IPCONFIG /all command to display all the TCP/IP configuration information for DHCP, DNS, and WINS server addresses.

Answer A is incorrect because the NETSTAT –a command allows you to view all the TCP/IP protocol statistics and the current state of the TCP/IP connections.

Answer C is incorrect because you use the PING command to send IP echo request packets to a destination host on a local or remote network to determine if a destination TCP/IP host is available, functional, or has a valid IP address.

Answer D is incorrect because the FTP utility is a connection utility application on your client computer that allows you to log onto a network and transfer files to or from an FTP enabled server and your computer.

9.   Answer D is correct. It is important to test with PING in the correct sequence. The loopback test is the first test you need to perform to determine that TCP/IP is on your computer and functional.

Answer A is incorrect because TRACERT is a diagnostic utility that traces the path of a packet as it travels in real time through routers (hops) from the local host to a remote host.

Answer B is incorrect because NETSTAT is a diagnostic utility that can display all TCP/IP protocol statistics and the current state of TCP/IP connections with the NETSTAT –a command.

Answer C is incorrect because IPCONFIG utility is a diagnostic utility you can configure to view the TCP/IP configuration of DHCP, DNS, and WINS server addresses.

10.   Answer B is correct. You can run the PING utility to test the IP address of the default gateway router to determine if the router is functional.

Answer A is incorrect because IPCONFIG allows you to view all the TCP/IP protocol statistics and the current state of the TCP/IP connections.

Answer C is incorrect because NETSTAT a diagnostic utility that allows you to view the TCP/IP configuration of DHCP, DNS, and WINS server addresses with the netstat –a command.

Answer D is incorrect because the ARP utility is a diagnostic tool that displays the cache of IP address to local hardware address resolutions that are on your local computer to allow you to find invalid entries or purge the cache.

Lesson 6: Connectivity

Connectivity describes the structure of communication between one computer and another computer that might be a server or host computer on the Internet. Connectivity includes a discussion of the type of circuitry, hardware, and software that aide in the transmission of data as well as the rate of speed at which data can be sent and received. Good connectivity supports the import and export of data from an array of programs in different formats. Poor connectivity limits the types of data that may be imported and exported.

After completing this lesson, you should have a better understanding of the following topics:

- Connectivity Concepts

- Modems

- Remote Connectivity

- Dial-Up Networking

- Proxy Server Connectivity

Connectivity Concepts

Successful connectivity with another computer, network, or the Internet requires selecting formats, devices, and standards, or protocols that enable linking to take place. You must choose protocols that accommodate your computer hardware and software as well as devices that support your computer.

Asynchronous Protocols

Digital asynchronous communication means that data transfers intermittently instead of in a smooth flow (Figure 6.1). Each character transmits a signal as it is being sent. This kind of communication is accomplished through a "start-bit," "stop-bit" process called "start-stop" transmission. Most communication between computers and devices is asynchronous. One of the problems involved in this kind of communication is that the computer must distinguish between legitimate data and background noise.

Figure 6.1 Asynchronous and Synchronous Data Flow

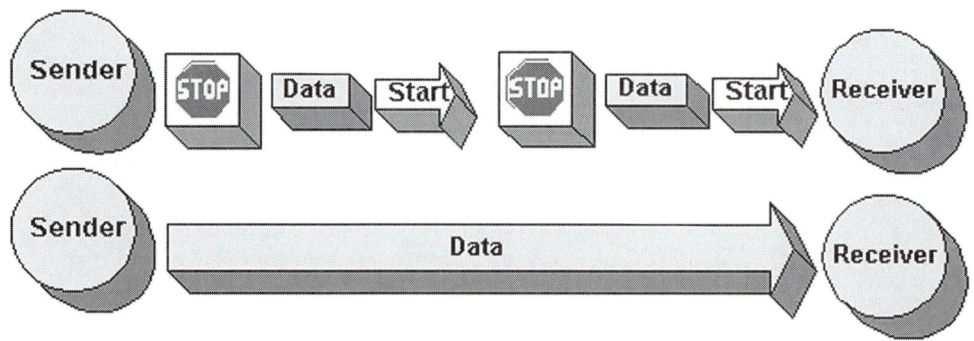

Packet-switching technology is a kind of asynchronous technology with protocols that divide messages into packets that travel separately along different routes to a final destination. Upon arrival, these packets reformat into their original form. WANs use this packet-switching technology.

The difference between normal telephone service and packet switching is that telephone service is circuit-switching technology where one line is dedicated for transmissions between parties. This provides a more assured delivery of data.

The benefit of using circuit-switching technology is that data transmits quickly and arrives in the same format in which it was sent. Packet switching is valuable because of its efficiency combined with the ability to bear up well with transmission delays.

Asynchronous Transfer Mode (ATM) technology merges the best features of packet switching and circuit-switching networks. ATM technology facilitates sending data packets at a high rate of speed over LANs and WANs.

Tip: ATM offers data transfer rates of up to 1.2 gigabits per second. ATM gauges transfer rates in terms of fiber optics that can go up to 622 megabits per second.

Synchronous Protocols

In opposition to the "start-bit," "stop-bit," asynchronous protocol communication, synchronous transmission flows smooth with a regular pulse where characters are sent at timed intervals. "Start-stop" is replaced by a "synch frame" methodology. There is an error-checking process embodied in synchronous protocols at higher speeds where interruption results in an automatic resending of data.

UARTs

Universal Asynchronous Receiver-Transmitter is a computer module that maintains both receiving and transmitting abilities, allowing asynchronous serial communication. Computers use UARTs to manage serial ports. Each modem has its own UART. Recent UARTs have the ability to send asynchronous data very rapidly. For example, the new 16550 UART contains a 16-byte buffer that allows rapid data transfer as compared to the older 8250 UART version.

Tip: As modem speed increases, UARTs have been found to be the cause of increased bottlenecks.

Serial Ports

RS-232C—the asynchronous data communication port—is the industry standard port used most frequently in today's computers. Frequently called the "COM port," this port can handle speeds from 110 to 115,200 bps. UART transforms and serializes data signals received by your computer and manages the COM port.

DB-25 or DB-9 connectors are the present standard connectors for the COM port. RS-232 has ten pins that function in transmitting signals valuable for the "data negotiation" process. Two pins function as grounders and two others are reserved to "receive" and "transmit." The functions of the remaining pins are as follows:

DTR—Data Terminal Ready Data Set Ready (DSR)
RTS—Request to Send
CTS—Clear to Send
CD—Carrier Detect
RI—Ring Indicator

Asymmetric Digital Subscriber Line (ADSL)

Asymmetric digital subscriber line technology presents unlimited opportunities for high-quality videoconferencing. In addition, the medical and educational fields will be able to share information and technology though separated by continents. The following are potential beneficiaries for ADSL technology:

- Movies

- Television

- CD-ROMs

- Video catalogs

- Distance learning

- Videoconferencing

- Medical community
- Interactive games

Virtual Private Network (VPN)

A VPN is a private network on a Point-to-Point Protocol (PPP) connection that is formed upon a public system. VPNs allow its restricted users to communicate through encryption, keeping data secure from public access. A VPN functions the same whether established across the Internet or a private line. Only those authorized for access can participate in a VPN where communication is encrypted.

Virtual Private Networks can be formed upon the Internet, or participants in a VPN require only modem and client software to establish a VPN. VPNs allow customers or employees to access a private organization through remote-access for secured communication. This can be accomplished through an Internet Service Provider (ISP) or by direct connection with the Internet.

Tip: Using VPN technology allows transferring secure information as well as reducing telephone costs.

Modems

A modem (modulator/demodulator) is a device that facilitates transmitting information over telephone lines (Figure 6.2). Modems perform the role of being an intermediary between the computer that sends out digital (electronic) signals, and a telephone line that sends out analog (sound) pulses. The modem converts the computer's digital signals into analog pulses that are in turn converted by the receiving computer back to electronic signals. Therefore, a modem has the capability for both voice and data transmission.

Figure 6.2 Modem Connections

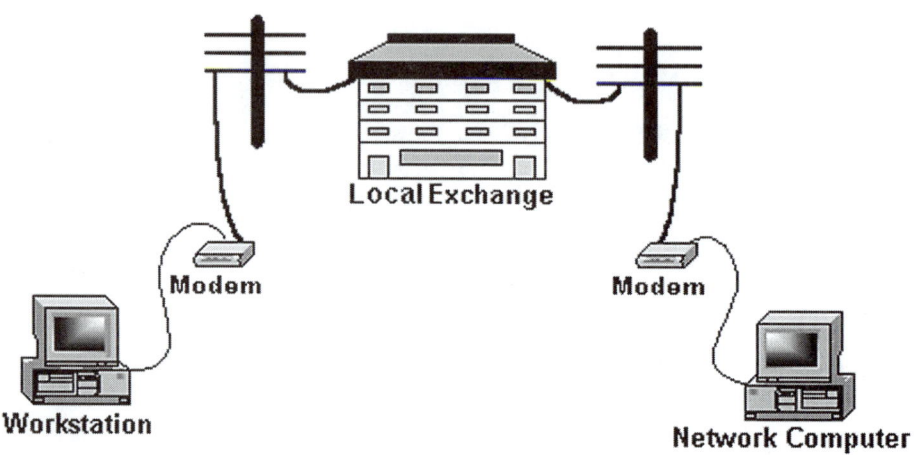

Modem Technology

Your modem is a device connected to a port. The speed at which a modem transmits messages is measured in bps (bits per second). The baud rate for the modem, such as 28.8K or 56K, measures the number of signals that occur per second during transmission. The baud rate and the bps rate of your modem will differ. Currently, modems can run at 56,000 bps. Modem speed has increased over the years with advancing technology. The evolution of modem speeds is presented in Table 6.1, according to data compiled from Networking Essentials, Self-Paced Training.

Table 6.1 Evolution of Modem Speeds

Bps Rate	Year Developed
2,400 to 9,600 bps	1984
14,400 bps	1991
19,200 to 28,800 bps	1993
28,800 bps	1993
57,600 bps	1995

Tip: If your modem runs at 28,000 bps but the receiving modem accommodates only 19,200 bps, your data will be received at the lower rate of transmission.

Modem Standards

Before two modems can talk to each other, there needs to be a process of negotiation to determine at what speed they will communicate. A variety of standards or "protocols" has been agreed upon internationally for modem negotiation. The basic standards were developed by Bell Laboratories who developed the modem. Those standards are the Bell 103 (300 bps), and Bell 212A (1,200 bps). The international equivalent is called the v.22.

Modem Lines

The options for data communication lines range from basic use of copper-wire telephone lines to high-speed connections. Your choice will be either analog or digital lines. Some of those options are provided in this section.

Public Switched Telephone Network (PSTN) (POTS)

PSTN, frequently referred to as "plain old telephone service," (POTS), is the most available and economical means for networking and transmitting data. PSTN requires only the use of a normal telephone line that carries voice data, but offers limited add-on services and features.

V.34

V.34 is the most common standard for today's modems that sends data through phone lines. V.34 uses an analog connection and transmits up to 33.6 Kbps. V.34 modems have the ability to adapt their speed of transmission according to the quality of the lines on which they transfer data.

Integrated Services Digital Network (ISDN)

Developers of ISDN envisioned it as a state-of-the-art universal, digital communication system to transfer voice, video, or data through telephone lines. ISDN conveys data through digital signals that can carry large quantities of voice, data, and video with exceptional speed and clarity. However, the acronym became tagged "It Still Does Not Work," as developers gave up working on ISDN technology as other technologies, such as ADSL, surpassed ISDN technology.

 Note: An ISDN line transfers data at three times the normal rate of telephone line service.

T-1 and T-3 Carriers

T-1 and T-3 carriers are dedicated phone lines that support data at differing rates of speed and with unique functions. T-1 lines, also called DS1 lines, contain 24 channels that transmit data at 64 Kbps per channel. T-1 lines have the potential to transmit both voice and data. T-1 lines are frequently leased and used by organizations and ISPs to connect to the Internet. T-1 transmits a signal over twisted-pair copper cables with a bandwidth that is nearly the same as its data rate,

1.5 MHz

Fast T-3 carrier lines, also called DS3 lines, are frequently referred to as "the Internet backbone" because they are frequently used to connect with the Internet. T-3 lines transfer data at the rate of approximately 43 Mbps. They have 672 channels that can support 64 Kbps per channel.

Modem Configuration

Your computer can be equipped with an internal or an external modem. An external modem connects to the computer through its serial port. An internal modem is a modem card that is seated into a slot on your computer's motherboard. An external and internal modem can both operate at the same speed. The only advantage with an internal modem is that it takes less room at your workstation.

Data transmission speeds depend upon the bits per second (bps) of your modem. For example, a 56K modem sends and receives data faster than a 28.8K modem. In addition, several modem types can compress data, which decreases the time it takes to transmit data.

In dial-up networking, the modem on your computer actually dials the modem on the connecting network. Most modems have an auto answer option that enables your computer to automatically answer as well as send and receive facsimiles in your absence.

Serial Port Interrupt Request Line (IRQ)

There are as many as sixteen IRQ lines built into the internal hardware of your computer. Many of these lines are in use and assigned by the system for specific functions by default. They transmit requests made by devices such as input/output (I/O) ports, disk drives, the keyboard, and so forth to the computer's microprocessor to prioritize and complete the requests.

The serial port IRQ is set to indicate which line your modem can use. When a peripheral device, such as a modem, is added to a computer, the Interrupt Request Line must be set with the number of the peripheral. Table 6.2 lists the default IRQ levels for the serial ports.

Table 6.2 Serial Port IRQs

Port Name	Interrupt
COM1	4
COM2	3
COM3	4
COM4	3

Warning: If two separate peripheral devices use the same IRQ numerical setting, a conflict will result that may interfere with the transmittal of data or even cause a complete transmission failure.

Input/Output (I/O) Addresses

When you gather data, you do so through devices such as the keyboard, the mouse, or disk files. This process is called "input." When you display or print data, the process is called "output."

Information is either entered or extracted from your computer. This data travels through conduits in and out of your computer called "ports." These dedicated I/O ports can send and receive data. The base I/O port that transmits data shows up in the Central Processing Unit (CPU) as an "address." Each device connected to your computer has an I/O address. Table 6.3 lists the default I/O port addresses.

Table 6.3 I/O Port Addresses

Port Name	Base Address
COM1	03F8(Hex)
COM2	02F8(Hex)
COM3	03E8(Hex)
COM4	02E8(Hex)

Port Speeds

Your modem is assigned to a COM port, also called a serial port, on your computer. A "bit rate" is the speed at which serial data is exchanged. A port speed must accommodate the modem to which it is connected for data transmission to take place. The speed cannot be slower than the speed at which your modem operates. A typical port speed is 115K bits.

Remote Connectivity

Remote connectivity is the ability to log onto a network, another computer, or the Internet from a distant location through means of a dial-up connection. Once you achieve remote access, you become a full-fledged "host" on the network and can directly send and receive files and other data.

Remote access from your computer to another computer, a network, or the Internet requires software and hardware to establish the connection. You will have the same access to data from your remote site as a "host" at workstations directly connected to the network, except you will receive the transmission of data at a slower transfer speed. As a remote host, you will typically dial-in to a server-based networking station that contains a communications server, mail and file servers, a print server, and onsite workstations (Figure 6.3).

Figure 6.3 Server-Based Networking

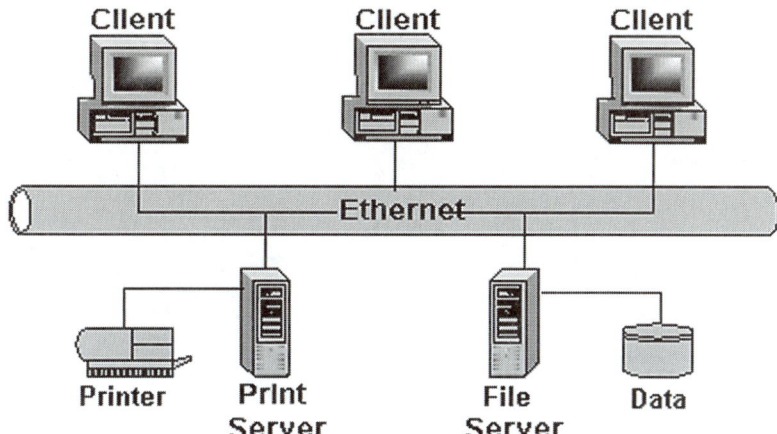

Your computer and a modem or other device are required to establish this connection as well as an appropriate set of industry-established protocols that serve as the "physical connection" to accomplish connectivity. Protocols are a uniform standard of rules and procedures that have been formulated for accomplishing remote connectivity. These settings establish the parameters of your connection and are essential before you can become a full-fledged host on the network. The kind of network—peer-to-peer, server-based—with which you want to connect will determine the preferred protocol used. Microsoft® Windows NT has the ability to support a variety of protocol settings (Figure 6.4).

Figure 6.4 Remote Access Protocols on Windows NT

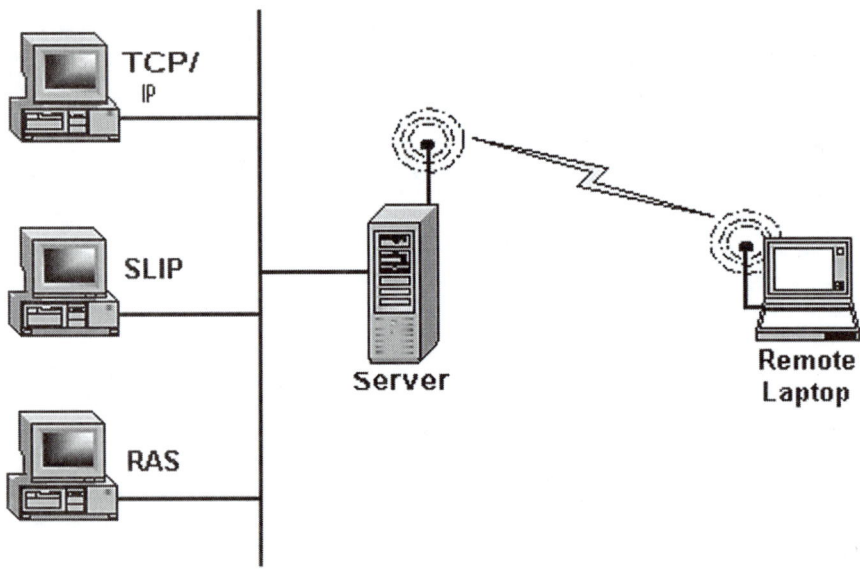

Dial-Up Access

Devices, such as a modem, can connect you with another computer, a network, or the Internet through a telephone network. One modem calls another over normal phone lines to make that connection. However, the quality and speed of data transmissions vary according to the telephone lines. New dial-up access technology provides faster rates of data transmission from the once maximum of 56,000 bps that normal telephone lines have provided. Some of the options are discussed in this section.

ISDN

Integrated Services Digital Network technology, although no longer state of the art, allows you to use two lines (B Channels) that separate voice and data transmittals. However, you may combine use of both to speed up the transmittal rates. Sending voice and data over can provide a data rate of 128 Kbps. A newer version of ISDN technology may allow faster transmission rates through use of fiber-optic cable.

ISDN provides the following advantages over POTS:

- Faster data transmission

- More reliable data transmission

- Two B channels, each with 64-Kbps transmission speed for data (a total of 138 Kbps per line), and one A channel for administration. POTS lines operate at one-fifth the speed

- Simultaneous voice and data transmission

- Superior technology

- More efficient RAS network connections

T-1 Lines

T-1 digital connection lines are used by organizations, such as large businesses, because these dedicated phone lines offer great flexibility for data support. They transmit both voice and data and contain 24 channels that send data at a rate of 64 Kbps x 24. These lines have the ability to support multiple simultaneous transmissions. T-1 transmits a signal over twisted-pair copper cables with a bandwidth that is nearly the same as its data rate, 1.5 MHz.

Fractional T-1 Lines

Normal T-1 carrier lines contain 24 channels. However, "fractional T-1" lines are available. The difference in fractional lines is that they provide less bandwidth—they transmit less data during a set ratio of time than normal T-1 lines. The advantage of using fractional lines is that they are less expensive. These lines send data at 56 Kbps rather than 64 Kbps.

Leased Lines

A viable alternative for an organization with remote sites or an individual who wants exclusive and readily available dial-up connections is leasing a permanent telephone connection. A telecommunications carrier can set up this connection between two points for a set fee that reserves the line. This arrangement provides immediate access to the "active" line, and assures a pre-arranged level of line quality. The lessee has the option to "multiplex" the connection—separate the single connection into voice, data, or high-speed use. Leased lines offer cost-effective and fast data transfer with the assurance that service is available whenever the lessee needs it.

Point-to-Point Protocol (PPP)

Point-to-Point Protocol (PPP) is an industry-standard set of Internet protocols that facilitates dial-up Internet and network connections for sending and receiving data. PPP is a more complex protocol than the older Serial Line Internet Protocol (SLIP). It has been enhanced to provide increased flexibility of service as well as added security features. The following are added functions available using PPP:

- Encrypted password control for log-in

- Ability to support more than one protocol during a connection

- Dynamic Internet protocol (IP) addressing

- Improved error control

- Facilitates data compression

Serial Line Internet Protocol (SLIP)

SLIP is an older industry-standard set of protocols that facilitates dial-up Internet and network connections for sending and receiving data. Whichever protocol you choose, PPP or SLIP, will depend on the preferred level of service you want for your organization. The following are disadvantages of using SLIP as compared with PPP:

- Inability to encrypt logon information

- Inability to support more than a single protocol (TCP/IP)

- Static IP addressing

- No error control

- Lack of data compression support

Tip: A dial-up networking client using PPP can only connect to a server that is PPP-enabled.

Point-to-Point Tunneling Protocol (PPTP)

Point-to-Point Tunneling Protocol (PPTP) is a new technology used in conjunction with the Transmission Control Protocol/Internet Protocol (TCP/IP). TCP/IP is the main protocol for the World Wide Web and the Internet.

PPP and PPTP protocols form the basic framework for connecting with Internet services. PPTP enables creation of a virtual private network (VPN) that transmits secure messages through the Internet among VPN nodes similar to branches of a tree. You dial the desired network through the Internet to use PPTP.

PSTN (POTS)

Using Public Switched Telephone Network or "plain old telephone service" is the most common type of line used in dial-up access. PSTN uses copper wiring that carries voice data. PSTN is universally accessible and economical but is based upon older analog technology. Analog transmissions are continuous. Newer technologies are based on digital communication that allows more options for the type of data transmitted and the speed at which data can be sent.

Remote Access Service (RAS)

Remote Access Service (RAS) is dial-up service that allows clients to configure Windows workstations for remote access to a LAN (local area network) or a WAN (wide area network) through a modem and telephone lines or an ISDN line. Once the connection is established with the RAS server, a "gateway" is established between the dial-up client and the network. RAS is a popular service used by telecommuters who can access all resources and services as if they were directly attached to the network.

Tip: RAS is flexible and can function with a variety of protocols, including PPP and TCP/IP.

To use RAS software, you will need a computer running a Windows operating system, a communications line (PSTN), and a modem. In addition, installation requires a compatible modem or ISDN adapter and 2 to 4 MB of hard disk space for drivers. After you have set up the connection to the server with the appropriate protocols, you can dial-in on an as-needed basis to obtain files, electronic mail, printing capabilities, database information, scheduling, and so forth. Installation of additional software and appropriate protocol configuration can further ensure security of data transmitted and received during RAS telecommunication.

Note: Before you attempt to remotely dial into your network server through RAS, make sure both your computer and the receiving network that you are dialing into have RAS software already installed. No networking can be performed without RAS installation on both ends.

UNIX and Connectivity

UNIX is a multi-tasking operating system developed by AT&T. UNIX manages computer hardware (CPU processing, video, hard disk, and so on). UNIX also provides networking capabilities between clients and servers. Networking functionality, however, is built into the system as the operating system allows for multiple protocols. Networking in UNIX is accomplished through inter-process communication (IPC)—the ability of one process to communicate with another. IPC functionality operates between applications and networking protocols.

UNIX and Remote Connectivity

UNIX facilitates dial-up service during the process of account set up. On the New Account page under Dial setup, an administrator enters your name and phone number as well as the name and telephone of your ISP. After this information is entered, the Authentication menu automatically selects the connection to your ISP. When these preliminaries are complete, you can begin to configure IP settings. To complete the connection, you need to configure the following:

- IP settings (a 32-bit IP address, a subnet mask, and a default gateway)

- DNS servers

- The gateway

- Logon script

- The modem

Internet Protocols (IPs) are usually "dynamically assigned" by your Internet provider. This means that the action (protocol assignment) takes place on an as-needed basis rather than in advance of your need to use the service. Therefore, when you dial-up to your ISP, an address is assigned at the same time the connection is made. The PPP protocol recognizes dynamically assigned addresses. If you prefer to have a static IP address, select the "Static IP Address" button in IP Setup on the New Account page. If you want dynamically assigned addresses, select the "Dynamic IP Address" button in IP Setup on the New Account page.

Domain Name System (DNS) servers are dedicated servers that track names and IPs assigned to computers. You make "requests" to these servers when you enter an address. Your request provides information about your ISP, such as domain name, that helps establish the connection. Specific information on your DNS servers, such as domain name, address, and list, are found under DNS on the New Account page.

A "gateway" allows you to connect to networks outside of your ISP. A gateway can be either a special computer or a router that connects these other computers to your ISP network. Selecting a default gateway, static gateway, or assigning a gateway can be done in Gateway Setup on the New Account page.

Before you can access your ISP or another network, you must have a logon script that will automate your dial-in process. These scripts may be simple or complex. Consult your ISP for specifics on how to generate the script that is known as a "chat script." The script can be selected and edited in "Edit Script" on the Edit Account page.

After the settings and protocols have been completed, you will need to configure a modem based upon the parameters of the configurations you have selected. Modem mapping, flow control, line termination, connection speed, and timeout can be set in "Device" on the kppp configuration page. Novell® Netware and Connectivity NetWare facilitates connecting to larger host machines through a multiprotocol management process called Open Data-link Interface (ODI). ODI is an industry standard that supports multiple protocols—IPX, TCP/IP, and Macintosh® AppleTalk. ODI was developed by Novell and Apple and for use on workstations, file servers, and on non-DOS platforms. Once the architectural ODI model has been implemented, it provides the following:

- Connectivity with computers, file servers, and mainframe computers

- Connectivity with LAN adapters compatible with ODI

- Flexible configuration

Novell's Netware and Remote Connectivity

Novell facilitates dial-up networking through its remote communication platform Novell® NetWare Connect. Off-site users on Microsoft® Windows, Macintosh, or DOS platforms use this product to perform all the functions on a network that include file review, e-mail access, printing and so forth.

NetWare Connect contains the network shell as well as the dial-out enabling software. Dedicated servers are set up with NetWare Connect that has enhanced NetWare Loadable Modules (NLMs) required for a large number of connections. NLMs provide added features that facilitate network protocols, workstations, servers, and so forth.

NetWare Connect provides the following:

- Remote security

- Shared use of modems, communication lines, and other dial-in/dial-out resources—this consolidates resources, allowing more users access to the network

- Simplified remote dial-in access from remote nodes

- The ability to support different kinds of connection types

- An industry-standard platform that supports a wide variety of hardware and software.

NetWare Connect security can be set up from a variety of options: Novell Directory Services (NDS) based user passwords, Password Authentication Protocol (PAP), Challenge Handshake Authentication Protocol (CHAP), or Third-Party Challenge and Response. In addition, dial-back options for remote users check the number and call the party back to ensure identity and reduces telephone costs. In addition, shared use of modems, communication lines, and other services frees up unused lines to support more users.

Windows and DOS users of NetWare can use PPP dialer to establish IP connections from their off-site computers. Microsoft® Windows 95 and Microsoft© 98 facilitate Novell network connections with no additional software installation. Macintosh users require an AppleTalk connection.

NetWare Connect supports a variety of connectivity methodology—ISDN, high-speed modems, X.25. The system is flexible and "scalable" in that it can expand from connecting a few nodes to many. In addition, NetWare adapts to support hardware and software of varying specifications that include, but are not limited to, LAN and WAN hardware specifications, Windows and Macintosh requirements.

Dial-Up Networking

Protocol "negotiation" is the process of one modem calling another and matching protocols to begin the process of linking (Figure 6.5.). To be ready for dial-up networking, you need to set the configuration parameters for the serial port, interrupt request line (IRQ), input/output I/O address, and maximum port speed.

 Note: Windows 95 and 98, as well as Windows NT, provide a screen wizard that simplifies the process of configuring your modem and setting your parameters.

Figure 6.5 Dial-Up Networking

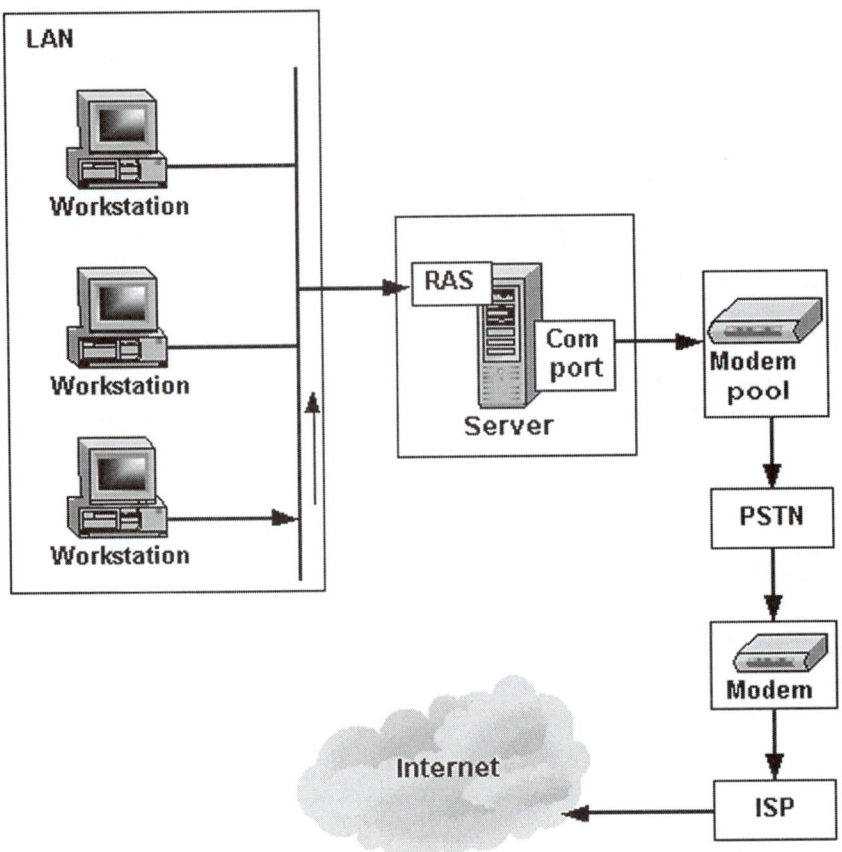

To prepare your computer for dial-up networking, you will need the following:

- 2 MB of free disk space on your computer.

- A telephone or communication line (ISDN or PSDN).

- An installed and configured modem.

- An Internet Service Provider (ISP) or direct connect between two computers.

- The appropriate Internet tool, such as a Web browser, FTP (File Transfer Protocol, or Telnet).

You will need to set your protocols for Internet or a network dial-up service. All computers on the Internet use the Transmission Control Protocol/Internet Protocol (TCP/IP). You can install dial-up networking and TCP/IP through the Dial-up Networking wizard in My Computer.

You must also set configuration parameters for the serial port, interrupt request line (IRQ), input/output (I/O) address, and maximum port speed. Your ISP will help facilitate the PPP or SLIP settings.

Post Office Protocol (POP)

Post Office Protocol is the standard protocol used to receive e-mail from a server dedicated to mail. POP is the most popular protocol for e-mail, although a more recently developed application is Internet Message Access Protocol (IMAP). POP comes in two versions, POP2, the standard for the 1980s that must be used with Simple Mail Transfer Protocol SMTP to send mail. POP3, the more recent application, functions with our without SMTP protocol. POP is also known as "point of presence," indicating a location on a WAN where a user can connect with a telephone carrier.

Windows 95/98 Dial-Up Networking

The set up and functions of dial-up networking in Windows 95/98 are similar to those for RAS. Adding dial-up networking components to Windows can be accomplished by accessing the Control Panel, opening the Add/Remove Programs Properties page, and selecting Communications from the Components list. Then select the Details button. The Dial-Up Networking checkbox appears. This option allows you to copy networking files to your local computer. After the files are copied to your computer, system configuration can begin through the Dial-Up Networking Connection wizard.

Windows NT Dial-Up Networking

The set up and functions of dial-up networking in NT are similar to those for RAS. The TCP/IP protocol or the NetBEUI protocol must be installed on your workstation for RAS communication. The PPTP protocol can provide security when you access a corporate network through Internet.

Installing dial-up networking in Windows NT can be accomplished from the Start menu by selecting Programs, Accessories, Dial-Up Networking, and installing the program files. The system automatically searches for any existing RAS-capable devices (modem or ISDN adapter). If none are available, the Install New Modem page appears, giving you the option to manually set up the devices. The Network Configuration page allows for setting of network protocols. After installation, you will be asked to set up a phonebook through the New Phonebook Entry Wizard. This wizard leads you through the last steps of configuring your dial-up connection.

Configuring a Dial-Up Connection

Dial-up networking requires that you configure your computer in order to connect to another computer, another network, or the Internet through a modem. Most operating systems provide easy setup process through use of a "wizard" that leads you step-by-step through the process of up software configuration for dial-up connection. Activate the setup wizard through the My Computer icon on your desktop.

 Note: If you use SLIP protocols for dial-up Internet and network connections, you must make sure it is installed on your computer.

Regardless of whether you are accessing your office, another computer, or a network, both your computer and the one to which you are trying to connect must have modems installed. In addition, the computer into which you are dialing must have a network server set up in order for you to access files and have full host privileges. In addition, Your POP protocol must be set and your ISP must provide both a gateway and your computer's IP address and DNS server address.

Access Dial-up Networking set up and reconfiguration options through the My Computer icon on your desktop. You have the option of placing a shortcut for dial-up networking on your desktop.

 **Tip:** If you want to access an online bulletin board, you must use HyperTerminal, not Dial-up Networking.

Proxy Server Connectivity

A proxy server is kind of an "understudy" for an actual server that a client may access. This intermediary proxy server is located between a client program and the actual server being contacted (Figure 6.6). It intercepts requests to see if it can accommodate them. If it cannot, it sends them on to the actual server for processing.

Figure 6.6 Proxy Server Connectivity

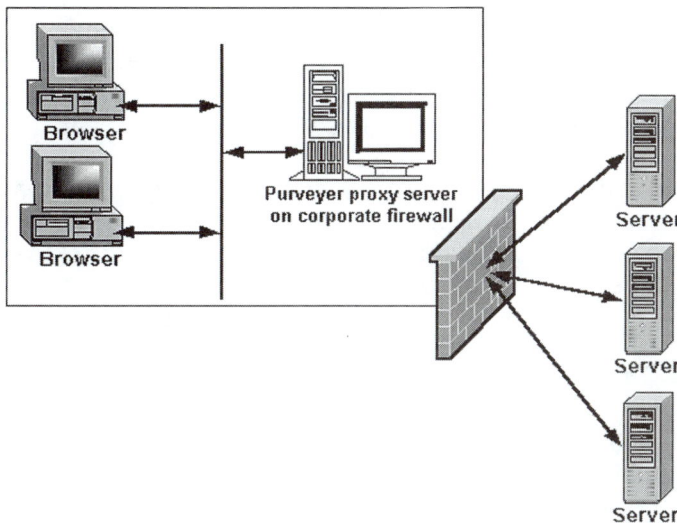

A proxy server makes the process of data transfer efficient by streamlining user requests as well as the response time. Proxy servers accommodate thousands of users by channeling and saving requests and accepting or denying requests according the protocol checks. They capture and retain these requests, sending out replies at the appropriate time. Large online services, such as AOL and CompuServe, maintain numerous proxy servers to facilitate thousands of customers.

The Microsoft Proxy Server 2.0 acts as a firewall—a device that functions as a kind of security blanket to prevent unauthorized access to or from a private network. Proxy Server 2.0 performs packet filtering by sifting through Internet data, pulling out data, holding it, and answering those requests that it can fulfill. This process prevents access to unauthorized information or Web pages along with restricted access to company employees.

Web Proxy Server

A Web proxy server is a dedicated Hypertext Transfer Protocol (HTTP) server. HTTP is the protocol used for the Web. This allows any client behind a firewall inside an organization to become a full host on the Internet. The Web proxy server scans all external servers and sends responses to internal clients. Normally, one proxy server is used for an entire subnet of users, increasing efficiency. The proxy server allows internal clients to "cache" documents, meaning that you can keep a copy of the Internet document.

A Web proxy server can access Internet service when your local network might not have direct access to the Web. The browser you are using may not be able to retrieve the file you want, and the proxy server can perform this role.

For example, through using software available on a firewall, the proxy server may have Internet access. From that position, it takes requests from your browser in the form of a Uniform Resource Locator (URL) and translates it into HTML, the language of the Web. Caching also serves to allow you to access a Web page even if the network is down.

Microsoft® Proxy Server

The Microsoft Proxy Server 2.0 serves as a cache server, is fast, provides firewall service, and has good Internet security features. The uniqueness of this proxy server is that it combines firewall and high-performance caching, also known as "active" caching. It is especially popular with large organizations and Internet content providers because of its caching ability, scalability, and fault tolerance through cascading proxy server. "Blocking" unwanted URLs and filtering data packets is also a feature of the Microsoft Proxy Server that functions upon existing networks and can be supported by a variety of protocols.

Some of the features and advantages of Microsoft Proxy Server are as follows:

User Access Control—A highly selective system to manage employee Internet access. This service can be applied to an entire organization or selected employees.

Security—Dynamic packet filtering allows closure of all firewall ports between requests. In addition, only selected packets are admitted and interchanged, according to the protocols exchanged. Other security features include an alerting system for rejected packets that could signal a potential intruder, and domain filters that limit or deny a small business network to certain Internet or Intranet sites.

Local Address Tables (LATs)—Stores records of IP addresses used by the DHCP server. This information indicates whether IP addresses are on the internal or external network for correct routing.

Internet session viewing—An administrator has the option to monitor an actual Internet session.

Caching—Storing URLs is done by default but may be manually modified as desired. The advantage of locally storing URLs is that it decreases the need to access the external network and the Internet. Caching keeps the frequently used Internet pages in the proxy server. Therefore, the client retrieves the page from the proxy, and does not have to contact an ISP or go online.

Vocabulary

Review the following terms in preparation for the certification exam.

Term	Description
ADSL	Asymmetric digital subscriber line is a new technology that facilitates high-speed, rapid transfers of data over POTS lines
asynchronous	The transfer of data intermittently in a "start-stop" process
ATM	Asynchronous Transfer Mode merges the best features of packet switching and circuit-switching networks. ATM technology facilitates sending packets at a high rate of speed in packets over LANs and WANs
baud rate	The pulse rate at which signals are transmitted through a modem. Baud is not the same as bps
bit rate	The speed at which data is exchanged
bps	Bits per second or the speed at which a modem can transfer data. The bps is not the same as baud rate
cache	Storing URLs
COM port	Also known as a serial port. The port to which a modem is connected
cps	Characters per second or the speed at which printers that are not lasers (dot-matrix or ink-jet) print. Cps can also measure the rate at which a device, such as a modem, transmits data

CPU	Central processing unit contains the "brain" of the computer. This is the device that processes and transmits data
DNS	Domain Name System are servers dedicated to track names and IPs assigned to computers
FTP	File Transfer Protocol is a protocol that allows you to download files from the Internet. The process usually requires a "client" and a "server"
gateway	A gateway is a device that facilitates changes of protocols or topography
host	A host is just another node computer
I/O	Input/output refers to the process of gathering data and distributing the information. Gathering is accomplished with devices such as a mouse and disks. Output is performed through display and printing features
IRQs	Sixteen interrupt request lines are built into the internal hardware of a computer that transmit requests made by devices
ISP	Internet service provider is a private concern that offers connectivity to the Internet to individuals or organizations
LAN	A local area network connects computers over a limited geographic area to share resources and data
modem	A device that allows data to be transferred over telephone lines
network	A group of connected computers that can serve a large area (WAN) or a smaller area (LAN)

node	A location with an address that might be another computer, a printer, or another device
ODI	Open Data-Link Interface is an industry standard developed by Apple and Novel that supports multiple protocols and allows connectivity
POP	Post Office Protocol is the standard protocol used to receive e-mail
port	A connection on a computer into which peripherals are attached
POTS	Plain Old Telephone Service refers to normal telephone lines that connect telephones
PPP	Point-to-Point Protocol is an industry-standard set of Internet protocols that facilitates dial-up net-working
PPTP	Point-to-Point-Tunneling Protocol is a newer proto-col used in conjunction with TCP/IP which is the main protocol for the WWW and Internet
protocol	Uniform industry standards of rules and procedures established for accomplishing remote connectivity
Proxy Server	A server that serves as an intermediary server and a firewall to intercept and answer requests
PSTN	Public Switched Telephone Network is frequently referred to as "plain old telephone service" or POTS. This is the most available and economical means for networking
RAS	Remote Access Service is a Windows service that allows clients to configure Windows workstations for remote access to a LAN or WAN

SLIP	Serial Line Internet Protocol is an older industry-standard set of protocols that facilitates dial-up Internet and network connections
synchronous	The smooth transfer of data where characters are sent at regular timed intervals
TCP/IP	Transmission Control Protocol/Internet Protocol is a protocol invented by the U.S. Government as the "standard" for transmitting data over the Internet
UART	Universal Asynchronous Receiver-Transmitter is a computer module that contains both receiving and transmitting abilities that allow asynchronous serial communication
UNIX	UNIX is a multi-tasking operating system developed by AT&T. Unix manages computer hardware
URL	A Uniform Resource Locator is an Internet address that can be located through use of a browser
VPN	A Virtual Private Network is a private network on a Point-to-Point Protocol (PPP) connection that is formed upon a public system that allows its restricted users to communication through encryption
WAN	A wide area network connects computers over an extended geographic area to share resources and data

In Brief

If you want to...	Then do this...
Use the best technology to facilitate videoconferencing	Use ADSL
Use the most economical point-to-point connection method	Use POTS
Send data at the rate of 56,000 bps	Make sure the recipient has a modem with the same or higher bps rate capability
Use RAS service for telecommuting	Install RAS software on the server and on the remote computer
Have Dynamic IP address capability	Use the PPP protocol rather than SLIP
Transfer state-of-the-art video images	Use the PPP protocol for best results
Have help with setting your PPP or SLIP settings for dial-up networking	Contact your Internet Service Provider
Encrypt your data before sending it over the Internet	Use the PPTP protocol
Follow the easiest process for setting up dial-up networking	Use the Dial-Up Networking wizard to set your modem and networking parameters
Have private access to information within an organization such as a bank	Join a Virtual Private Network (VPN)
Use the standard e-mail protocol	Use POP

Use an asynchronous technology that merges the best features of packet-switching and circuit-switching	Use ATM
Configure your computer for a dial-up connection in the easiest way possible	Use the step-by-step "wizard" option

Lesson 6 Activities

Complete the following activities to better prepare you for the certification exam.

1.   Define the three major functions of remote connectivity.

2.   Describe the term "protocol."

3.   Identify some of the advantages of using the PPP protocol over SLIP.

4.   Identify the most common communications format for transmitting data from modem to modem.

5.   Define the process of "protocol negotiation."

6.   Name some of the advantages of using leased lines.

7.   Explain how and why you would use RAS.

8.   Explain the value of using a VPN.

9.   Describe the purpose of a Proxy Server.

10.   Describe how UARTS transmit data.

Answers to Lesson 6 Activities

1. Remote connectivity allows you to log on to a network, another computer, or the Internet from a distant location. Once you are granted access to the network, depending on the level of access you are granted, you can send and receive files and other data.

2. Protocols are industry-established standards that facilitate remote connectivity.

3. PPP allows for greater security with password log-on procedures. PPP can support more than one protocol during a connection and supports dynamic IP addressing. It can also support data compression for faster transmission. SLIP does not provide any of these features.

4. PSTN or telephonic lines are the most common means for transmitting data. This method offers limited services and features but is accessible and economical.

5. "Protocol negotiation" is the process of one modem calling another modem to initiate checking and matching protocols for completing the dial-up network connection.

6. Leased lines allow an individual or a large organization with off-site locations exclusive and available permanent telephone connection for dial-up access. In addition, leased lines allow for "multiplexing" that separates connections into voice, data, or high-speed use. Leased lines are cost-effective and transfer data fast.

7. Remote Access Service (RAS) is a popular and flexible service that you can use with Windows workstations for dial-in networking. RAS allows remote access to a LAN or WAN to obtain files, e-mail, database information, or for printing and scheduling, and so forth.

8. A Virtual Private Network is a private network that allows users to communicate with organizations, like banks and corporations, through encryption that ensures safe and private transmission of data.

9.   A proxy server is located between a client program and an actual server to intercept requests to see if it can accommodate them. It streamlines the process of data transfer.

10.   UARTS are modules that contain both receiving and transmitting abilities that allow asynchronous communication. Asynchronous means that each character sends a signal in a start-stop type of transmission.

Lesson 6 Quiz

These questions test your knowledge of features, vocabulary, procedures, and syntax.

1. Which of the following functions accurately describe the word "protocols"?

    A. Facilitate modem configuration
    B. Establish parameters for your remote connection
    C. Aid in data compression
    D. Are not required for "full-fledged host" status

2. Which of the following is true of the PPP protocol?

    A. It is outdated but facilitates dial-up networking
    B. Provides enhanced security features like password control
    C. Cannot facilitate data compression for faster transmission
    D. Does not function well with Internet service

3. What is the only difference between a network "host" and an on-site user?

    A. A host has a unique encrypted log on procedure
    B. A host can never alter network directory files
    C. More than one protocol cannot be supported during a connection
    D. The rate of speed at which data is transferred

4. Which factors are important when choosing either a SLIP or PPP protocol for your network?

    A. The bps speed of your modem
    B. The I/O port set up
    C. The type of operating system you are using
    D. The preferred level of service required

5.   Why do many administrators prefer an ISDN line rather than POTS?

      A.   It transfers video images
      B.   Is less expensive to use
      C.   Supports only one protocol
      D.   Transfers data at the twice the speed of POTS

6.   Which choice best describes a modem's function?

      A.   Transfers analog signals over telephone lines
      B.   Transfers digital signals over telephone lines
      C.   Cannot transfer both voice and data transmission
      D.   Works better when it is an internal rather than an external device

7.   Which type of conflict can occur with a serial port IRQ?

      A.   Cause your operating system to crash
      B.   Corrupt files on the network
      C.   Interfere with transmittal of data
      D.   Change the settings on your modem

8.   Which of the following best describes a proxy server?

      A.   It is an inefficient but necessary device
      B.   It is located on the client end of communication
      C.   It sends out responses at a uniform rate
      D.   It filters incoming requests

9.   When is RAS service not accessible?

      A.   With a LAN server
      B.   With a SLIP protocol
      C.   On a computer not running a Windows program
      D.   With POTS lines

10.   Which communication option is best suited for computers connecting to the
       Internet?

       A.   File Transfer Protocol
       B.   Telnet
       C.   RAS
       D.   TCP/IP

Answers to Lesson 6 Quiz

1.   Answer B is correct. The most important function of protocol selection is to
      establish the parameters of your connection to become a full-fledged host on
      the network.

      Answer A is incorrect. A modem is a device for connecting to the network.

      Answer C is incorrect. PPP supports data compression.

      Answer D is incorrect. A "full-fledged" host status is an access level granted by a
      network administrator.

2.   Answer B is correct. PPP is a state-of-the-art protocol allowing great flexibility
      and increased security features over similar protocols.

      Answer A is incorrect. PPP is a new protocol.

      Answer C is incorrect. PPP facilitates data compression.

      Answer D is incorrect. PPP functions exceptionally well with the Internet.

3.   Answer D is correct. Once a dial-up connection is made, you can become a
      "full-fledged host" on the network with ability to send and receive files, except
      you will receive the transmission of data at a slower rate of speed.

      Answer A is incorrect. Host logon is no different from anyone else's logon.

      Answer B is incorrect. A host with "Full-Access" rights can alter a file.

      Answer D is incorrect. PPP is flexible and supports more than one protocol.

4.   Answer D is correct. The preferred level of service required is the main factor
      that determines the kind of protocol you choose. PPP is a newer set of protocols
      that allows you increased security as well as increased flexibility of service.

      Answer A is incorrect. The bps modem speed determines the rate of data transmittal.

Answer B is incorrect. Devices connect to ports.

Answer C is incorrect. SLIP and PPP support all operating systems.

5.   Answer A is correct. ISDN can transfer video, voice, and data through telephone lines.

Answer B is incorrect. ISDN is more expensive than POTS.

Answer C is incorrect. ISDN supports more than one protocol.

Answer D is incorrect. ISDN transfers data at three times the speed of POTS.

6.   Answer A is correct. A modem transfers computer digital signals into analog pulses that are converted back into digital signals at the receiving end.

Answer B is incorrect. The computer sends digital signals.

Answer C is incorrect. A modem can transfer both video and voice data transmission.

Answer D is incorrect. The functioning of a modem is not determined by whether the modem is internal or external to the computer.

7.   Answer C is correct. If the Interrupt Request Line is not set with the correct number of the peripheral being served, the conflict may interfere with transmittal of data.

Answer A is incorrect. The settings may interfere with performance but will not necessary cause a crash.

Answer B is incorrect. It does not act like a virus, but just interferes with the speed of data transmittal.

Answer D is incorrect. The settings on the modem were configured before attachment.

8.   Answer D is correct. A proxy server acts as a filter for incoming requests.

      Answer A is incorrect. A proxy server is not a device and it increases efficiency
      because it screens incoming requests and tries to answer them.

      Answer B is incorrect. The proxy server is located between the client program
      and the actual server being contacted.

      Answer C is incorrect. A proxy server saves requests and sends out replies at an
      appropriate time.

9.   Answer C is correct. RAS must be configured on Windows workstations.

      Answer A is incorrect. RAS is accessible on a LAN.

      Answer B is incorrect. RAS is accessible with SLIP.

      Answer D is incorrect. RAS is accessible through POTS.

10.   Answer D is correct. Transmission Control Protocol/Internet Protocol (TCP/IP)
      must be used for the Internet and can be set up through the wizard or by
      accessing the My Computer icon on your workstation.

      Answer A is incorrect. FTP is a choice of protocols.

      Answer B is incorrect. Telnet is a choice of protocols.

      Answer C is incorrect. RAS is a remote connectivity service of choice.

Lesson 7: Network Security

Operating a secure network requires more than just security products. It requires architecture, planning, and a strategy for communicating your rules to everyone who uses your systems. You need rules and guidelines in place before implementing a security infrastructure.

After completing this lesson, you should have a better understanding of the following topics:

- Security Basics
- Security Models
- Passwords
- Permissions
- Cryptography
- Data Encryption
- Firewalls
- Virus Protection
- Security Myths

Security Basics

The first element of network security involves the creation of user accounts and assignment of passwords that provide permission to log on to the network. Limiting the number of authorized users as well as what they can access enhances security over data stored on the network. To log onto a network, you type a password, and then the server validates your name and password entry, giving or denying access by checking a database on the server. If you have user rights to access the server's data, permission is granted. This process of password checking is called "authentication."

Once you have access to a system, your ability to perform specified actions on the system is called "rights." Rights apply to the entire system and are assigned to groups or users by the system administrator. Rights give users access to services, such as backing up files and directories, shutting down the computer, logging on interactively or changing system times which normal discretionary access controls do not provide.

Authentication

Authentication is the process of a user request to use a server and the server's response to that request. Gaining access to computer and network files is based upon authenticating the identity of the user. When a user logs on a user name and password into the computer, the process of "authentication" begins as the system checks this information against an authorized list of user names and passwords. A user chooses a user name and password, which is a series of secret characters that prevents unauthorized individuals from accessing a computer. Once the user name and password are confirmed by the system, the user is granted access to the system.

 Tip: Authentication takes place on the "Session Layer" of the Open Systems Interconnection (OSI) model. This layer determines where two computers begin, use, and end a communication session.

Security Models

A security model is an established policy that users can easily understand and follow that helps protect system and individual user data stored on a network. Security models form a framework for the appropriate use of a system and define who can do what and when. Most organizations have developed a security model that ensures data safety from the desktop to the network through user-level and share-level protection levels. User-level is more specific and more complex to define; share-level is simpler. A user-level of security requires a centralization of user accounts with access rights given to individual files and directories. A share-level of security has broader access through network folders and single-password access.

 Tip: The best security model combines a share-level of security with a user-level security.

However, regardless of the model chosen, there should be a plan that defines how you want to secure your data prior to implementing a security plan. This plan must be documented and understood by users within your network infrastructure. A security model should include some or all of the following security measures:

- Access control
- Password policies
- Permissions
- Data encryption
- Firewalls
- Virus protection

Microsoft® Windows NT Security Model

The Windows NT security model is based on the following components:

- Local Security Authority (LSA)

- Security Account Manager (SAM)

- Security Reference Monitor (SRM)

In addition, Windows NT security includes logon processing, access control, and object security services that form the foundation of security in the Windows NT operating system, known as the security subsystem. This subsystem affects the entire operating system.

Local Security Authority (LSA)—The LSA is the heart of the security subsystem that validates local and remote logons for all types of accounts. It accomplishes this by verifying the logon information from the SAM database, and also provides the following services:

- Checks user access permissions to the system

- Generates access tokens during the logon process

- Manages local security policies

- Provides user validation and authentication

- Controls auditing policies

- Logs auditing messages generated by the SRM

A graphic representation of the Windows NT security model is shown in Figure 7.1.

Figure 7.1 Windows NT Security Model

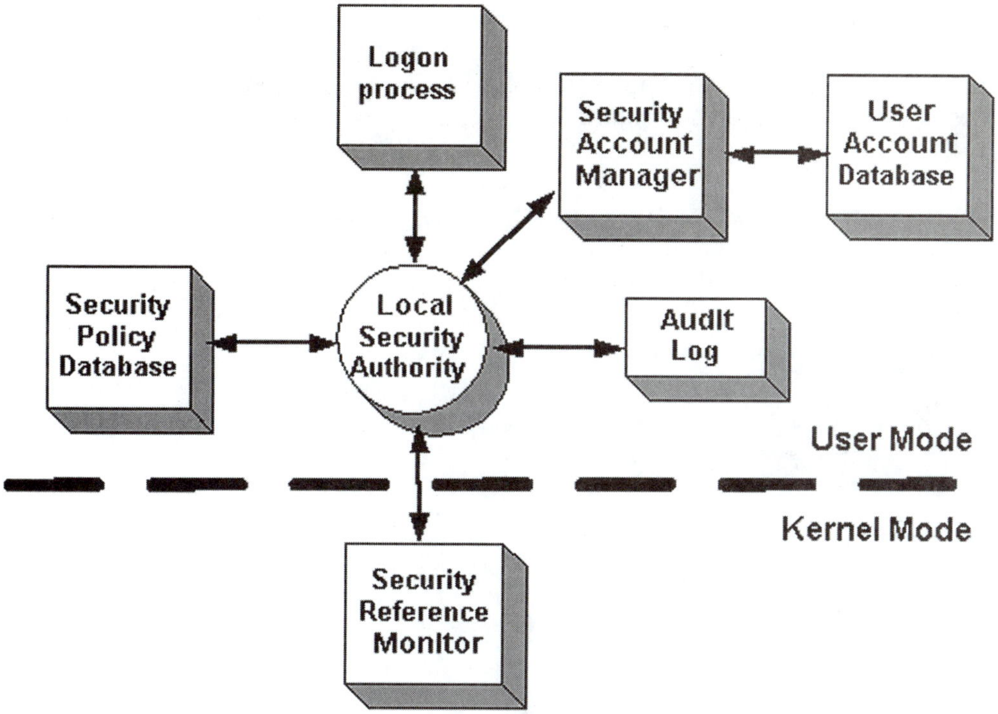

Security Account Manager (SAM)—The SAM manages a database that contains all user and group account information. SAM provides user validation services that are used by the LSA but are transparent to the user. SAM checks logon input against the SAM database and returns a secure identifier (SID) for the user, as well as a SID for each group to which the user belongs.

When a user logs on, the LSA creates an access token that includes the SID information along with the user's name and associated groups. Every process that runs under this user's account will have a copy of the access token. When a user requests access to an object, a comparison is made between the SID from the access token and the object's access permissions list to validate that the user has the correct permissions to access the object.

Security Reference Monitor (SRM)—The SRM runs in kernel mode and is a Windows NT Executive component that enforces access validation and audit generation policies required by the LSA. SRM provides services for access validation to objects and access privileges to user accounts. The SRM maintains only one copy of the access validation code on the system to ensure that objects are protected, regardless of type. To determine user access to objects, the SRM adheres to the steps below, and as shown in Figure 7.2.

Figure 7.2 SRM Access Validation Process

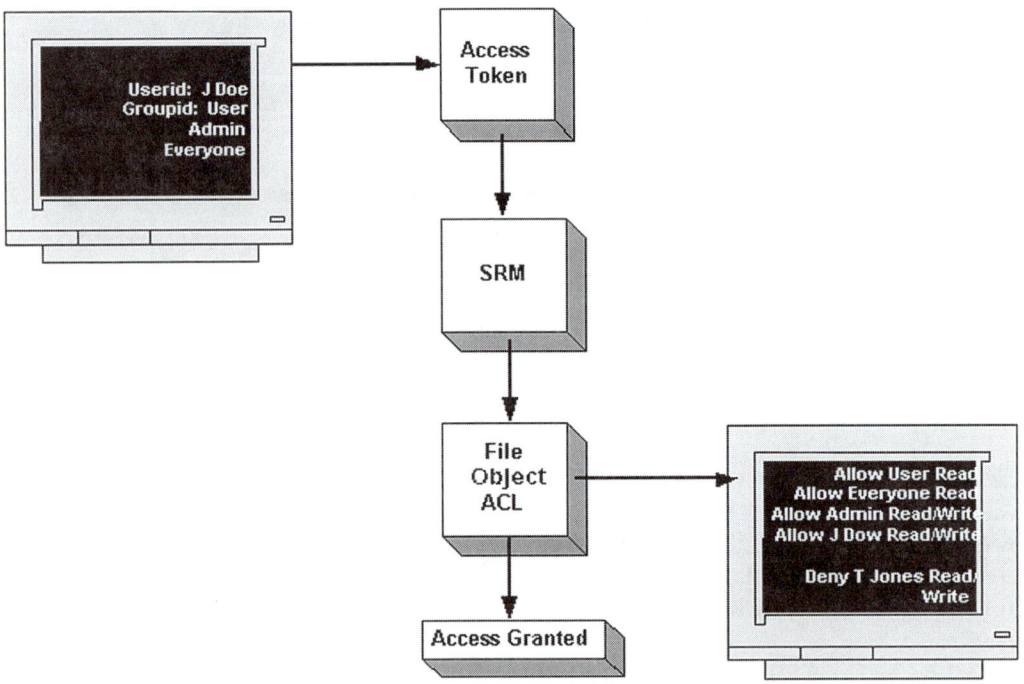

When data access is requested, a comparison is made between the file's security descriptor and the SID information stored in the user's access token. If you have sufficient rights, you can obtain access. The security descriptor comprises all of the Access Control Entries (ACE) included in the object's Access Control List (ACL).

When the object has an ACL, the SRM checks each ACE in the ACL to determine if access to the data is granted. If the object does not have an ACL associated with it, SRM automatically allows access to everyone. If the object has an ACL with no ACEs, then all access requests to that object are denied.

After the SRM grants access to the object, continued validation checks are not needed when accessing the object. However, any future access to the object is obtained by using a handle that was created when the access was initially validated.

Novell Security Model

Novell's multi-layered security model is called Novell® NetWare. This layered security model consists of the following:

- Login/password security

- Access rights

- File attributes

Users must pass through this hierarchy of security before they can access directories and data. In addition, access restrictions, privileges granted, and file conditions further restrict users from data. Furthermore, users are limited by access restrictions, assigned privileges, and specified file conditions.

Login/password embodies the authentication process and initial access to a system. However, even authenticated users may be restricted from certain workstations or network access during certain timeframes. Access rights set down privileges users have once they access the system. Seven kinds of access rights can be assigned in NetWare. File attributes, or attribute security, deals with file sharing, reading, writing, executing, and so on. The difference between access rights and attribute security is that access rights grant users access to files, and file attributes tell users what they can do with the file once they have access. More details on these NetWare security layers are discussed in the next paragraphs.

Login/Password Security

This first layer of NetWare security controls system access through requiring two pieces of information—logon name and password. When the system checks these two separate IDs against each other and they do not match, log on to the system is denied. If access is allowed, the user passes through another level of security called "login restrictions" that checks the following, among other factors:

■   Does the user have access rights to the computer on which logon took place?

■   Is the user restricted from entering the system on the date logon took place?

■   Is the user restricted from entering the system during the timeframe that logon took place?

In addition, another security feature called "Intruder Detection/Lockout" monitors unauthorized logons. When this feature detects violations, it automatically locks accounts that reach a threshold of violations.

An enhanced logon procedure available in Novell that eliminates some steps required in general login-password security, while allowing use of all Novell applications, is called "single sign-on." Single sign-on is a security feature where the user requests access to Novell Directory Services (NDS) by entering a single password. The system checks the single sign on for authentication. The requested application or system loads upon authentication. If the user is denied authentication, the system asks for submission of a password appropriate for the application or system. This process eliminates the need for multiple passwords and repeatedly logging on to system applications during a session.

When the system requests that a user change passwords and a new password is entered, the system immediately registers the change in NDS. The system further secures user profile information through encryption and storage of authentication credentials in NDS and not on user computers.

Access Rights

The following table lists the seven types of NetWare access privileges.

Table 7.1 Novell Permissions Levels

Privilege	Symbol	Definition
Access Control	(A)	Permission to determine access rights
Create	(C)	Permission to create and write new files or subdirectories
Erase	(E)	Permission to delete files or subdirectories
File Scan	(A)	Permission to search a directory or subdirectory
Modify	(M)	Permission to modify file names or attributes
Read	(R)	Permission to read an existing file
Supervisor	(S)	Permission to all rights of a file or folder
Write	(W)	Permission to edit an existing file

Access rights are assigned to users according to the functions they perform in directories. You must understand which rights (W, A, C, M, F, and so on) are associated with NetWare activities.

Tip: Access rights change according to the version of NetWare. Therefore, check the specific access rights privileges given with each version of NetWare you use.

Directory rights in NetWare are assigned by default. Directories have rights that are independent and do not transfer down. "Locks" are placed upon directories that can be accessed through a kind of "trustee assignment" that allows access with a user key. Only those directories that are considered sensitive need have locks. NetWare directory rights are classified Maximum Rights Mask (MRM). Directory users have "effective rights" to a directory that take into consideration the actual privileges when using a directory and the directory MRM.

File Attributes

File attributes pertain to what a user can do with a file once access has been made. This affects all file users and is a separate issue from access rights. You may have Write (W) access to a file, but if the administrator has assigned the file READ-ONLY status, this level of NetWare security overrides your Write status. File attributes can be divided into "security" and "feature."

Security Attributes

The following are file sharing and file alteration attributes:

Non-Sharable (NS)—Access is limited to one user at a time.
Sharable (S)—Access by simultaneous multiple users is allowed.
Read/Write (RW)—Users can view and alter a file.
Read/Only (RO)—Users can view only the file.
Execute Only (X)—Users cannot copy or delete the file.
Hidden (H)—Users cannot view, write, delete, or copy the file.

The hidden attribute applies to archived files that cannot be seen, written to, deleted, or copied to.

Feature Attributes

The following are feature attributes that provide information on the following features of a file:

Activate TTS (TTS)—Transactional tracking
Indexed (I)—Turbo FAT indexing
System (S)—A system-owned file
Not Yet Archived (A)—File has been modified since last backup

The combined power of these file attributes lays the base for an effective security system that controls user access to specialized NetWare files.

 Tip: The default attribute for all system files is "non-shareable read/write" (NSRW).

UNIX Security Model

UNIX supports a large number of simultaneous users. Therefore, UNIX tools for adding, removing, and managing users are flexible and strategic in maintaining a good security model. This layered security model consists of the following:

- Login/password security
- Access rights
- Permissions

Logon/Password Security

The UNIX first-line password security uses the DES encryption algorithm to encrypt passwords. Passwords are a minimum of six characters and a maximum of eight characters. User and group encrypted passwords and related information are stored in databases named `/etc/passwd`, `/etc/group`, and `/etc/shadow`. When you log on, the system encrypts your password and compares it with the information stored in one of these two files. If the two match, access is allowed. In addition to encrypted passwords, the `/etc/passwd` database contains the following:

- List of user logon names
- User IDs (UIDs)—a number that represents log on
- Group IDs (GIDs)—the group log on default membership
- Location of home directories

In addition to encrypted passwords, the `/etc/group` database contains the following:

- List of user groups
- Group passwords
- GID number
- List of all usernames that belong to a group

The `/etc/shadow` database contains "shadow" passwords that can be read only by root. Shadow passwords keep encrypted passwords private because passwords are normally stored in the `/etc/passwd` file that can be accessed by all users.

The `/etc/shadow` file can only be accessed by privileged users. Shadow passwords are installed on the system by default and are important for protecting system passwords because when this file is in place, all passwords contained in the `/etc/passwd` file can only be read by root.

 Tip: UNIX has a Network Information Service (NIS) that shares a server `passwd` file containing usernames, groups, and passwords with UNIX computers.

Access Rights

Because UNIX is a multi-user operating system, sharing files requires a strong security mechanism to protect those files. To support this need, UNIX has built permissions mechanism into the UNIX file system. When a user is added to the system, a home directory is created, appropriate database entries are made for user files, and the user is added to a user group. To protect personal or group files from unwanted access, the UNIX "permissions" system assigns each user to an appropriate level of access upon account creation.

Access level for a user is contingent upon the user role to a file. Each file contained in the system is "owned" by one user and related to one group. An administrator assigns and changes these memberships as appropriate. When you log on, the system matches your identity with a permissions file. At this point, the system decides what files you can access and what you can do with those files. Each file contains nine "switches" controlling the file actions of read, write, or execute.

Users can perform one or all of the following roles regarding a specific file:

Owner/User—The creator of the file can assign file permissions.
Group—The user is a member of a group that has access to a file.
Other—The user is not the creator, the "owner" of the file, or a member of the group that has access to the file.

Permissions

Three permission "switches" for the owner/user, group, and other roles can be defined. These permissions for files are listed in the following table:

Table 7.2 UNIX Permissions Levels

Switch	Symbol	Definition
Read	(R)	Permission to read the file contents
Write	(W)	Permission to change file content or delete file
Execute	(E)	Permission to run a file change directory

Some files contain permissions established by default. In general, users first have an established role—owner, group, other—and then permission level according to a specific file or directory.

As an extra security level, entire UNIX directories and subdirectories can be protected by a blanket denial of access. In addition to read, write, and execute, there is a permission level entitled "Access Denied," indicating that the directory or subdirectory cannot be read, written, or executed.

Directories have additional numerical permission bits called SETUID, SETGID, and STICKY. The SETUID bit allows users to read and write to files that are normally not accessible. The SETGID bit, when used with a directory, sets group ownership of a file. The sticky bit protects directory files.

Tip: UNIX permission levels can also be set by the following hierarchy of numbers:
4 = Read permission
2 = Write permission
1 = Execute permission
0 = No permission

Passwords

A password is an unspaced sequence of characters that determines if a computer user requesting access to a computer system is really that particular user. Typically, users of a multi-user or securely protected network system claim a unique name, often called a user ID.

To verify that someone entering that user ID really is that person, a second identification, the password, known only to that person and to the system itself, is entered by the user. A password typically contains between 4 and 16 characters, depending on how the computer system is set up. When a password is entered, the computer system is careful not to display the characters on the computer screen, in case others might see it.

 Note: A "strong password" cannot be easily guessed and should be eight characters or longer that combines letters and numerals. Unique characters may also be chosen for complexity. The word should be difficult to remember. It can be symbolically easy to remember (ez4u2no).

Password Creation Guidelines

Good criteria when choosing a password or setting up password guidelines include the following:

- Do not create a password that someone can easily guess who you are, such as: Your Social Security number, birthday, maiden name, or your pet's name.

- Don't use a word that can be found in the dictionary (since programs exist that can rapidly try every word in the dictionary!).

- Don't use a word that is currently newsworthy.

- Don't use a password that is similar to your previous password.

- Be sure to create a password that contains multiples of alpha and numeric characters.

- Choose a password that you can remember, because you don't want to write your password on a sticky note and attach it to your computer monitor so everyone can gain access to your network's data.

- Many networks require that you change your password on a periodic basis.

Your unique user name and password give you permission to enter the network. The password you choose should not be easily identifiable or guessed, and shouldn't be left where another can find it. Passwords should be changed frequently, approximately every 30 to 60 days, in case your password has been compromised.

Permissions

Your ability to access a resource on a network directory is determined by the level of "permission" that has been assigned to your user account. Permissions can be set up on an individual or a group basis. At the top of the hierarchy of permissions is "full access" that grants you permission to read, modify, delete, save, copy, or generally alter a file. Other permissions include "read-only access" and "no access" to network directories.

Tip: "Read-only access" to network files prohibits modifying, deleting, or altering files. The lowest level of permission is "no access" to network directories. In addition, some directories may be restricted from certain groups or individuals.

Network administrators assign access permissions to individuals or groups who share network directories or files. The levels of accessibility can range from "read-only," to "read, execute, write, delete" that allows opening, reading, creating, editing or altering files, and removing files from a directory. The lowest level is "no access."

Network administrators can control and limit the number of individuals or groups who can enter the network through assignment of access rights. In addition, software and database systems provide tools that allow network administrators to follow an audit trail that lists the names of those who have entered a system, when they used it, and what functions they performed.

Cryptography

Cryptography is the process of securing private information passed through public networks by mathematically scrambling (encrypting) data, making it unreadable to anyone except the person or persons holding the mathematical "key" that can unscramble (decrypt) it.

 Tip: Many network operating systems automatically encrypt user names and passwords.

Cryptography is no longer relegated to the fields of espionage, government, the military, or wartime societies. The global proliferation of information and advancing technologies has warranted use of cryptography as a "tool" to ensure security and privacy when sharing data. Cryptography is the skill of encrypting messages and cryptanalysis is the skill of decrypting messages. Cryptographers and cryptanalysts must have knowledge of message securing, authentication, digital signatures, electronic money, and a number of other related programs.

Many methods of cryptography rely on algorithms for creating cipher text or encrypted messages. In this reference, an algorithm is a logical sequence of steps that can translate a kind of encryption and decryption called a "cipher." Cryptography is the art of breaking a cipher without having a key—algorithms require use of a "key" to encrypt and decrypt text. The two types of key-based algorithms are "symmetric" and "asymmetric."

The two most common types of cryptography are "same-key" and "public-key." In same-key cryptography, a message is encrypted and decrypted using the same key, which is passed along from one party to another in a separate transmission. A more secure method is public-key cryptography. This method uses a pair of different keys (one public, one private) that have a particular relationship to one another, such that any message encrypted with one key can only be decrypted with the other key and vice versa.

As methods for cryptography become more sophisticated, the challenge for deciphering text has increased. Depending on the situation, this decryption skill is called either "cryptanalysis" or a "cryptographic attack." The following is a list of some kinds of deciphering techniques:

Chosen plain-text attack—The cryptanalyst or attacker can have any text encrypted with an unknown key. The challenge is to pinpoint the key that was used for encrypting the text.

Cipher text-only attack—The cryptanalyst or attacker has no knowledge about the text and must guess about its content. Precedent about how text and correspondence is formatted (heading, body, closing) can aid in this random approach.

Known plain-text attack—The cryptanalyst or attacker has knowledge or has the ability to guess parts of the text. The challenge is to determine the unknown parts of the text by a comparison with the known part. Guessing the key used in the cipher text could complete this process.

Man-in-the-middle attack—During the process where two users exchange keys to prepare for private communication, the cryptanalyst or attacker takes a position between the parties. In this central location, the intervener switches keys and obtains knowledge of both keys that allows eavesdropping on all communication between the two parties who think they communicate in private.

Timing attack—This method requires a sophisticated process of measuring modular exponentiation. This is used in deciphering complex algorithms as RSA (Rivest, Shamir, Adelman) and Diffie-Hellman.

Encryption

Encryption is the process of translating data into a code that cannot be interpreted by anyone who does not have the key or password to read the encryption. Data that can be read without translation is called "plain text." Data that has been encrypted is called "cipher text." Cipher text cannot be understood by unauthorized people.

Encrypting data by translating it into a secret code is the most effective means of ensuring that information stored on a network or data being transferred during networking cannot be compromised. Scrambling data in this manner makes it unreadable. You can choose from a variety of ways to encrypt data to make it indecipherable to an unauthorized individual.

 Tip: Some operating systems have the ability to automatically encrypt internal and external communication. Passwords and server data are especially sensitive and must be protected.

The use of encryption is as old as the art of communication. In wartime, a cipher, often incorrectly called a "code," can be employed to keep the enemy from obtaining the contents of transmissions. Technically, a code is a means of representing a signal without the intent of keeping it secret; examples are Morse code and ASCII.

The Data Encryption Standard (DES) is the U.S. Government standard for encryption developed in cooperation with IBM. It is the encryption algorithm of choice for private commercial electronic communication. DES protects data through using what is called a 56-bit encryption key. This 56-bit key allows the use of as many as 72,057,594,037,927,936 possible keys.

Simple ciphers include the substitution of letters for numbers, the rotation of letters in the alphabet, and the "scrambling" of voice signals by inverting the sideband frequencies. Ciphers that are more complicated work according to sophisticated computer algorithms that rearrange the data bits in digital signals.

To easily recover the contents of an encrypted signal, you need the correct decryption key. The key is an algorithm that "undoes" the work of the encryption algorithm. Alternatively, a computer can be used in an attempt to "break" the cipher. The more complex the encryption algorithm, the more difficult it becomes to eavesdrop on the communication without access to the key.

Public Key

Asymmetric and symmetric are the two types of encryption. Private keys use symmetric encryption. Symmetric encryption uses the same key to both encrypt and decrypt data. Therefore, senders and receivers of messages have the same key that encrypts and decrypts. Public key technology requires the use of two keys—a public key that encrypts messages and a private key that decrypts messages received.

 Tip: Public key technology is also called Diffie-Hellman algorithm.

Kerberos

MIT developed a network authentication protocol that identifies a client from a user name and a password. Kerberos provides the security measures that are not found in many Internet protocols, making the Internet data transfer especially insecure. Kerberos provides both tools of authentication and powerful cryptography that can be used on a network. This technology is useful for systems that send unencrypted passwords over the network, client/server programs where client identity is unchecked, and programs that put the responsibility on the client to restrict network activity. Through assignment of a key called a "ticket," Kerberos allows two clients private interchange upon a network. Users must log on to the network using the ticket that inserts itself into the messages in a format that identifies the sender.

Secure Sockets Layer (SSL)

SSL is a Netscape-developed protocol that allows sending private data through the Internet. SSL requires the use of a "private key" to encrypt data that is transferred in this method—making a secure client/server connection. SSL provides a valuable service to those Web sites that require user input that request confidential and financial information. The SSL protocol does not limit the amount of data that can be sent through this secure channel and it supports multi-browser access.

Although SSL is presently the more popular protocol for securing Web connections, a second protocol called Secure Hypertext Transfer Protocol (SHTTP) is more powerful and flexible than SSL. This protocol offers a choice of key options and cryptographic algorithms through a type of "open negotiation" between the sender and receiver.

Certificate Authority (CA)

A CA guarantees the identity of two individuals exchanging online information. This generally requires pre-arranged authentication by a third-party organization (bank, financial institution, credit card concern) that issues the certificate. These organizations issue digital certificates and digital signatures with two keys—one public, one private—that facilitate secure online interchange.

A digital signature is a code of data created using a secret key that attaches to an electronic message and identifies the sender. Encryption secures the electronic signature, which guarantees the identity of the individual issuing the digital signature. It also can be used as a "timestamp" on documents to verify that the document was signed at a specific timeframe.

A digital certificate is an attachment to an electronic message that also verifies the identification of the message sender but provides the receiver the means to encode a reply.

If you want to send an encrypted message, you can obtain a digital certificate from a CA. The CA gives you an encrypted certificate that contains a public key. The CA keeps a copy of the public key that it makes available, along with your identification certificate, to the individual who receives your encrypted message.

 Tip: The most common standard for digital certificates is X.509.

Data Decryption

Decryption is the process of converting encrypted data back into its original form so it can be understood. As ciphers become more sophisticated, the challenge of decryption increases. However, wireless communications are easier to "tap" and decrypt than hard-wired counterparts. The increase of online credit-card purchases and sensitive organization online communication has encouraged a demand for the development of stronger ciphers to make it more difficult for unauthorized people to break the codes. As organizations seek more secure codes, the cost for developing this security measure also increases.

In recent years, a controversy has arisen over so-called strong encryption. This refers to ciphers that are essentially unbreakable without decryption keys. While most companies and their customers view stronger decryption methodology as a means of keeping secrets and minimizing fraud, some governments view it as a potential vehicle by which terrorists might evade authorities. These governments, including that of the United States, want to set up a key-escrow arrangement. This means that everyone using a cipher would be required to provide the government with a copy of the decryption key. These decryption keys would be stored in a supposedly secure place, used only by authorities, and used only if backed up by a court order.

Opponents of this scheme argue that criminals could hack into the key-escrow database and illegally obtain, steal, or alter the keys. Supporters claim that while this is a possibility, implementing the key escrow scheme would be better than doing nothing to prevent criminals secure and evasive online access.

Recent concessions have been made by the U.S. Government, and especially the Office of the Attorney General, to allow U.S. companies to sell powerful data-scrambling technology overseas. Companies have pressured the government to allow them to export these encryption-capable applications because many overseas companies are marketing these programs. In response, the U.S. Congress will appropriate monies for the government to develop techniques to unscramble messages encrypted by terrorists and criminals.

Firewalls

A firewall is considered a first line of defense in protecting private information. It is a device that provides a type of security blanket that prevents unauthorized access to or from a private network. A proxy server sometimes serves as a firewall, filtering and recreating data received. Firewalls can be implemented in both hardware and software or a combination of both (Figure 7.4).

Figure 7.4 Intranet Firewall

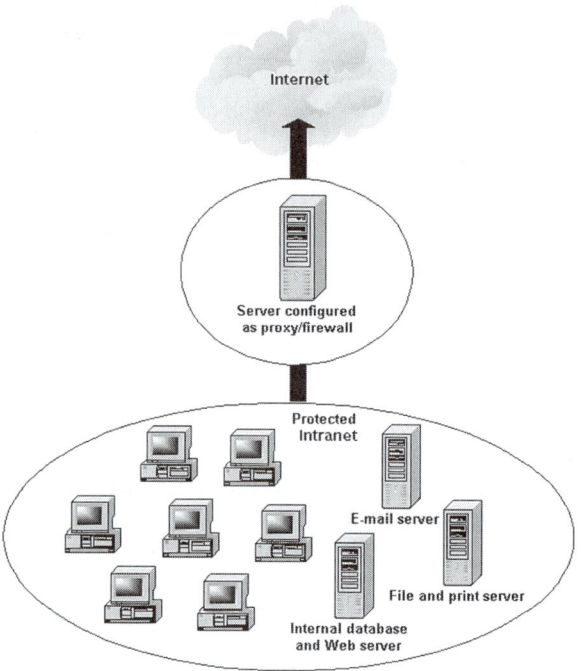

Firewalls are frequently used to prevent unauthorized Internet users from accessing private networks—especially Intranets—connected to the Internet. Generally, an Intranet is a network used by an organization exclusively for its members. Firewalls can also be used to prevent employees on a private network from accessing all or certain locations on the Internet.

All messages entering or leaving the Intranet pass through the firewall, which examines each message and blocks those that do not meet the specified security criteria. Both the Internet and an Intranet use the same protocols and technology, but while information on the Internet is universally available, information on Intranets is limited to authorized personnel within an organization.

The following paragraphs define several types of firewall techniques:

Packet filter—Examines each packet entering or leaving the network and accepts or rejects it based on user-defined rules. Packet filtering is fairly effective and transparent to users, but it is difficult to configure. In addition, it is susceptible to IP spoofing.

Application gateway—Applies security mechanisms to specific applications, such as FTP and Telnet servers. This is very effective, but can impose performance degradation.

Circuit-level gateway—Applies security mechanisms when a TCP or User Datagram Protocol (UDP) connection is established. UDP is a connectionless protocol normally used for broadcast messages. Once the connection has been made, packets can flow between the hosts without further checking.

Proxy server—Intercepts all messages entering and leaving the network. The proxy server effectively hides the true network addresses.

In practice, many firewalls use two or more of these techniques in concert.

Firewall Evaluation Checklist

When considering whether to add a firewall to your network security plan, you will want to review the following firewall evaluation checklist:

■ Can the firewall protect against these security problems?

Node spoofing attacks—Someone outside a network posing as an authorized user with an IP address within your network range uses a trusted external address to gain access. There is only one legitimate IP address within a firewall—the address of the external server.

TCP sequence number prediction attacks—Transmission Control Protocol sequence number prediction means that someone has guessed the TCP number sequence and has inserted packets into the system using the correct number that can execute commands.

Session hijacking attacks—"Denial of service" is the term used when someone outside your system disrupts, jams, floods, or crashes your service. Hackers and crackers always have the advantage of being able to disrupt service somewhere in the long stream of data flow. Network administrators, firewall administrators, or ISPs can only maintain control over some elements in the flow of information along a network, Intranet, or Internet.

Source routing attacks—Source-routed traffic from a firewall is vulnerable to a hacker who masks generating traffic from inside a firewall. During construction of the firewall, this kind of illegal access can be limited by building "block source routing" into the version.

DNS attacks—Two basic methods can help you avoid serious Domain Name Server (DNS) compromise. The first is to configure Domain Name Servers to perform reverse lookups. This limits someone who may have a system's trusted name from gaining access. In addition, implementing a "split brain" DNS approach allows computers inside the firewall to perform DNS Internet lookups but prohibits outsiders to enter. This configuration requires that there be two primary DNS servers—one inside the firewall and one outside. The inside DNS server has full access to all workstations inside the firewall, but obtains information about anything outside from the server on the other side of the firewall. This allows internal computers to perform global DNS lookups, but stops hackers from penetrating the firewall.

RIP attacks—The Routing Information Protocol, which classifies routers as active or passive, can be compromised when a hacker breaks the authentication key or password. RIP is a commonly used gateway protocol.

ICMP attacks—An Internet Control Message Protocol attack by a hacker can take the form of an "ICMP redirect" that has the power to override or reroute elements on a targeted system's routing table. Firewalls can be built to screen ICMP traffic from the network.

- **Has your existing firewall ever been breached?** If so, most of the time the problem is with the configuration of the proxy server that functions as the firewall. Reconfiguring the server to meet the crises should eliminate the problem.

- **What types of data do I need to protect?** A good security plan includes a review of all sensitive documentation in your system as well as where it is stored. Highly classified material that could be considered secret or top secret should be isolated from network access. Determine which elements and data in your system should be secured by a firewall.

■   **How likely is it that someone will want to break, steal, or alter my organization's data?** Hackers have the luxury of free time on their hands. Many of their attacks are random and high-profile sites are not their only targets. They may want to access your site to get to another site to which you are connected. The National Center for Computer Crime Data has stated that U.S. businesses lose on the average of $550 million from unauthorized access to their networks.

■   **Have we incorporated a good backup policy for important files in the event a server failure?** Most files are backed up on tape so that data can be restored if your file server fails. The following is a description of the five basic kinds of backup methods:
    Full backup—Files are backed up and marked "backed up."
    Copy—Files are backed up but not marked "backed up."
    Incremental backup—Files are backed up and marked "backed up" only if they have been altered since the last backup.
    Daily copy—Files are backed up that have been changed on a specific day but are not marked "backed up."
    Differential copy—Files are backed up if they have been altered since the last full backup but are not marked "backed up."

■   **How badly will it hurt our organization if a hacker succeeds in accessing or destroying our data?** Think of the consequences of someone destroying your Web site or obtaining strategic financial information or a marketing plan. Perhaps distributing or publishing some sensitive information could have serious consequences for your financial and long-term success.

Most important, the major question you should ask when considering the implementation of a firewall is: "If something happens to our network, will it put us out of business?" Connecting to public networks greatly increases that chance of something happening, and you must be prepared to weight that in your design plans. Regardless of whether or not a security problem could put you out of business, try to estimate the kinds of business damage that downtime or system clean up might cost.

Note: An attacker choosing a network target normally doesn't
                  bother to research the target to see if the data is
                  valuable. It is much easier to hack in and look around.
                  Most of these types of attacks can be classified as
                  random.

Firewall Types

A firewall can be thought of as a gap between two networks filled with something
that lets only a few selected forms of traffic through. The designers of the firewall
can explain the mechanism that enforces the separation. However, the firewall itself
should not be easy to break into, since breaking into a firewall gives an attacker a
foothold on your network. This section reviews the following four types of firewalls:

- Router-screening firewall

- Dual-homed gateway firewall

- Proxy firewall

- Dynamic packet filtering firewall

Router-Screening

The simplest and most popular form of firewalls is called router screening. Since
most commercial routers can restrict traffic between destinations while permitting
other traffic, screening routers operate only at the network level and can make all
permit or deny decisions based on the contents of the TCP/IP packet header. Router
screening is fast, flexible, and quite inexpensive, but lacks the ability to provide
detailed audit information about transmitted traffic.

Tip: Screening routers have often proven vulnerable to attack since they
             rely on software being correctly configured on the hosts behind
             them. Many experts prefer to avoid screening routers as a sole
             defense.

Dual-Homed Gateway

"Dual-homed gateway" is a system with two network interfaces on both the protected network and on the public network. Since the gateway communicates with both networks, you can install security software for carrying data back and forth.

This type of software agent is called a proxy. A proxy is customized for the service it provides. For example, a dual-homed gateway that has a proxy for WWW traffic will have some form of agent running on it that manages making requests to the remote networks on behalf of the user.

Proxy

Proxy firewalls, also known as application firewalls, are attractive to many sites because the proxies perform detailed auditing of data passing through them. Some experts consider proxies more secure since software proxies can be customized to specifically deflect known attacks to which host software behind the firewall might be vulnerable.

The main disadvantage of proxy firewalls is that they are sometimes not completely transparent, and they do not support protocols for which a proxy has not been developed.

Dynamic Packet Filtering

A dynamic packet filtering firewall is a cross between a proxy firewall and a screening router. It appears to the end-user as if it is operating only at a network level, but the firewall actually examines traffic as it passes by, just like a proxy firewall's proxy application does.

When a user connects out through the firewall, it records that fact and allows data to come back to the user for the duration of that session. Dynamic packet-screening firewalls provide an attractive technology that shows a lot of promise for securing network data.

Firewall Trade-Offs

Like other security systems, firewalls are not perfect. The trade-off usually represents a choice between ease of use and security. The more rigorously the firewall checks the user's identity and activity, the more likely the user is to feel interrupted and frustrated with the process.

Firewalls cannot solve problems with internal violations of security. Information can leak from your organization telephonically and most likely through use of disks. Additionally, firewalls don't ensure your system against viruses because of the variety of encoding on files transferred through networks and the fact most computer viruses are carried on disks.

When you choose a firewall, don't discount user resentment as a factor in your decision. Many organizations with firewalls have internal networks that users ignore in preference for installing their own dial-in or dial-out modems that bypass the firewall.

Virus Protection

Computer viruses infect programs with a damaging effect that can range from file corruption to destruction of a hard disk. A virus is a piece of program code that embeds itself into one of your programs or files and "self-replicates and propagates." This means that computer viruses hide and reproduce in computer code so that when a program with a virus is executed, the virus spreads throughout the program and computer.

Warning: Viruses existing in a program can only be activated by "execution." Opening the attachment can only activate a virus present in an e-mail attachment. Do not open an attachment to an e-mail that you suspect may contain a virus.

Computer viruses share the following characteristics:

- They reproduce similar to the biological bacteria.

- They "ride" on other programs.

- They infect and damage a computer system.

Computer viruses are invisible but can be programmed to activate. When you open an infected file or start your computer with a disk that might be infected, you are actually "executing" or "activating" the virus itself. As it spreads by attaching with other programs and files, you may not even be aware that a virus is active within your system. Viruses can be classified as "benign," having no adverse presence in your computer. They can also be just annoying, disappearing after rebooting. However, they can also be categorized as "malignant" with an ability to corrupt programs, erase files, and destroy hard drives.

If a malignant virus strikes one of your programs, it may not function normally. The program might totally stop working, it might write incorrect information in documents, or you might not be able to open other programs or locate existing documents. You need to be alert for both subtle and drastic changes in programs and documentation to detect existing virus attacks. These next paragraphs define some of the more common types of computer viruses.

Boot Sector Infector

A drive—hard or floppy—contains a section called a "boot sector" that maintains information regarding disk formatting, data stored, and a program for booting. This program can become infected by a boot sector virus. When a disk containing the virus is inserted in a drive and the machine is rebooted, the virus replicates onto the hard drive. These viruses may also be downloaded from the Internet.

Warning: Use caution when you download a program from the Internet. It may contain a boot sector virus.

File Infector

In general, this virus attaches or replaces .EXE or .COM files. However, it can infect other file extensions. When you open a program that contains a file infector virus, it can infect your hard drive and spread to any other program. Directory files and entire directories can be contaminated with this virus.

Macro

This virus is a deviation of program viruses that attack Microsoft® Excel and Microsoft® Word templates that can proliferate in all other files. Macro is a very common cross-platform virus.

Meta

This virus is frequently carried through Microsoft Word and Lotus® AmiPro documents, and can infect multiple platforms.

Multi-Partite

This insidious virus is a blend of several viruses—polymorphic, boot sector, stealth—that infects programs, files, and the master boot sector (MBR). MBR is a program that opens upon computer bootup. This is not a common virus.

Polymorphic

This virus frequently alters appearance, making it especially hard to detect through virus scanning. Polymorphic infects boot sectors and files.

Stealth

This virus is the most difficult to detect. When it infects a file, it adds bytes. However, at the same time it takes away an equal number of bytes from the file directory entry, giving the impression no change has taken place. This virus remains in the computer memory to hide file size change.

Viruses cannot infect write-protected disks, many documents (except documents and templates in Word 6.0 or above), or compressed files that were virus free before compression. In addition, computer hardware, monitors, and chips are immune from virus contamination.

Measures are being taken to improve and test software to ensure it is virus free before it is put on a network, the Internet, and bulletin boards before users download. The federal government has formed a SWAT team (Computer Emergency Response Team) that investigates complaints from major computer networks about threats imposed against them. In addition, the Software Publishers Association is reviewing the adoption of guidelines and measures that will alleviate the problem of computer virus proliferation. However, the bottom line is that no one who uses computers is safe from virus attacks.

The best course of "treatment" for any of these viruses is installation of computer anti-virus software before networking occurs. Two of the most powerful and popular anti-virus software products are manufactured by McAfee and Norton. These virus control packages detect any present viruses upon computer boot-up.

Security Myths

The following statements are based upon false assumptions and are discussed in the next paragraphs:

- My private password ensures that no one else can gain access to my PC.

- Intranet firewalls protect organizational networks from outside penetration.

- Data sent over the Internet should never be considered secure.

- Technology can solve all security problems.

- Security only applies to the present, future security problems can be addressed when they arise.

My private password ensures that no one else can gain access to my PC.

Passwords only allow you access to the network. They are the first-measure of authentication when you log on to a system. However, this process does not make your computer secure from outside penetration to your system from the outside (LANS, modems). In addition, passwords do not protect your outgoing data nor make them secure once they leave your machine.

Intranet firewalls protect organizational networks from outside penetration.

Firewalls are one aspect of an organization's security plan and can be effective in keeping outside intruders from entering an organization's system. The degree of sophistication of the firewall configuration will determine the level of security it can provide. However, regardless of the adequacy of an organization's firewall, many breaches of security are internal in nature. A firewall cannot address security problems that arise inside an organization. Complacency about internal security procedures when a firewall is existent can be detrimental to an organization's overall security plan.

Data sent over the Internet should never be considered secure.

Despite worldwide access to the Web, data sent over the Internet can be made secure through utilization of various means of encryption. Sensitive banking, financial, and credit transactions take place on the Internet every day. The acceleration of the need for private transmittal of data through the Internet has led to the development of more sophisticated ciphers with secure decryption techniques.

Technology can solve all security problems.

Technology is only as good as the people who use the technology are. A fail-safe security policy should include ongoing training for all users. Despite the fact though most companies have built-in security features in their systems and applications, user knowledge about how these features can be best used and applied is equally as important. A major factor in security incidents is the human factor. Many of these problems are not recognized until a sizeable amount of damage has already been done. Technology is not a substitute for good training and security awareness by system users and administrators.

Security only applies to the present, future security problems can be addressed when they arise.

Waiting to address matters of security breach or serious problems until they happen can disrupt business, be costly, and even seriously damage an organization. Constant surveillance of the security measures that have been implemented as well as keeping abreast of state-of-the-art security technology can save time, money, and devastating setbacks when internal or external security incidents have to be addressed.

Vocabulary

Review the following terms in preparation for the certification exam.

Term	Description
Access Control List	A list that contains information that specifies user and group permissions levels
access rights	Assignment of the appropriate permission level when sharing directories or files
algorithm	A logical sequence of steps. Ciphers work according to sophisticated computer algorithms that rearrange the data bits in digital signals
audit trail	Software or tools that allow reviewing a list of names of those who have entered a network
CA	A Certificate Authority guarantees the identity of two individuals exchanging online information
cipher text	Text translated into a secret code that is unreadable by an unauthorized person who accesses a directory or file
DES	The Data Encryption Standard is the U.S. Government standard for encryption
digital certificate	A digital certificate is an attachment to an electronic message that also verifies the identification of the message sender but provides the receiver the means to encode a reply
digital signature	A digital signature is a code that attaches to an electronic message and identifies the sender

encryption	Translating data into secret code
firewall	A firewall is a device designed to prevent unauthorized access to or from a private network
gateway	Software or hardware that unites incompatible programs or networks to facilitate the data transfer
LSA	Local Security Authority is the Windows NT security subsystem that validates local and remote logons for all types of accounts
NIS	UNIX has a Network Information Service that shares a server `passwd` file containing usernames, groups, and passwords with UNIX computers
password	The unique identifier chosen by a computer user to securely access the network
permissions	The ability to read, execute, write, delete, or in general access a network directory or file
proxy server	A server that is located between a client workstation and a server that filters and recreates data received
public key	One of the keys used in public key encryption that encrypts and decrypts messages
public key encryption	The security process of using keys (an algorithm) to encrypt and decrypt messages
router	A piece of hardware or software that facilitates connecting two or more networks
SAM	In the Windows NT security model, the Security Account Manager manages a database which contains all user and group account information components

shadow passwords	In UNIX security, shadow passwords are installed on the system by default and are important for protecting system passwords because when this file is in place, all passwords contained in the `/etc/passwd` file can only be read by root
SRM	In the Windows NT security model, the Security Reference Monitor manages a database which contains all user and group account information
strong password	A strong password is complex and not easily guessed by another. They are preferable and should be eight or more characters and a combination of letters and numbers
UDP	A User Datagram Protocol is an Internet connectionless protocol normally used for broadcast messages
user access	Logon names and passwords provide users unique levels of system access
user accounts	The first element of network security involves the creation of user accounts and assignment of passwords that provide permission to log on to the network
user ID	Typically, users of a multi-user or securely protected network system claim a unique name, often called a user ID
virus	A potentially destructive program that spreads through a program or computer
weak password	A weak password is easily guessed because it has been chosen as a matter of simple association. These should be avoided

In Brief

If you want to...	Then do this...
Protect your files and hard drive against virus infection	Install a computer anti-virus program
Ensure that a virus doesn't spread from your e-mail system	Never execute (open) an attachment of a suspicious e-mail
Rid your computer of an annoying virus that isn't potentially destructive	Reboot your machine and check the program again or install an anti-virus program
Limit the tasks a user can perform in a network	Assign "access rights"
Check a list of names of those who have entered a network	Install audit trail software on your system
Make sure of password confidentiality	Change your password every 30 to 60 days
Make sure you can alter files once you access a network	Check with the system administrator to make sure you have been given a "Full-Access" permissions level
Enhance security control for a network once the basic features of security are in place	Limit the number of authorized users on a network
Implement an easy way to assign access rights	Assign blanket "group rights"
Use the official U.S. Government standard encryption process	Use Data Encryption Standard (DES)

Ensure the identity of another individual with whom you are exchanging online data	Obtain a Certificate Authority (CA)
Filter all messages entering and leaving your network	Install a proxy server as a firewall
Make sure the disk you are using does not have a virus	Use a write-protected disk

Lesson 7 Activities

Complete the following activities to better prepare you for the certification exam.

1.   Explain how "access rights" are a security element in dial-up networking.

2.   Describe the most effective and sophisticated method of ensuring data transferred during networking.

3.   Identify the purpose of a firewall.

4.   Explain the best defense against a computer virus.

5.   Explain why the U.S. Government has been reluctant to allow U.S. companies to export powerful data-scrambling technology.

6.   Describe cipher text and its purpose.

7.   Describe the effects that the least serious and most destructive kinds of computer viruses can have on your system.

8.   Define what full-access rights means to you as a user.

9.   Discuss two factors an organization needs to consider before good security practices are in place.

10.   Define how a proxy server can function as a firewall.

Answers to Lesson 7 Activities

1. The logon process where the user enters his/her name and password is the first level of security as the network server identifies and validates the name as authorized or unauthorized. Access rights determine the level of permission assigned to the user. The hierarchies of permission levels range from "full access" rights to "no access."

2. Encryption of data that turns information into a secret code is the most effective means of preventing compromise of information transferred during networking. To read encrypted text, you must know the decryption key or password that translates the code.

3. Firewalls are security systems that are used by organizations that have an Intranet for exclusive use by their employees.

4. Installation of anti-virus software.

5. The U.S. Government fears that this technology will be used by terrorists to evade authorities.

6. Cipher text is text translated into a secret code that is unreadable by an unauthorized person who accesses a directory or file. Encryption is a sophisticated security measure used by organizations to protect their data.

7. The least serious kinds of computer viruses present no adverse presence in your computer. They can be annoying but will disappear after rebooting. The most critical kinds of virus can erase files and even destroy entire hard drives.

8. Full-access rights is a level of "permission" that grants you the right to access files on a network and read, modify, delete, save, or copy that file. There is no higher level of permissions than "full access."

9. An organization needs to formulate and document a plan that does the following: 1. Targets the level of security needed, and 2. Evaluates the resources in place to actually implement that plan.

10. A proxy server can serve as a firewall by sitting outside a network, filtering and recreating data received.

Lesson 7 Quiz

These questions test your knowledge of features, vocabulary, procedures, and syntax.

1. Which of the following is the most detailed source of information on user access to the network?

    A.   Access control
    B.   Permissions
    C.   Audit trail
    D.   Passwords

2. How is the ability to access a resource on a network directory determined?

    A.   Through a password and the network response
    B.   By the user name entered before the password upon logon
    C.   From the number of authorized users accessing the network
    D.   Through the permissions level assigned

3. What must you know to read cipher text?

    A.   Know the code
    B.   Have "read-only" access
    C.   Know the key to press
    D.   Gain entry to the network

4. Which of the following most accurately describes who can use a firewall for network security?

    A.   An organization or corporation
    B.   Dial-up networks
    C.   The Internet with specific protocols
    D.   Telecommuters using RAS

5.   What does "full access" rights mean?

    A.   You can write to a file but not delete it
    B.   You can edit a file and delete the file
    C.   You can delete a directory on the network
    D.   You can enter the system without logging in

6.   What is the best method to rid your system of a computer virus that you know is contained in an e-mail attachment?

    A.   Install one of the anti-virus software programs immediately
    B.   Open the e-mail attachment and delete the contents
    C.   Delete the e-mail without reading it
    D.   Send the e-mail to your network administrator

7.   Which of the following best explains the reason behind the recent U.S. Government efforts to discourage "strong encryption?"

    A.   The government wants to monopolize sale of all export of software that uses this technology
    B.   The fear that drug cartels, criminals, and terrorists will use the technology to minimize fraud
    C.   This method violates government DES code
    D.   The U.S. Government has passed a law that allows strong encryption if the FBI can have access to all encryption keys

8.   What is the first-level network security that you should consider before implementing your network security plan?

    A.   Encryption
    B.   A firewall
    C.   An audit trail of network users
    D.   Creation of user accounts and assignment of passwords

9.   What occurs in the user logon process?

    A.   The server identifies and verifies the name as authorized or unauthorized
    B.   The permission level of the user is checked against a database

C.   The server checks your group status
D.   The server creates a user account under the password entry

10.   Whom are security permissions granted to?
A.   Intranets
B.   Groups or individuals
C.   Internet users only
D.   Only those holding an encrypt key decoder

Answers to Lesson 7 Quiz

1. Answer C is correct. An audit trail provides a listing of the specific name, date, and time of everyone entering the network.

   Answer A is incorrect. Access rights tell only what level of user right an individual or group has after connecting to a network.

   Answer B is incorrect. Permissions encompass the categories of "user rights."

   Answer D is incorrect. Passwords are the first-level of network security and restrict those without assigned passwords but do not provide information about the time, date, and events that took place after log on.

2. Answer D is correct. The permissions level assigned by the network administrator determines the level of accessibility you have on the network—"read only," write, execute, etc.

   Answer A is incorrect. Password and user name only allow access to the network during the log on process.

   Answer B is incorrect. A user name combined with the password only allows accessto the network during the log on process.

   Answer C is incorrect. The number of authorized users on a network does not affect user access but limiting the number of users is a good security practice.

3. Answer C is correct. You must know the correct key to press to decipher the text.

   Answer A is incorrect. You do not have to know the code to decipher text. You need to know which key "translates" the code.

   Answer B is incorrect. Access level does not affect whether or not you can read cipher text that is a security practice.

   Answer D is incorrect. Gaining entry to the network is dependent upon entering your user name and password that the system checks and either "authorizes" for entry or "unauthorizes" that prohibits entry.

4.   Answer A is correct. Organizations or corporations use a firewall around
       their Intranets.

       Answer B is incorrect. Dial-up networks call in from remote workstations that
       do not have a firewall.

       Answer C is incorrect. The Internet has worldwide access, and is not surrounded
       by a firewall.

       Answer D is incorrect. RAS is a dial-up networking service that accesses a network.

5.   Answer B is correct. Edit a file and delete a file as well as read, save, copy,
       or modify the file.

       Answer A is incorrect. "Full access" rights allow you both write to a file (edit or
       change the file) and to delete the file.

       Answer C is incorrect. Network directories are protected by security features and
       may not be deleted by users regardless of access levels.

       Answer D is incorrect. Everyone entering a system must use the log on system that
       is the first level of security.

6.   Answer C is correct. The virus in the attachment can only be activated by
       "executing" or opening the e-mail. Deleting it immediately without reading is the
       correct action.

       Answer A is incorrect. Installing an anti-virus after a virus is already in the
       system will not eradicate the problem.

       Answer B is incorrect. Opening the attachment will spread the virus throughout your
       system.

       Answer D is incorrect. You should notify your network administrator about the
       problem, but do not forward the infected e-mail to anyone.

7. Answer B is correct. The U.S. Government fears that strong encryption will be a tool that will be used by subversive elements to communicate in secret.

Answer A is incorrect. The Attorney General's Office has recently loosened its authority and granted permission for U.S. companies to sell powerful data scrambling technology overseas.

Answer C is incorrect. DES refers to the U.S. Government standard for encryption. It is the encryption algorithm of choice for private commercial electronic communication.

Answer D is incorrect. Privacy groups will not agree to a having a government agency hold all encryption keys and there is no such law.

8. Answer D is correct. Creation of user accounts and password implementation is the initial step in controlling the number of network users having access to data stored on the network.

Answer A is incorrect. Encryption is a very sophisticated and expensive method of security used by some organizations.

Answer B is incorrect. Firewalls are used only by organizations with Intranets.

Answer C is incorrect. An audit trail of network users is software that can be used totrace the names, time, and dates of those entering the network.

9. Answer A is correct. At log on process, the server identifies and validates the name entered as authorized or unauthorized. Only verified names will be allowed access to the network.

Answer B is incorrect. At log on, only the name and password are checked to authorize network entry.

Answer C is incorrect. Your group status will not be checked until you have been authorized to log on to the network.

Answer D is incorrect because the server does not create a user account; a network administrator creates a user account.

10.   Answer B is correct. Permissions can be granted to groups or individuals by
      network administrators.

      Answer A is incorrect. Permissions are given to groups of people or individuals, but
      not to an Intranet.

      Answer C is incorrect. Permissions are given to groups of people or individuals, not
      to an Internet.

      Answer D is incorrect. Having access to an encryption decode key does not
      have anything to do with granting of user access permissions.

Lesson 8: Network Requirements

Implementing a functional, high performance network depends on several critical installation factors, which all focus on defining your requirements in advance.

After completing this lesson, you should have a better understanding of the following topics:

- Network Implementation Overview

- Safety and Preventative Maintenance

- Administrative and User Accounts

- Network Components

- Network Connectors and Ports

- Network Troubleshooting Tools

- Network Scenarios

Network Implementation Overview

A network allows computers to access information from other devices without the need for physically walking to other devices to retrieve the information. To reach the goal of developing and implementing an efficient network, excessive planning must take place.

The network administrator must be ready to build, maintain, and support all aspects of a network to keep the network running smoothly. The safety of both the administrator and the computer equipment should be seriously considered during implementation and maintenance of the network.

Accounts for both users and administrators must be planned and tested to verify that the accounts will allow the users to access only the information needed to complete their work and the information they are allowed to access on the network. You will need to verify that all network components, connectors, and port styles and types are compatible with each other and that they are available by the implementation date.

In addition, you must prepare for the inevitable troubleshooting that will take place during the installation process and after the network is fully functioning.

Safety and Preventive Maintenance

While setting up your network, take into consideration environmental conditions. The physical environment of an office and the placement of devices in the office can affect network performance.

Electrostatic Discharge (ESD)

ESD occurs when two objects of dissimilar charge come in contact with each other. Each object exchanges electrons to standardize the electrostatic charge between the two objects. This charge can damage electronic components and is the most severe form of Electromagnet Interference (EMI).

Tip: EMI is the disruption of electrical equipment performance.

These electrostatic build-ups discharge very rapidly into an electrically grounded body or device. Placing a 25,000-volt surge through any electronic device is potentially damaging. However, the human body can build up static charges that range up to 25,000 volts.

When the charge reaches 2,500 volts, it could discharge to grounded metal parts. ESD is the most damaging form of electrical interference associated with digital equipment.

The most common causes of ESD are:

- Movement

- Low humidity

- Improper grounding

- Unshielded cables

- Poor connections

Movement causes friction that produces large amounts of electrostatic charge on their bodies. For example, walking across carpeting can create charges in excess of 1,000 volts.

ESD frequently occurs during periods of low humidity. If the relative humidity is below 50%, static charges accumulate easily. ESD generally does not occur when the humidity is above 50%. We advise the use of a room humidifier to maintain the humidity level at 50% or above in an area where computers are maintained.

Anti-static mixtures applied with a soft cloth to the following items can decrease static electricity:

- Floors

- Carpets

- Desks

- Computer equipment

In addition, you can use an anti-static mat or a grounding wrist strap (Figure 8.1) while working on computer equipment to decrease the risk of static shock and harm to the computer equipment from ESD.

Figure 8.1 Grounding Wrist Strap

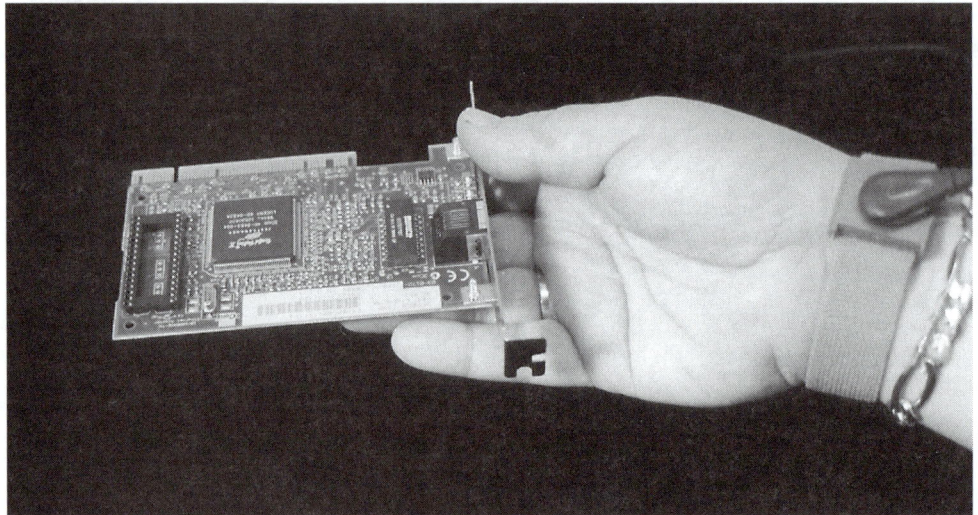

Computer Equipment

Computer chips have tightly packed circuits that heat up during use. Heat produced by the processor can corrupt the internal signals of the chip or even destroy its crystal structure. The faster a chip runs, the more heat is produced by the processor.

To prevent computer chips from failing or destruction due to excess heat, chip manufacturers have developed heatsinks to defer the heat.

 Note: A heatsink has rows of pins, called fins, on its surface to facilitate heat disbursement.

A heatsink is a piece of aluminum placed on the processor that allows the heat to spread across the heatsink and dissipate from the chip.

Most computers are also equipped with a cooling fan (Figure 8.2) to diffuse the heat that the heatsink is unable to disperse. Many computers have their own active cooling fan within the chip. If the chip did not have a cooling fan or a heatsink, the chip would melt from the heat it generates.

Figure 8.2 Cooling Fan for a Microprocessor

Excessively high temperatures can cause your server to overheat, which is why the ideal environment for servers is a well-ventilated room, preferably air-conditioned around 60-65°F.

If your server room is controlled by an air-conditioning schedule, and if the air is turned off at night and on weekends, you should think about purchasing a climate control. A climate control system could be expensive, but it is less expensive than replacing your server and network data.

 Note: A technology advantage is to have a message sent to your pager when your server room temperature exceeds the maximum conditions.

Network Wiring

Occasionally, rushed installers will push wiring out of the way and consequently break or short an existing wire. Coaxial and UTP connectors tend to go bad if they are moved around. These types of errors can cause delays in restoring normal operations.

For example, a client hired a group to install wiring in an old building that had been remodeled a number of times and contained a false ceiling. While installing the cable, the installers encountered 20- to 30-year-old telephone wiring. When this wire was moved to make room for new cables, it disrupted service for an entire floor of that building. The phone company was called to repair this problem at great expense to the company.

 Note: Many suppliers offer bulk-pack wiring. If it is mislabeled, the problem may not be discovered until after the wiring is installed.

The wiring you choose for your network connections is critical. The outside casing of shielded twisted-pair (STP) and unshielded twisted-pair (UTP) cables look identical (Figure 8.3). If the network is installed with the wrong wiring, it may work fine for a while before a problem occurs. Verify that the wiring meets the specifications of the network by checking both the box or spool and the stenciling on the wire before you install it.

Figure 8.3 STP and UTP Cables

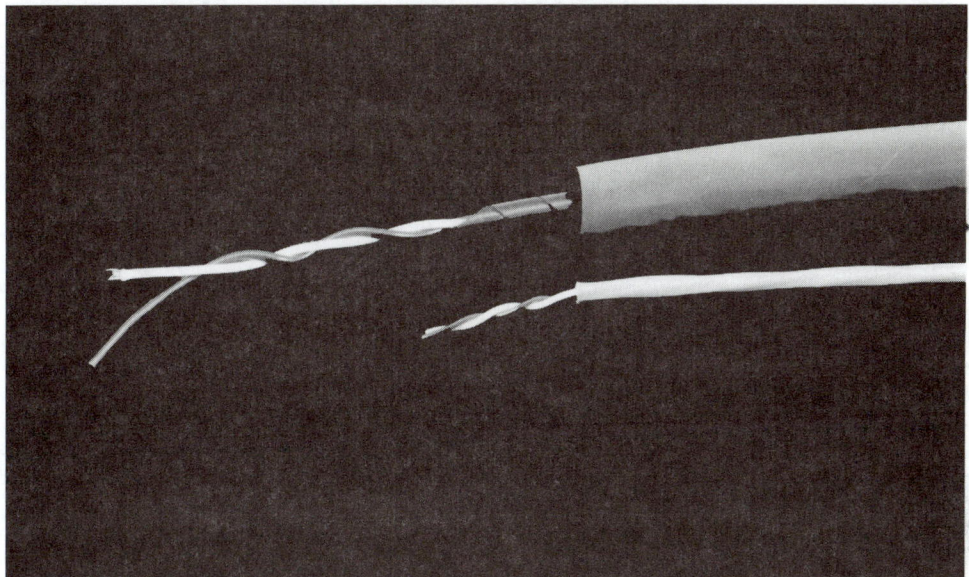

Environmental considerations can create wiring problems. It is critical that you use wiring which is appropriate to the environment. If wiring will be exposed to outside temperature and moisture, you'll want to use wiring rated for that purpose to improve network longevity. Check the wire manufacturer specifications to verify the type of suitable environment.

Workstation Errors

Workstation errors include network adapter configuration, logon, and cable connections. If, for some reason, the network adapter is reconfigured, it acts as if the network is no longer present, which confuses users who must then call customer support. Electrical devices that produce magnetic waves can corrupt network signals if electrical devices are located near the network cables. Since the main server and other key devices on the network are usually stored in secure areas, electrical devices that interfere with the network usually affect data in the cables.

Data transmitted through an unshielded twisted-pair cable can be corrupted if the cable is installed too close to ceiling fluorescent lights. Space heaters, microwave ovens, and even elevators can also cause data corruption if they are located too close to cable runs or electronic equipment. In addition, if the electrical device is plugged into the same outlet as the computer equipment, the circuit can overload.

Administrative and User Accounts

Prior to installing a network, you must design the structure based on the following factors:

- Administrative accounts

- Test accounts

- User accounts

- IP addresses

Administrative Accounts

Administrative accounts allow you to change and maintain system settings, such as formatting and partitioning network drives, system backups, monitoring performance, and using Dynamic Host Configuration Protocol (DHCP). DHCP is a TCP/IP protocol that allocates and manages IP addresses. Prior to network installation, appoint someone as network administrator to manage administrative accounts.

To secure the network, the administrator creates two accounts, an administrative account, and a user account. If you log on as an administrator and access non-administrative network features, such as word processing and e-mail, you could change entire network features by accident. Having both administrative and user accounts makes more work for an administrator, but the benefits outweigh the time it will take to maintain both accounts.

Warning: If a virus is activated in an administrative account, it
can do much more serious damage than if activated
in a user account.

Test Accounts

Test accounts are non-administrator accounts set up by the administrator to test the network. The network should be thoroughly tested to verify it works correctly before allowing user access.

User Accounts

The administrator must assign a user ID and password for each user account. The ID and password allows access to the system at a designated security level set by the network administrator.

Note: Administrators are responsible for adding and
maintaining user accounts.

Internet Protocol (IP) Addresses

After the installation of the network on the server is complete, each of the assigned workstations must be assigned an (IP) address to uniquely identify each workstation on the network. Most systems will automatically assign an IP address to a workstation during configuration. However, the administrator must verify that the IP address has been properly set up and be prepared to configure the address manually in Administrative Tools, if needed.

Network Components

During the planning stage of the network, peripherals, connectors, and network components must be included in the overall design. You will need to evaluate peripherals and components for the types of ports and connectors required to verify that the parts are available for the physical setup of the network.

Peripherals

A peripheral is a device that connects to a computer or network, such as printers, modems, and keyboards. Each peripheral has a port to physically connect the peripheral to the network.

There are various types of ports found on peripherals available for your network. Serial ports are used with a modem and mouse. The port name for a serial port is represented as COMx ("x" signifies ports from 1 to 4). COM ports can be viewed in Device Manager in the Windows operating system.

Parallel ports are used for high-speed devices such as disk drives and printers. Parallel ports send and receive data 8 bits at a time. The port name for a parallel port is represented as LPTx ("x" signifies the ports from 1 to 4). Parallel ports can be viewed in the Device Manager of the Windows operating system.

A Universal Serial Bus (USB) is a port that connects peripherals to your computer. Rather than a simple point-to-point port, the USB acts as an actual bus that allows you to connect up to 127 devices, such as keyboards and mice, into one jack on your computer. All the linked devices on the USB share the same signals. A USB transfers data from 1.5 to 12 megabits per second (Mbps).

Print Server

A print server is a computer designed to accept print jobs as they are generated from computers on the network. The print server is responsible for keeping track of the incoming print jobs in the Print Queue until the printer is available to print the next print job.

 Note: To print a number of documents, the operating system (or a special print spooler) queues the documents by placing them in a special area called a print buffer or print queue. The printer then pulls the documents off the queue one at a time. Another term for this is print spooling.

Usually, printers are connected directly to the network. However, you can also connect a print server to a Small Computer Systems Interface (SCSI) connection. You will need to install a SCSI interface card in an expansion slot in the print server on your network. This type of SCSI cable has a DB-25 connector that attaches the printer to the SCSI card. Figure 8.4 displays a SCSI connector.

Figure 8.4 SCSI Connector

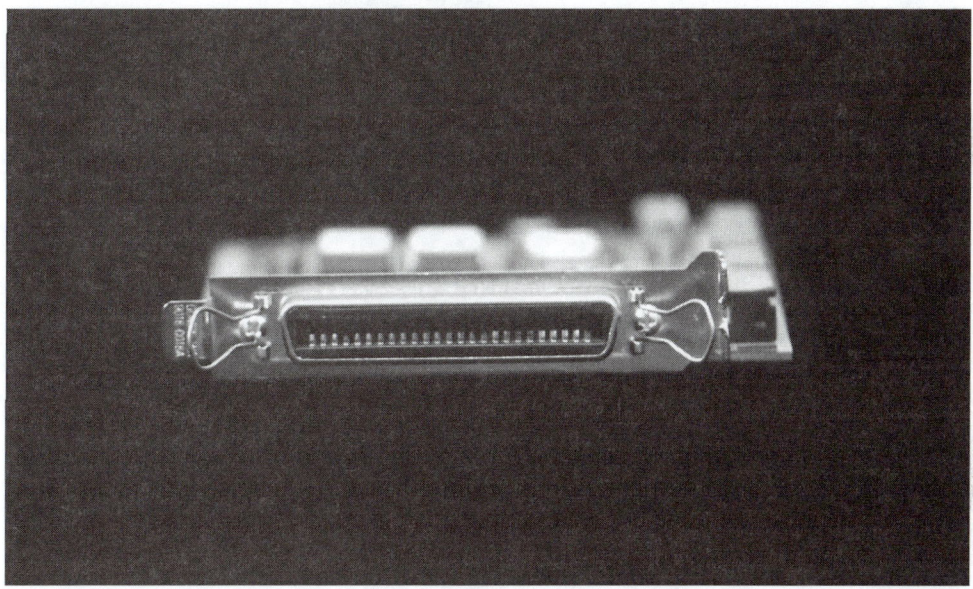

Uninterruptible Power Supply (UPS)

A UPS is a device that provides continuous power to a network to prevent data loss during a power outage or fluctuation in voltage. A UPS monitors line voltages to verify that adequate voltage is sent to the network. If the voltage extends above or below the normal preset standards, the UPS changes to battery back-up mode and supplies the network with the correct voltage (Figure 8.5).

Figure 8.5 UPS

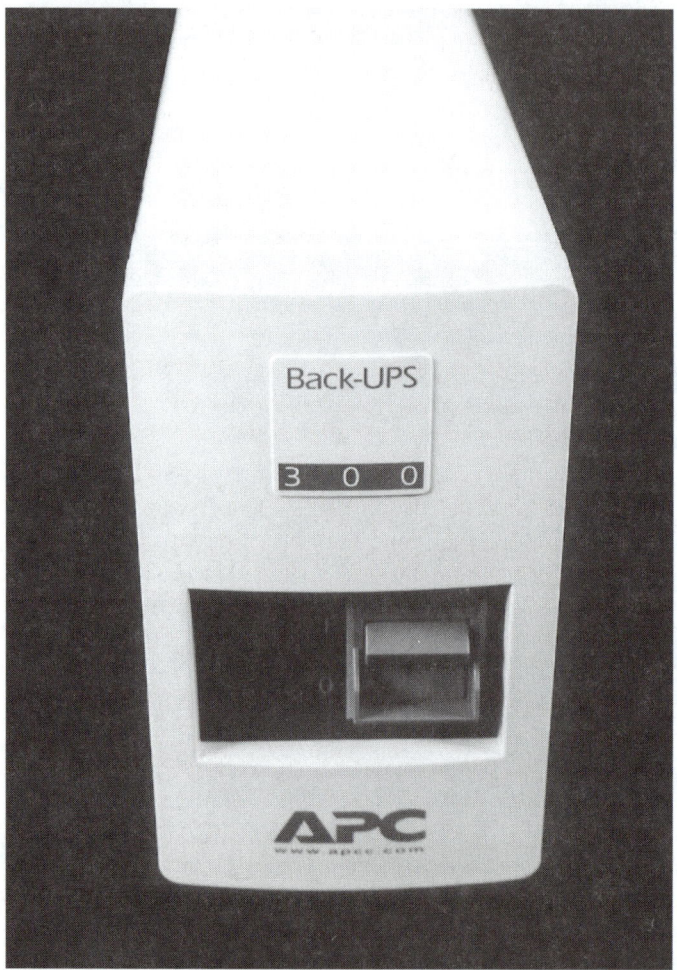

A UPS should be installed on all devices that need to be protected from power failure, including hubs, routers, and bridges.

Tip: For maximum protection against power failure on your network, install a UPS on each cable that connects your computer to modems, telephones, or other network equipment.

Network Interface Card

A Network Interface Card (NIC) is a device that allows the network cable to connect the computer to the network. A NIC can contain both an external and an internal location to connect the cable (Figure 8.6).

Figure 8.6 NIC

The type of connection you choose will be determined by the type of cable you have.

Tip: If a network card has both internal and external connection options
       and you select one setting, you must disable the unused
       connection. This can be accomplished by using jumpers. Jumpers
       are small plastic connectors placed around two pins to
       communicate with an expansion card.

Token-Ring Media Filter

A token-ring media filter is an adapter that allows UTP cables in a token-ring
network to connect to the network devices. The following are examples of token-ring
media filters:

TokenEase NIC Adapter—Used with a token-ring device that has a standard 9-pin
DB connector.

TokenEase MAU Adapter—Used with a token-ring device, which has an IBM
MF41601 (MIC) connector.

Tip: A TokenEase Relay setup tool is required to reset the ports of the
       TokenEase MSAU or the TokenEase Slim MAU once you have
       installed either device.

Network Connectors and Ports

A connector is a device that connects cables to nodes on your network.

Hub

A hub is a device that connects nodes on a network into a central location. You can
add hubs to a network to increase connectivity.

If a hub contains eight ports, then you can connect only a maximum of eight devices to that hub (Figure 8.7). You must determine how many devices you want to connect to each hub before designing your network.

Figure 8.7 Hub

You can use one port on a hub to connect an additional hub, which will support more devices. For example, if you need to connect fourteen devices, you can purchase two, 8-port hubs and use one of the ports to connect to the additional hub.

Routers

A router is a device that connects segments of a network together. A router operates at the network layer (Layer 3) of the OSI model. Routers read and process data on a network. Routers also maintain information about all known network addresses, connections between routers, and information about path alternatives.

The type of network you are installing will determine the type of router your network requires. Visit your local computer store or online computer manufacturer for the best router type for your network.

Bridges

A bridge is a device that transmits data over the network by viewing the address on the data unit, and then referring to the information table stored at the bridge. A bridge operates at the Data Link Layer of the OSI model. When the proper destination is detected, data frame routes to the receiving node.

Tip: A learning table is a list of addresses of nodes on the network. A bridge develops a learning table as it sends the data frames to the network to locate the nodes for further transfers.

Brouters

A bridge router, referred to as a brouter, is a device that incorporates both a bridge and a router in a single unit. A brouter connects segments of a local area network (LAN) to transfer data packets through the most efficient route on the network.

Patch Panels

A patch panel is a centrally located device that links device cables to additional network devices. The patch panel is usually stored in an isolated location, such as a closet or the back of an office, so the cables will not affect other devices or interfere with office traffic.

Wires inside a patch panel are electrically connected to set up the connection for the devices. Cables connected to the patch panel can be designed to accept any type of connector.

Network Troubleshooting Tools

The following tools can help you troubleshoot a network:

Digital Volt Meter (DVM)

A DVM (Figure 8.8) measures ohm resistance, AC and DC voltage, and in some cases low frequencies. Most of the time, a DVM measures a wire to determine that the connection is continuous and unbroken. In resistance mode, the DVM sends a small voltage down the wire and measures the drop between the two ends. This voltage drop translates into resistance. If the line is open, a high-resistance state exists, which means the signals cannot pass through. Adjacent wires can also be tested for shorts, which would ground out the signal. With practice, most people only need the DVM to troubleshoot simple network problems.

Figure 8.8 Digital Volt Meter

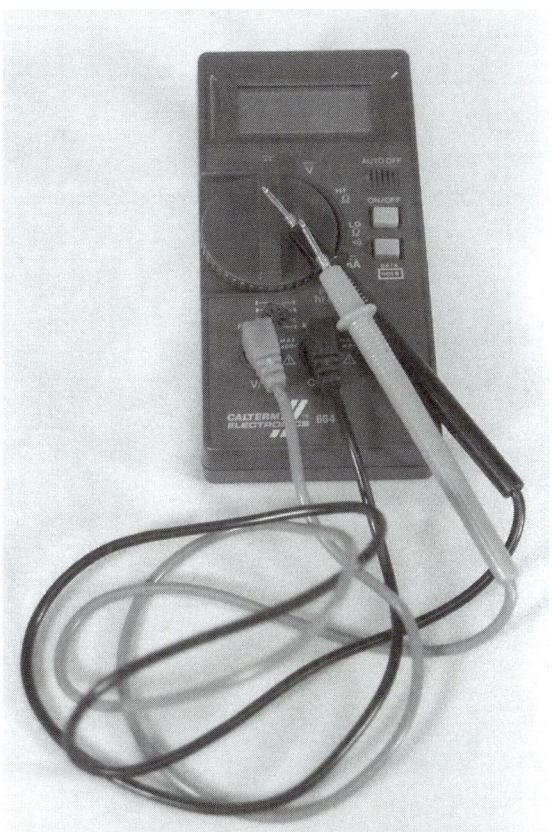

Time-Domain Reflectometers (TDRs)

A TDR is a device that sends pulses down a cable and bounces the signal back to where any impedance difference occurs. The time measured between the transmission and return can be translated into distance. Cable breaks or shorts can be identified to within a few feet with most TDRs

Oscilloscopes

An oscilloscope is the primary tool for most electronic technicians. The oscilloscope presents the signals it probes in a graphical format that is time based (Figure 8.9). With an o-scope, you can watch voltage changes over a network line and see exactly what is occurring. When used in conjunction with a TDR, an o-scope graphically shows you where line characteristics change.

Figure 8.9 Oscilloscope

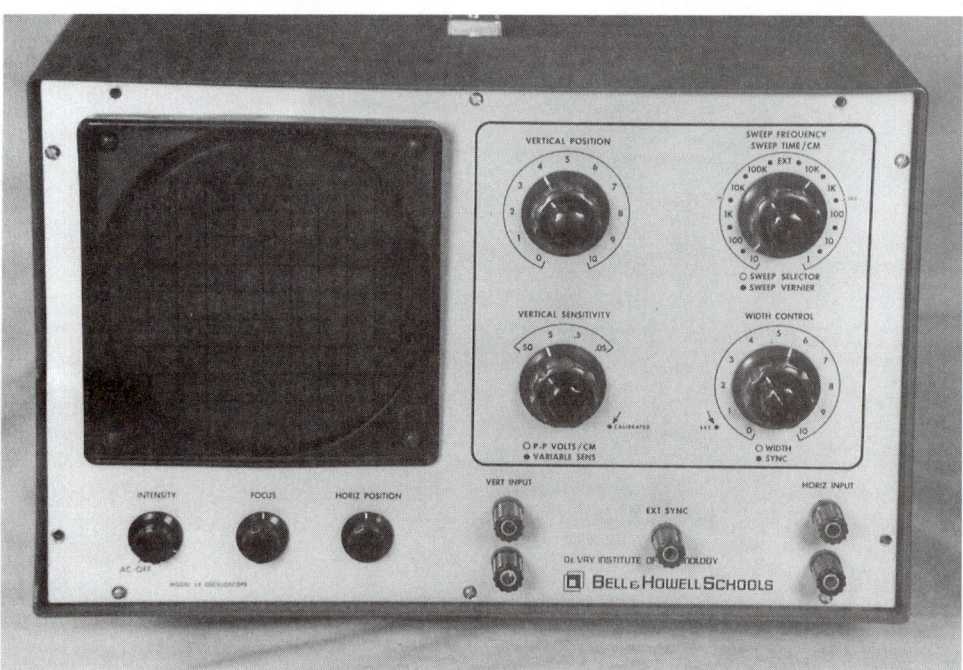

Advanced Cable Testers

An Advanced cable tester is a device that checks the smart wiring at the OSI physical layer in layers 2, 3, and 4 and you can collect information about transmission characteristics. Newer models can be plugged into a computer to graphically display information on the monitor. Cable testers can check the physical condition of the cable by measuring the types of activities on the cable.

Network Monitors

Network monitors are software programs that track and measure network activity. Typically, network monitor software is installed on a computer dedicated to this purpose. The network monitor stores and records information on virtually every aspect of the network, including individual nodes, devices or workstations.

Network Installation Scenarios

There are a variety of cables available on the market to meet the needs of the extensive amount of peripherals for computers and networks. This section will give three scenarios of typical situations that you may encounter when you are planning and installing a network.

Installing an Analog Modem on a Digital Jack

If you are implementing a network in a newer building using older equipment, you may encounter the need to install an analog modem on a digital telephone jack. A modem is a device that connects your computer to a phone line allowing you access to sending a receiving data over distances. Phone lines transfer information as analog signals, and sounds are converted into electrical signals as they are sent into the phone line.

The data signals produced by a computer must also be converted to analog signals to be transferred over the phone line. A modem converts between the digital signals produced by a computer and the analog signals required by an analog interface. The digital lines in the new building are communication lines that transfer data in binary form. Digital signals are created to have a set number of values.

Note: Modern modems not only convert signals to analog and
                     digital, but they can also correct errors existing in the
                     signals passed.

Many modern phone lines use digital signals and do not require a conversion to
analog signals, which may be the case with your installation. However, there are
many existing plain old telephone service (POTS) that require conversion from digital
to analog signals.

Modulation is the process of transmitting digital signals through an analog phone
line. Demodulation is the process of translating the analog signals back in to digital
signals. An internal or external modem must be physically connected to a phone
line.

 Tip: The word Modem comes from MOdulator/DEModulator.

Internal modems are usually placed in an expansion slot inside the computer and
the phone line connects to the section of the modem card exposed outside the
computer (Figure 8.10).

 Tip: Older modems have jumpers that require settings to properly
                     communicate with the interface or operation of the modem.
                     Jumpers are small plastic connectors around two pins that
                     communicate with an expansion card.

Figure 8.10 Internal Modem

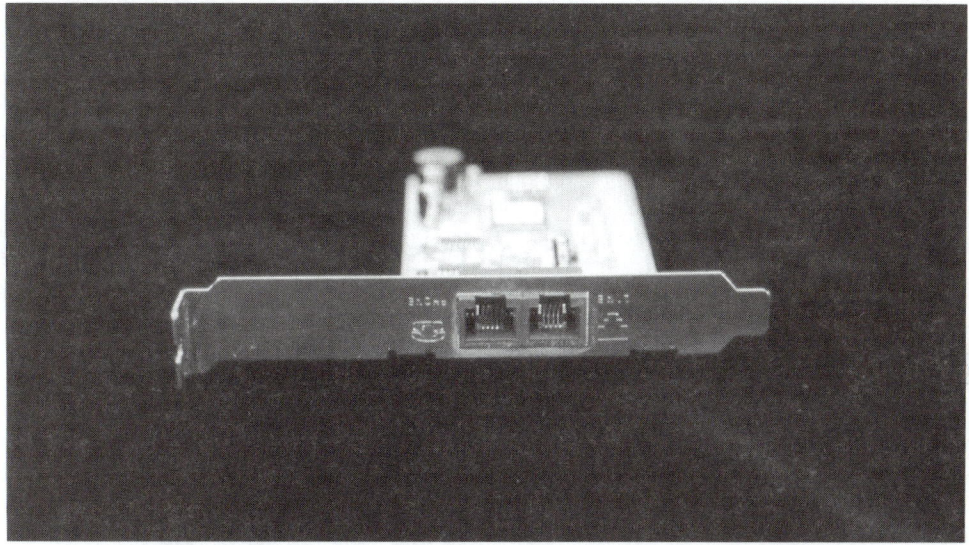

External analog modems are connected by a cable that connects to your computer and require a phone line connection on the back of the modem (Figure 8.11). The connector on your modem that attaches to your computer is usually a female D-shell 25 pin. The connector on your computer is usually a male DB 9 pin.

Figure 8.11 External Modem

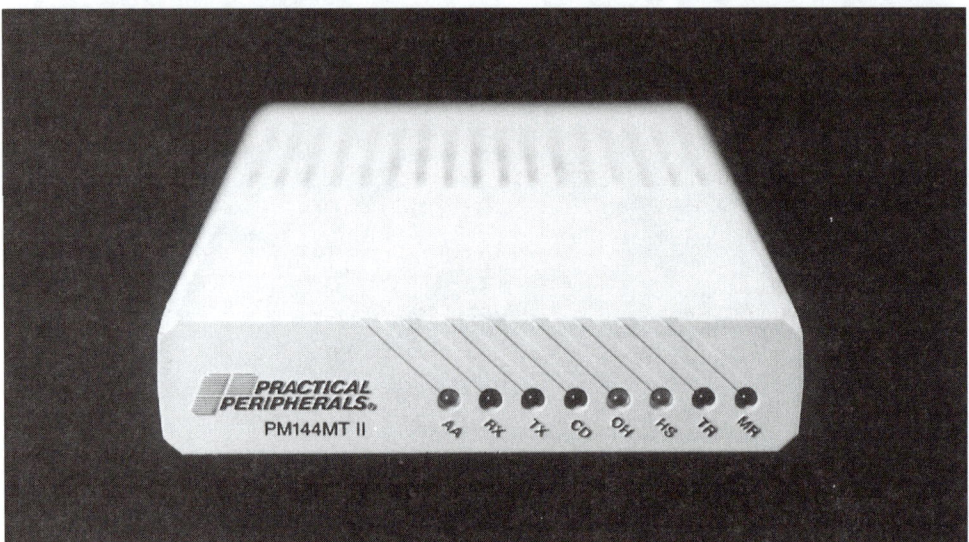

The cable you use for your modem must have a male end connection and a female end connection to properly install the modem. In addition, external modems require a separate power source and require a connection into an electrical outlet. Both internal and external modems use a standard telephone and cable to connect the modem to the modular telephone jack.

A null modem environment allows two computers to connect together to share hard drive information without modems. The computers are connected through serial ports by a single cable, called a null modem cable. The null modem cable connects the line that transmits data one computer to the line that receives data on the other computer so data can be transferred.

Connecting a 10BaseT Cable Using Registered Jack-45 (RJ-45) Connectors

During a 10BaseT network installation using Cat3 cables you will need to become knowledgeable about RJ-45 connectors. An RJ-45 connector is a device that connects computers to a variety of devices related to a LAN. Similar in appearance to a normal phone cord (RJ-11), an RJ-45 connector has eight cable connections—twice as many as a normal phone connector (Figure 8.12). Therefore, it is wider than a phone cord and will not fit into a normal phone jack.

Figure 8.12 RJ-45 and RJ-11 Connector and Jack

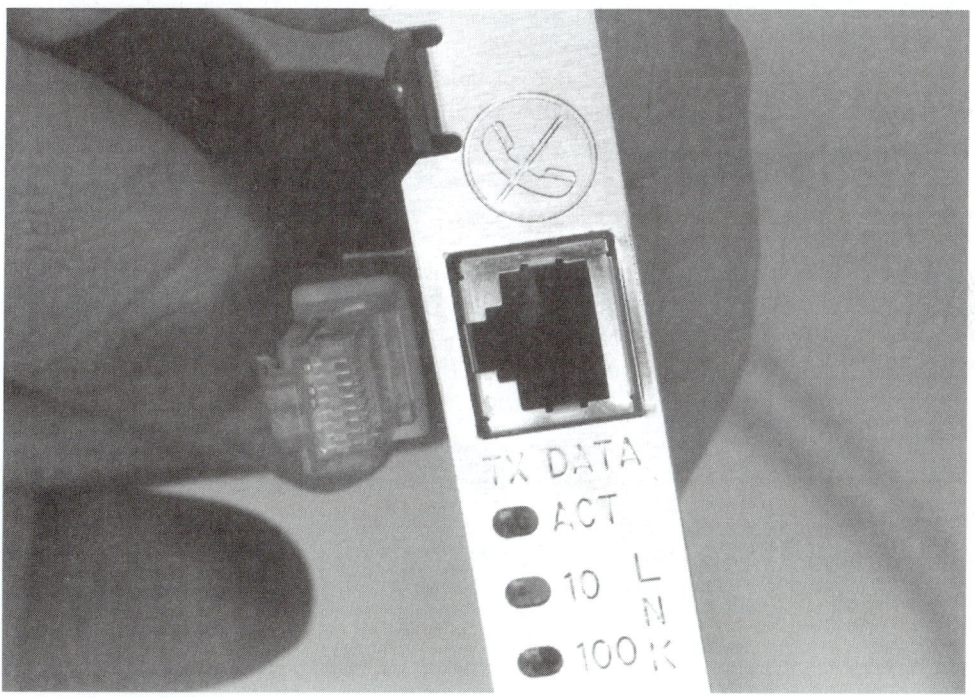

Note: The eight wires or connections inside the RJ-45
                  connector are color-coded.

There are two different types of RJ-45 connectors, straight and rolled. You should
become familiar with RJ-45 cables and understand the difference between the two.

■   **Straight RJ-45 connectors**—Color-coded wires travel straight through the
     cable.

■   **Rolled RJ-45 connectors**—Color-coded wires twisted throughout the cable.

Tip: The easiest way to determine if you have an RJ-45 straight or
                  rolled connector is to hold both ends of the cable next to each
                  other. If the colors fall in the same order, the connector is a
                  straight RJ-45 connector. If the colors are opposite, the
                  connector is a rolled RJ-45 connector.

Refer to Lesson 2 for detailed information regarding 10BaseT and 100BaseTX
cables. 10BaseT is an Ethernet network that uses UTP cables to connect the nodes.
The 100BaseTX, also called Fast Ethernet consists of two data-grade pairs of twisted-
pair cables and transfers data at the rate of 100 Mbps. Both 10BaseT and
100BaseTX use standard RJ-45 connectors for Category 5 unshielded twisted-pair
(UTP) cabling, a rolled RJ-45.

 Note: UTP cables are often incorrectly referred to as "RJ-45
                  connectors." The RJ-45 refers to an 8-pin modular
                  connector wired for use with an AT&T Definity PBX.

Using Patch Cables to Extend the Network

After a network installation is complete, you may encounter a situation where you will need to add additional computers and devices to the network. Patch cables allow adding additional devices to increase the overall length of the cabling segment.

Generally, patch cables are used on token-ring networks, allowing the addition of a special-purpose hub, or Multistation Access Units (MAU) to increase the number of devices that can be connected to a network. In this lesson, we will discuss connecting a MAU to an existing token-ring network.

Tip: The difference between a MAU and a hub is that a MAU must be part of the network ring.

Generally, a MAU has ten connectors. Eight jacks are designed for connecting devices, the other two connection slots are for Ring In (RI) cables and Ring Out (RO) cables (Figure 8.13). To successfully connect additional MAUs into a network, unplug a patch cable from an existing MAU (MAU1) on the RO connection slot. The end of that patch cable plugs into the RI connection in MAU2.

Figure 8.13 Network Configuration Including MAUs

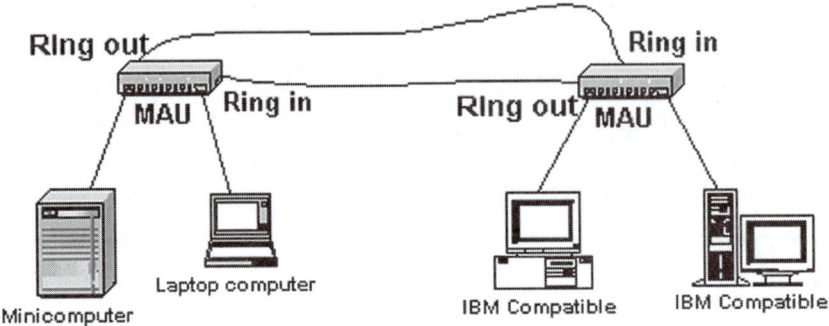

Plug the end of the first patch cable into MAU3. Then, plug the patch cable into the RO connection slot of MAU3. Plug the second patch cable into the RI connection slot in MAU3 and then connect it to the MAU1—RO slot.

For more information regarding network connections, refer to the following Web sites:

http://technet.microsoft.com
http://www.microsoft.com
http://www.zdnet.com
http://news.cnet.com
http://www.3com.com
http://www.rad.com/networks.htm
http://www.connectworld.net

Vocabulary

Review the following terms to prepare for the certification exam.

Term	Description
analog phone lines	Standard phone lines that transmit continuous signals
bridge	A device that transmits data over the network by viewing the address on the data unit
brouter	A device that incorporates both a bridge and a router in a single unit
DHCP	Dynamic Host Configuration Protocol is a TCP/IP protocol that allocates and manages IP addresses
digital phone lines	Communication lines that transfer data in intervals
hub	A device used to connect network nodes into a central location
IP address	An Internet Protocol address uniquely identifies the workstation on a network
jumpers	Small plastic connectors placed around two pins that communicate with an expansion card to determine which circuits will be used
NIC	A Network Interface Card is a device that allows the network cable to connect the computer to the network
patch cables	Cables that allow the addition of MAUs to continue the flow of data on token-ring networks

patch panel	A centrally located device linking cables from existing devices to additional devices on the network
peripheral	A device connected to a computer or network
print queue	The list of documents or print jobs waiting in a special part of computer memory to be printed
print server	A computer designated to accept print jobs generated from network computers
RJ-45	Similar to a phone cord (RJ-11) except it has eight cable connections
router	A device that connects subnets and different networks
TokenEase Relay Set-Up Tool	A tool that resets the ports of TokenEase MSAU or TokenEase Slim MAUs once you have installed either device
token-ring media filter	An adapter that allows UTP cables in a token ring network to connect properly to the devices on the network that require traditional token-ring connectors
UPS	A device that monitor electrical power to a network to prevent data loss during a power outage or power fluctuation
USB	A Universal Serial Bus is a successor to the RS-232C serial port and provides a basic mechanism for connecting peripherals to your computer

In Brief

If you want to...	Then do this...
Be prepared for installation of a network	Create administrative network accounts, test network accounts, user network accounts, and IP addresses before setup of your network
Further secure your network from entry errors or extreme virus attacks	Have a regular and an administrative account for a network administrator
Maintain a healthy server environment	Keep your server in a well-ventilated room
Connect devices together on a network	Use a hub
Provide a continuous power supply to a network	Install a Uninterruptible Power Supply (UPS) on the network

Lesson 8 Activities

Complete the following activities to better prepare you for the certification exam.

1.   Explain the purpose of having both a user account and administrative account for a network administrator.

2.   Explain the naming convention for serial and parallel ports.

3.   Define and explain the purpose of test accounts.

4.   Explain the purpose of a user ID and password.

5.   Define and explain the purpose of an IP address.

6.   Explain the types of information that can be found in the IPCONFIG utility program.

7.   Explain physical conditions that can harm your network.

8.   Describe nine network components and their connectors.

9.   Explain the difference between analog and digital lines.

10. Explain the difference between straight and rolled RJ-45 connectors.

Answers to Lesson 8 Activities

1. To secure your network, the network administrator should have two accounts, an administrative account and a user account. If you are logged on as an administrator and using normal features, there is a possibility that you could accidentally change network features. Having both an administrative and a user account makes more work for the administrator, but the benefits outweigh the amount of time it takes to maintain both accounts.

2. The port name for a serial port is represented as COMx, ("x" signifies port numbers from 1 to 4). The port name for a parallel port is represented as LPTx ("x" signifies port numbers 1 to 4).

3. Test accounts are normal user accounts set up by the administrator to test the network. The network should be thoroughly tested to verify that the system properly works when the users access the network the first time. An installation of a network system, a series of common procedures, such as e-mail and file manipulation, should be performed on workstations with the test accounts to verify the processes are properly performing.

4. Each user on the system should be assigned a user ID and a password for logon. The ID allows access to the system at the designated security level set by the network administrator. Passwords allow authorized users access to the network, but restricts others from accessing the network without and authorized ID.

5. After network installation on the server is complete, each workstation must be assigned an Internet Protocol (IP) address to uniquely identify the workstation. Most systems automatically assign an IP address during configuration. However, an administrator must verify that the IP address has been properly set up, and be prepared to manually configure it if needed.

6. You can view all TCP/IP information, such as the IP address and TCP/IP configuration in the IPCONFIG utility program.

7. Excessive heat can damage your server.

8.   **Printer Server**—A computer designated to accept print jobs generated
      from computers.

      Peripheral—A device connected to a computer or network. Typically, serial
      ports used with a modem and a mouse, and parallel ports are used for high-speed
      devices, such as disk drives and printers. The port name for a serial port
      is represented as COMx, where "x" signifies port numbers from 1 to 4. The port
      name for a parallel port is represented as LPTx, where "x" signifies port
      numbers from 1 to 4. A Universal Serial Bus (USB) is a successor to the RS-232C
      serial port and provides a basic mechanism for connecting peripherals to
      your computer.

      Hub—A device that connects other devices on a network into a central
      location.

      Router—A device that connects networks together. The router is responsible
      for reading and processing data to determine its destination network. Routers
      generally need a cable capable of connecting to a RJ-45 connection. An RJ-45
      connector is cable used to connect a telephone (RJ-11), but it is thicker, with
      eight conductors as opposed to an RJ-11 with only four conductors.

      Bridge router—Often referred to as a brouter, is a device that connects
      segments of a LAN to transfer data packets through the most efficient route on
      the network.

      Patch panel—A centrally located device linking cables from existing devices
      to additional devices on the network. Wires on the inside of a patch panel are
      electrically connected wires setting up the connection for the devices. Cables
      that travel to the connected devices attach to the patch panel with an RJ-45
      connector.

      UPS—A device that verifies continuous power to a network to prevent loss of
      data during a power outage power fluctuation.

      Network interface card (NIC)—A device that allows the network cable to
      connect the computer to the network. A NIC can contain both an external and
      an internal location to connect the cable. The type of connection you choose
      determines the type of cable you will need. If you choose to use a thinnet
      cable, you will need to use a coaxial BNC connector. If you choose a thicknet

connection, you will need to choose a 15-pin cable to connect the 15-pin (DB-15) connector to connect the network card to the network. If you choose an unshielded twisted-pair (UTP) cable, you will need to purchase RJ-45 connectors.

Token-ring media filler—An adapter that allows UTP cables in a token-ring network to properly connect to network devices on the network that require traditional token-ring connectors.
The following are two token-ring media fillers:

TokenEase NIC Adapter—Used when you have a token-ring device that has a standard nine-pin DB connector

TokenEase MAU Adapter—Used when you have a token-ring device that has a IBM MF41601 (MIC) connector.

9.   Analog phone lines are standard phone lines that transmit continuous signals. Digital phone lines are communications lines that transfer data in binary form.

10.   In Straight RJ-45 connectors, connections travel straight through the cable. In a rolled RJ-45 connectors, the connections travel through the twisted cable.

Lesson 8 Quiz

These questions test your knowledge of features, vocabulary, procedures, and syntax.

1. Why is it dangerous for a network administrator to complete daily tasks while logged on as an administrator?

    A.   The administrator can maintain the network system and test accounts
    B.   The administrator can locate an IP number that is faulty
    C.   The administrator may change network settings rather than the setting on the local computer and if a virus is activated there can be much more serious damage in an administrative account
    D.   The administrator can keep track of the incoming print jobs in the print queue

2. Which of the following allows a user personal access to the network?

    A.   IP address
    B.   User ID
    C.   IPCONFIG
    D.   Port

3. Which of the following verifies an IP number that is faulty?

    A.   Test account
    B.   Port
    C.   RJ-45
    D.   IPCONFIG

4. Which of the following defines a list of documents waiting to be printed?

    A.   Print queue
    B.   Print server
    C.   SCSI connection
    D.   USB

5.   Which of the following connectors can be used with a SCSI interface cable?

      A.   RJ-11
      B.   RJ-45
      C.   DB-25
      D.   COM2

6.   Which of the following defines a hub?

      A.   A device used to connect devices on a network in a central location
      B.   A device used to connect segments of a network
      C.   A device that connects segments of a LAN to transfer data packets
            through the most efficient route on a network
      D.   A centrally located device linking cables from existing devices to
            additional devices on the network

7.   Which of the following states the difference between an RJ-11 and a RJ-45
      connector?

      A.   An RJ-11 connector has two conductors and an RJ-45 has four
            connectors.
      B.   An RJ-11 connector has four conductors and an RJ-45 has eight
            conductors
      C.   An RJ-11 connector has eight conductors and an RJ-45 has sixteen
            conductors
      D.   An RJ-11 connector has sixteen conductors and an RJ-45 has thirty-two
            conductors

8.   Which of the following would you use to disable one of the connectors on a
      network interface card if there are two options built into the card?

      A.   TokenEase NIC adapter
      B.   IPCONFIG
      C.   Jumpers
      D.   Patch cable

9.   What is the ideal room temperature for a server?

      A.   30-35°F
      B.   50-55°F
      C.   60-65°F
      D.   70-75°

10.   Which of the following is considered the all-purpose electronic tool because it
      can measure resistance in ohms, AC and DC voltage?

      A.   Digital Volt Meter (DVM)
      B.   Time-Domain Reflectometers (TDRs)
      C.   Oscilloscope
      D.   Network Monitor

Answers to Lesson 8 Quiz

1. Answer C is correct. The administrator may change network settings rather than the setting on the local computer and if a virus is activated it can do much more serious damage in an administrative account.

    Answer A is incorrect. The administrator is responsible for maintaining the network system and test accounts.

    Answer B is incorrect. The administrator can locate an IP number that is faulty only in the IPCONFIG utility program.

    Answer D is incorrect. The print server keeps track of the incoming print jobs in the print queue.

2. Answer B is correct. A user ID will allow the users access the system at their designated security level set by the network administrator.

    Answer A is incorrect. An Internet Protocol address is used to uniquely identify the workstation for the other networked computers.

    Answer C is incorrect. The IPCONFIG utility program can locate an IP number that is faulty.

    Answer D is incorrect. A port is a place that the cable connects in a computer.

3. Answer D is correct. The IPCONFIG utility program can view all the TCP/IP information and can locate an IP number that is faulty.

    Answer A is incorrect. A test account is a normal user account set up by the administrator to test the network.

    Answer B is incorrect. A port is a place that the cable connects in your computer.

    Answer C is incorrect. An RJ-45 is a connector similar to a telephone wire except an RJ-45 has eight connectors rather than four.

4.   Answer A is correct. A print queue is a list of documents waiting to be printed that were sent to the printer by users on the network.

      Answer B is incorrect. A print server is a computer designed to accept print jobs as they are generate from computers on the network.

      Answer C is incorrect. A SCSI connection is a Small Computer Systems Interface that can transfer 8 bits at time.

      Answer D is incorrect. A USB is a universal serial bus that allows you to connect multiple peripheral into one jack on your PC.

5.   Answer C is correct. A SCSI cable has a DB-25 connector.

      Answer A is incorrect. An RJ-11 connector is a standard phone cord.

      Answer B is incorrect. An RJ-45 is similar to a standard phone cord except it has eight connectors rather than the standard four.

      Answer D is incorrect. A COM2 port is a serial port used a modem or a mouse.

6.   Answer A is correct. A hub is a device used to connect devices on a network in a central location.

      Answer B is incorrect. A router is a device used to connect segments of a network together.

      Answer C is incorrect. A brouter is a device that connects segments of a LAN to transfer data packets through the most efficient route on a network.

      Answer D is incorrect. A patch panel is a centrally located device linking cables from existing devices to additional devices on the network.

7.   Answer B is correct. An RJ-11 connector has four conductors and an RJ-45 has eight conductors.

      Answers A, C, and D are incorrect. An RJ-11 connector has four conductors and an RJ-45 has eight conductors.

8.   Answer C is correct. If a network card has both an external and an on-board connection, you will need to disable the type of connection that will not be used. This is done with jumpers. Jumpers are small plastic connectors that are placed around two pins to communicate with an expansion card about which circuits will be used.

      Answer A is incorrect. A Token Ease NIC Adapter is used when you have a Token Ring device that has a standard nine-pin DB connector.

      Answer B is incorrect. IPCONFIG is a utility program that can locate an IP number that is faulty.

      Answer D is incorrect. A patch cable is a cable designed to allow the addition of MAUs to continue the flow of data on a token ring network.

9.   Answer C is correct. The ideal room temperature for server storage is 60-65°F.

      Answer A, B, and D are incorrect because the ideal room temperature for a server is not 30-35°F, 50-55°F or 70-75°F.

10.   Answer A is correct. A Digital Volt Meter (DVM) is considered the all-purpose electronic tool. It can measure resistance in ohms, AC, and DC voltage, and in some cases low frequencies.

      Answer B is incorrect. Time-Domain Reflectometers (TDRs) send pulses down a cable and bounces the signal back to where any impedance difference occurs.

      Answer C is incorrect. The oscilloscope presents the signals it probes in a graphical format that is time based.

      Answer D is incorrect. Network monitors are software programs that track and measure network activity.

Lesson 9: Network Support and Maintenance

Ensure that your network functions at top performance by installing system upgrades and initiating regular preventative system maintenance. Updating your system with software and hardware adds functionality, features, and can remedy existing network problems. You can further enhance network devices and system or program applications with patches, fixes, and upgrades. Before implementing any upgrades to your system, always make a backup.

After completing this lesson, you should have a better understanding of the following topics:

- Network Support Concepts

- Patches and Fixes

- Test Documentation

- Upgrades

- Network Maintenance Concepts

- Monitoring Network Hardware, Network, Server and Workstations

- Implementing Patches, Fixes, and Upgrades

- Backups for Servers and Workstations

- Anti-virus Software

Network Support Concepts

Patches, fixes, and upgrades, sometimes called service packs, help keep your network systems, applications, and devices reliable. Monitor and test your system to prevent problems, and assess any new requirements. Patches, fixes, and upgrades are interim solutions to system problems until a new full version of the product is released.

Patches and Fixes

Patches are pieces of software code that are "patched" or added into a program. Fixes are grouped together into an executable program called a patch, as a temporary remedy for program or system problems. When problems with your network devices or operating system software and firmware (software routines stored in read-only memory) or bugs occur, your vendor can write a software fix for each problem that can be installed onto your system or network in the form of a patch.

Tip: A hot patch is code that fixes a program bug and does not require you to reboot the system for activation.

Test Documentation

Test documentation is the patch release notes that provide detailed information about the purpose of the fix. Sometimes a vendor will automatically notify you that a patch for your system is available, and distribute test documentation to you from a manufacturer's Web site, CD-ROM or DVD, or BBS.

There is a variety of online and hardcopy test documentation sources available for you to review regarding patches and fixes specific to your system. This information is accessible from:

- White papers

- Technical specifications

- README files
- Manuals
- News releases

White papers list solutions and test methods, explain the steps used to fix the problem, and define the theory of a technology. For example, a white paper can tell you how a particular patch could affect your system, and the reasons why you would configure your system in a particular way.

Technical specifications are a short listing of the specific components in a software or hardware product. You might see, for example, a list of a SCSI card specifications like the types of SCSI components that can be attached, the pins on the connectors, chip model, operating speed, number of devices supported, among other information.

README files are text files that contain information about the software patch that is not in any regular program documentation. Your test documentation for the patch should be sent to you on the same media as the patch release as a file named either README.TXT or README.DOC that can be opened from a word processor or text editor. Always read this test documentation before you install a patch, because it could contain information about your specific configuration.

Manuals are the hardcopy books that come with your hardware or software product. You can get or view manuals from Web sites and CDs. Use a text editor like Microsoft Word viewer or Adobe Acrobat to download and view online test documentation.

News releases on the manufacturers or vendors CD-ROM or Web site provide up-to-date information on new events. News releases are a quick way for you to check for upgrades to your products.

You can access Microsoft® TechNet Web site or CD-ROM for information on many technical issues.

Tip: Before you decide to apply a patch, you should read the
                    documentation first. For example, if the documentation tells
                    you that the patch fixes a problem with hardware that you
                    don't have, you could find out by reading the documenta-
                    tion that you don't need to install the patch.

Upgrades

Upgrades are releases of the latest version of your hardware or software. Upgrades
incorporate patches and are released on a regular basis to keep your technology
current. You might not want to upgrade to a new version every time your vendor
releases a new version. Assess what enhancements the new version provides.
Before you implement any upgrades to your system, read the test documentation,
backup your system, and test the upgrade.

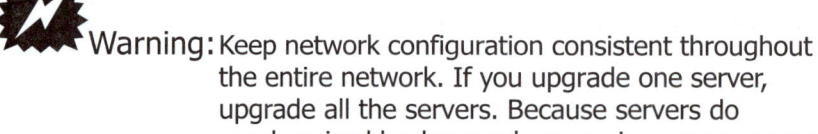Warning: Keep network configuration consistent throughout
                    the entire network. If you upgrade one server,
                    upgrade all the servers. Because servers do
                    synchronized background processing, errors occur if
                    they are not at the same patch or operating level.

Consider installing a software upgrade if:

■   The new software provides functionality the old one does not.

■   The enhancements provide a feature that those using the network need.

■   The vendor will not support your current software without the upgrade.

Consider network upgrades if:

■   The network slows.

■   Packets take too long to arrive at their destination.

■   Time-outs and slowdowns increase in frequency.

■   A new application requires additional or different requirements.

Consider hardware upgrades if you need to:

■   Increase memory

■   Increase hard disk capacity

■   Comply with standards such as Y2K

■   Provide more processing power

■   Replace obsolete technology

■   Comply with external changes

In summary, you should plan to install patches and fixes as part of your regular maintenance schedule, but read the test documentation to analyze the need for upgrading before you implement changes. Perform a system backup before making an upgrade.

Network Maintenance

You should develop a regular maintenance schedule for your hardware and software network systems, and monitor the server and network functions. Initiate preventative maintenance after a review of the conclusions and findings from your regular preventative maintenance exercises and reports generated from your system monitor. Before you implement any changes, make a backup, test the patch, or upgrade, and read the release notes. Moreover, you should perform regular system backups and make sure your anti-virus software is kept up to date.

Tip: Remember to upgrade anti-virus software.

Maintaining the Network

Your network needs constant attention to keep it running in optimal condition.
Network maintenance is less burdensome when you document the configuration and
characteristics of your network, and keep track of the changes in the network
configuration in a central log. Network documentation allows you to perform
maintenance tasks faster. Following is important information about your network
that you should keep in your central network log:

- Changes to the original network configuration

- Devices and shared devices like printers and faxes

- Documentation locations

- Installation directions, processes, and results

- Item descriptions and purchase dates

- Listings of critical files such as config.sys

- Network diagram

- Network drive assignments

- Network licenses for each application and where they are kept

- Network original configuration

- Network problem history and solutions documentation

- Network software and shared software

- Serial numbers

- Warranty and vendor information

Schedule preventative maintenance at least once every six months. Break down
your maintenance tasks into categories and rotate them in your schedule, so the
tasks aren't done at the same time. Your preventative maintenance plan should
include everything associated with your network, but not services outside of your
control, such as electrical services to your building and wiring, and the inside of
any devices.

Include the following tasks in your preventative maintenance rotation:

- Make sure all computers and devices are operational.

- Make sure all the servers, computers, and devices are connected to the network.

- Monitor the network and server to make sure your network is within the tolerance of the hardware.

- Back up the server and the workstations.

- Update antivirus software.

These tasks are included in the maintenance categories in the following information. You can rotate the following categories in your 6-month preventative maintenance, or maintain your system as problems arise.

Monitor the Network Hardware

Make sure all the servers, computers, and devices are operational and connected to the network. As a network administrator, you should know what the potential network bottlenecks are and which tools to use to monitor them. Bottlenecks are device problems that slow down the network.

The following tools can help you monitor the integrity of the network and its device components:

- The loop-back connector

- The cable tester

- The network management console

The loop-back connector tests the transmit and receive functions of a network interface card (NIC). You can use cable testers to identify and check tolerances of a particular cable to detect whether or not a cable is damaged and needs to be replaced.

In addition, you can set your network management console to monitor devices and report messages about events on those devices as they approach or reach your pre-set minimum limits or thresholds for hardware and electronic activities.

Tip: The network management console works with the Simple
            Network Management Protocol (SNMP) to trap your
            statistics in a database from devices on the network.

You will see a network alert when your pre-set thresholds are exceeded, and you
can gain a network baseline for use in future network trend analysis. Some common
thresholds or elements you may want your network management console to
monitor are:

- Device failure

- Network saturation

- Port malfunction

- Temperature

- Packet loss

- Excessive packet collisions

Monitor the Network, Server, and Workstations

You should monitor the network and server to make sure your network is within the
threshold of the hardware. You can monitor the server and workstations with a tool
that comes with the Network Operating System (NOS) designed to monitor errors.
The Microsoft tool to monitor the entire network, workstations, and servers from a
single computer, in local and area wide networks is called the Systems Management
Server (SMS).

The performance monitor in Windows NT® advises you about system performance,
and whether configuration or hardware changes have had any affect. Performance
information can give you early warning about potential problems. Windows NT
server installs a common set of performance counters for all of things you want to
monitor such as cache, disk usage, processor, processes, and threads. Network
counters are installed according to the network software at the system.

The performance monitor allows four view options to display server and system information—chart, log view, alert view, or report view. The following is a description of each view:

- Chart view—displays current data counters as a line or histogram, making it easy to see activity and utilization levels of resources or bottlenecks.

- Log view—displays data counters captured over a period of time from a log file. You can select objects (categories or areas) to monitor and hold the data in a log file, and view them in the chart view or report view.

- Alert view—displays an alert log under the alert view. When a specified threshold value is reached, an alert generates and an entry is made into the alert or event log.

- Report view—displays a spreadsheet type of display with the status of monitored resources.

In summary, the hardware tools and performance monitor tools allow you to observe trends, bottlenecks, and thresholds. As a network administrator, you can monitor categories and areas of computer performance and then decide whether to further tune your network or research the need for patches, fixes, and upgrades.

Steps to Implementing Patches, Fixes, and Upgrades

You should always approach the implementation of patches, fixes, and upgrades with a systematic testing plan in an isolated environment. Although you have determined the need to implement an interim solution, the production on your network system could be shut down. You can determine at this time whether or not the impact on your network production is worth an upgrade implementation.

Warning: Before you begin to test the upgrade, do a backup.

The following are four phases with steps you should use to implement an upgrade:

First do preliminary analysis, and test the upgrade in an isolated environment that include these steps:

1.   Research the test documentation

2.   Download the patch and related documentation into an isolated test network.

3.   Read the README files and manuals.

4.   Document the changes and define a test plan.

5.   Install the patch on a test workstation.

6.   Select the installation method that allows saving previous configuration to allow an uninstall.

7.   Record your option selections, such as driver configuration.

8.   Reboot the computer.

9.   If the operating system does not load or work properly, start over with a clean test machine.

10.   Test the new features. Test all security patches to see if they perform as required.

11.   Run the test workstation for two weeks. Reboot and try different tasks during this time.

If you find the test workstation environment tests acceptable, then test the upgrade on a server environment and do these steps:

1.   Install the patch on a test server.

2.   Repeat workstation test steps 1 through 11 on the server.

When you want to test the upgrade in a production environment, perform limited rollouts. A rollout is the implementation of the upgrade from the test environment to the production environment. If the server test runs are acceptable, perform these steps for the limited rollout.

1.   Update your information service staff's personal computers, and have them test the patch.

2.   Do a limited rollout to your user workstations if the staff testers determine the product is acceptable.

If your patch testing is successful, and you want to implement the patch, follow these steps:

1.   Roll out the patch to your production servers and all workstations by an automated procedure.

2.   Ensure proper revision control. Make sure that all equipment has the same approved patch.

In summary, with a systematic approach to implementing patches, fixes, and upgrades in an isolated environment, you have the opportunity to determine whether or not the impact of an upgrade is worthwhile. Once your upgrade is implemented, you can back up your system with the upgrade.

Backup the Server and Workstations

Your system of preventative maintenance and implementing upgrades should be based on an appropriate backup system and strategy. Your backup strategy should meet the needs of your organization and guarantee recovery of lost data. You should always backup your system before implementing any patches, fixes or upgrades to removable media such as tapes, floppies, disks, or zip drives.

 Note: Microsoft Windows NT backs up and restores data to and from tape drives and to files and folders on the Windows NT file system (NTFS) and the File Allocation Table (FAT) volumes. You should ensure that the tape drive device is on your hardware configuration list (HCL).

Consider which files and folders to back up, whether or not to perform network backup or multiple local backups, and how often to back up. Back up using these criteria:

■   Back up critical files you can't get along without

■   Back up workstations or servers where your critical data resides

■   Back up as often as necessary to ensure the recovery of your organization's critical data

To protect your data, choose from one of these backup types: full, differential, or incremental.

Full backup—Allows you to daily copy all files from server hard disks to tape, as well as clear archive bits. Full backups are the easiest to set up because there are no configuration, special settings, or databases to search. You can restore data quickly because you have a complete copy of information. Full backups take up the most storage space and take the longest time to complete.

Differential backup—Allows you to backup files that have changed since the last full backup. You need to perform two backup sessions per week—a full backup at the beginning of each week, and a differential at the end of each week. It takes less time to do a differential backup than a full backup, but the restore time requires two backup tapes and two backup databases.

Incremental backup—Captures only the files that have been changed or added since the last backup, and work with a full backup session. Incremental backup is the fastest method, but the restores take the longest. To restore, the incremental backup requires the complete backup tape and every tape from the full back up to the day from which you wish to restore data.

After you analyze, upgrade, and back up your system to ensure against data loss, you should also further protect your data from destruction with anti-virus software.

Install and Update the Anti-virus Software

A virus is a small bit of computer code that changes and destroys data. Viruses are self-replicating and hide inside of programs. Their effects can range from trivial to catastrophic on your system. Viruses are classified by the system they target.

Warning: There are thousands of computer viruses with
                         hundreds of new viruses appearing each month.
                         Your system can be contaminated by these
                         destructive pieces of code unless you protect your
                         system with a current anti-virus software program.

Viruses are often concealed in e-mail attachments or part of a regular file. If you
open these infected files, they can infect a workstation, and are spread from
computer to computer when a file or e-mail is transferred. Some viruses can infect
the e-mail lists, and the addresses contained in each e-mail list they encounter.
Following are some virus types that target different parts of a computer system:

File-infecting virus—Attaches to executable files of other programs. When the
program is executed, the virus attaches to your computer's memory and infects any
other executed program.

Macro virus—A variation of the file-infecting program viruses, this type of virus
infects templates (usually the normal.dot file) for Microsoft Word and Excel
programs. Once a template is infected, every document or spreadsheet that is
opened with the infected program is corrupted.

Polymorphic virus—Infects boot sectors, files, or both. This virus changes its
signature with encryption variables, and escapes signature-scanning detection.

Only an anti-virus program with an up-to-date virus signature file can stop these
malicious virus programs. Anti-virus software programs contain a file of known virus
signatures or definitions. You can set the software to run in the background to
inspect files for known and suspected new viruses. An anti-virus program isolates
the known or potential virus, analyzes it, and destroys it.

Many anti-virus program manufacturers maintain Web sites that allow you to
regularly download the most current virus definition files. You should download
these files at least once or twice each month.

Tip: Run your anti-virus program daily to check for viruses, and have it run in the background.

You can use a server to download and distribute the signature updates or allow individuals to install and update their anti-virus software. Put anti-virus programs on every workstation, server, and proxy server/firewalls with the ability to check inbound e-mail is the best defense against catastrophe.

In summary, it is wise to develop a regular maintenance schedule for your hardware and software network systems, monitor the server, and monitor network functions. Perform preventative maintenance from the conclusions and findings from your regular preventative maintenance exercises and from the reports generated from your system monitor.

Before you implement any changes, make a backup, test the patch, or upgrade, and read the release notes. Moreover, you should perform regular system backups and make sure your anti-virus software is kept up to date to prevent catastrophe.

Vocabulary

Review the following terms in preparation for the certification exam.

Term	Description
anti-virus	Software that runs scans and cleans viruses by checking the computer signature or definition files
bottlenecks	Device performance problems that slow down the network
CD-ROM	Compact Disc-Read-Only Memory, a high-capacity, optical storage media that can hold up to 650 megabytes.
DVD	Digital Video Disc, a high-capacity, optical and video storage media that holds a minimum of 4.7 gigabytes which is enough for a full-length movie
fix	Solution to a software problem
hot patch	A fix that doesn't require rebooting the system for activation
patch	A group of fixes written into an executable program as a temporary remedy for program or system problems
removable media	Tapes, floppies, disks, and zip drives
rollout	Implementation of an upgrade from the test environment to the production environment
SNMP	Simple Network Management Protocol traps statistics in a database from devices on the network
test documentation	Patch release notes that give you the information about the purpose of the fix

upgrades	Version releases of software that incorporate patches and are released on a regular basis to keep your technology current
virus	A small bit of computer code that changes and destroys data
virus signature files	Definition files that list any existing viruses in the system and how they can be cleaned from the system

In Brief

If you want to...	Then do this...
Remedy a software or system problem	Install a patch, fix, or upgrade
Find out more about a patch, fix, or upgrade	Read the test documentation
See the newest, additional information about a specific patch	Look at the README file
Enhance your system with patches, fixes, or upgrades	Do a backup first
Gain more functionality, provide enhancements, or keep vendor support	Consider an upgrade to your software
Gain network speed or have different requirements	Consider an upgrade to your network
Increase memory, hard disk capacity, processing power, or replace old technology	Consider an upgrade to your hardware
Perform network maintenance tasks faster	Keep a central network log
Make backup and system maintenance easier	Rotate preventative maintenance tasks
Monitor the network hardware	Use the loop-back connector, cable tester and network management console to analyze problem areas
Monitor the server	Use tool or performance monitor that comes with your network operating system

Implement patches, fixes and upgrades	Use a systematic approach that includes backup
Ensure the safety of your organization's data	Choose to backup with full, differential, or incremental types
Protect your system from viruses and catastrophic disaster	Install and run an anti-virus program

Lesson 9 Activities

Complete the following activities to better prepare your for the certification exam.

1.   Describe some uses of patches, fixes, and upgrades

2.   Discuss the use of test documentation.

3.   Describe when you should consider a software upgrade.

4.   Explain the importance of a central network log.

5.   List five important tasks to include in your preventative maintenance rotation.

6.   Describe how you can use the performance monitor.

7.   Explain the three major phases to implement an upgrade.

8.   Describe the three types of backups.

9.   Describe how viruses work.

10.   Explain the need to update anti-virus software.

Answers to Lesson 9 Activities

1. Patches, fixes, and upgrades help you keep your network systems, applications, and devices reliable. Patches, fixes, and upgrades act as interim solutions to system problems until a new full version of the product is released.

2. Test documentation is the patch release notes that give you the information about the purpose of the fix. You can find the information in white papers, technical specifications, README files, manuals, and news releases. Before you decide to apply a patch, you should read the documentation first. If you do not have, for example, other software or devices that depend on the patch, you could find that you don't have to install it.

3. You should consider a software upgrade if the new operating system or application software provides functionality that the old one does not, the enhancements provide something that the people on the network need, and the vendor will not support your current software if you don't upgrade.

4. Your network system needs constant attention to keep it in optimal condition, and network maintenance should be less burdensome if you document your network, and keep track of the changes in the network configuration in a central log. Network documentation allows you to perform maintenance tasks faster.

5. Five important tasks you should include in your preventative maintenance rotation are:

    a. Make sure all the computers and devices are operational.
    b. Make sure all the servers, computers, and devices are connected to the network.
    c. Monitor the network and server to make sure your network is within the tolerance of the hardware.
    d. Back up the server and the workstations.
    e. Update the antivirus software.

6. The performance monitor has four display options so you can view information about your server and system in a chart view, a log view, an alerts view, or a report view. In the chart view, you can see counters displayed as a line or

histogram, which allows you to see resources and bottlenecks. The log file gathers data from selected counters over a period of time, so you can analyze them later. Files are kept to analyze later. In addition, you can create an alert log under the alert view. When the threshold value you specify is reached, an alert notifies you that and an entry was made in the alert log. The report view provides a spreadsheet type of display with the status of monitored resources.

7. To implement an upgrade, do preliminary analysis, and test the upgrade in an isolated environment. If the tests are acceptable, then test the upgrade on a server environment. If the server tests are acceptable, implement a limited rollout. Then, if your patch testing is successful, roll out the upgrade.

8. Three types of backups are: Full, Differential, and Incremental. Full backups allow you to copy all files from server hard disks to tape every day, and set the archive bits. Differential backups allow you to backup files that have changed since the last full backup. Incremental backups also work with a full backup session. You back up only the files that have been changed or added since the last backup. Incremental backup is the fastest method, but the restore process takes more time than the other backup types.

9. Viruses often are concealed in e-mail attachments, or regular files. If you open these infected files, a virus can infect a workstation, and spread from computer to computer. Some viruses can infect e-mail lists thereby infecting every message that is opened. Only an anti-virus program with an up-to-date virus signature file can stop these malicious programs.

10. Thousands of computer viruses run rampant during the course of a year, and hundreds of new viruses appear each month. Computer systems have the potential to contract any of these destructive pieces of code, unless you protect your system with a current anti-virus software program. Anti-virus software programs contain a file of known virus signatures or definitions. You can set the software to run in the background to inspect files for known and suspected new viruses. The anti-virus program can isolate the known or potential virus, analyze it, and destroy it.

Lesson 9 Quiz

These questions test your knowledge of features, vocabulary, procedures, and syntax.

1.   Which types of software code are designed to act as interim solutions to system problems?

      A.   Fixes
      B.   Patches
      C.   Upgrades
      D.   Backups

2.   Which type of patch doesn't require you to reboot the system to activate it?

      A.   Patches
      B.   Hot patch
      C.   Fixes
      D.   Upgrades

3.   Which of the following are test documents?

      A.   Patch release notes
      B.   README files
      C.   White papers
      D.   Technical specifications

4.   Why should you read the test documentation first before you install an upgrade?

      A.   You may not need to install it because you don't have the software or devices it is designed to fix
      B.   The people using the current software may not need the enhancements that the upgrade provides
      C.   You want to know what functionality the new release will provide
      D.   You will find out what enhancements the upgrade provides

5. What should you do to make your network maintenance easier?

   A. Document the network
   B. Keep track of configuration changes in the network
   C. Rotate the maintenance tasks
   D. Schedule preventative maintenance at least once every two years

6. Which tools allow you to monitor for performance on cache, disk usage, processor, processes, and threads?

   A. Loop-back connector
   B. Cable tester
   C. Network management console
   D. SNMP

7. Which of the following indicate that you need to upgrade your hardware?

   A. The manufacturer just released a new model
   B. You need to comply with changing standards
   C. Your company needs to use electronic data transfer with another company
   D. The newer models are faster

8. How often should you back up your system?

   A. Nightly
   B. Every two months
   C. After every upgrade
   D. As often as necessary

9. What type of anti-virus protection should when downloading information from the Internet?

   A. Anti-virus software on the workstations, servers, and proxy servers/firewalls with the functionality to check e-mail and downloads
   B. Anti-virus software only on the workstations
   C. Anti-virus software only on the servers
   D. Encryption

10.   What functions does an anti-virus program perform?

    A.   Automatically updates itself
    B.   Isolate a potential virus
    C.   Analyzes a virus against a file of known virus signatures or definitions
    D.   Destroys unknown viruses

Answers to Lesson 9 Quiz

1.   Answer A is correct because a patch is a grouping of software code called fixes, designed to remedy software problems.

      Answer B is correct because fixes are small pieces of software code designed to remedy problems.

      Answer C is correct because upgrades are version releases of software that incorporates fixes and patches.

      Answer D is incorrect because backups do not remedy software problems.

2.   Answer B is correct because a hot patch doesn't require you to reboot the system to activate it.

      Answer A is incorrect because you must reboot patches to activate them.

      Answer C is incorrect because you must reboot fixes to activate them.

      Answer D is incorrect because you must reboot upgrades to activate them.

3.   Answer A is correct because patch release notes give you information about the purpose of the software fix.

      Answer B is correct because README files come with the patch or fix, and describe any new information or tips about the fix since the last time you saw the test documentation.

      Answer C is correct because white papers list solutions, test methods, and explain the steps used to fix the problem.

      Answer D is correct because technical specifications list the specific components in the software or hardware product.

4.   Answer A is correct because if you read the test documentation before you install a fix, you may find that you don't need it because you don't have the

software or devices it is designed for.

Answer B is correct because if you read the test documentation before you install a fix, you could find that the people who use the current software don't need the enhancements.

Answer C is correct because if you read the test documentation before you install a fix, you should find out how it functions.

Answer D is correct because if you read test documentation before you install an upgrade, you'll discover the enhancements it provides.

5.   Answer A is correct because if you document your network, it will take you less to time to restore or recover if updates don't function.

Answer B is correct because if you keep track of configuration changes in your network, others can use your information to work with the system.

Answer C is correct, because if you rotate your maintenance tasks, your network maintenance occurs on a periodic basis.

Answer D is incorrect because scheduled preventative maintenance at least once every two years is too long. Preventative maintenance should be rotated and scheduled at least every six months.

6.   Answer C is correct because a network management console reports limits reached according to the counters you set.

Answer D is correct because SNMP traps your statistics in a database from devices on the network.

Answer A is incorrect because a loop-back connector is a connector that tests the transmit and receive functions of a network interface card.

Answer B is incorrect because a cable tester identifies and checks tolerances on cables to detect if the cable is damaged.

7.   Answer B is correct because standards changes such as Y2K compliance are valid reasons for upgrading hardware.

      Answer C is correct because it is a viable business reason to upgrade to comply with changes in your customer's business practices.

      Answer A is incorrect because you may find through analysis that newer models do not add functionality or meet your needs.

      Answer D is incorrect because through analysis you may find you don't need faster equipment.

8.   Answer D is correct because to insure the recovery of your organization's critical data, you should backup as often as necessary.

      Answer A is incorrect because a nightly back up may not insure the recovery of your organization's critical data. Nightly backups are a typical method of operation, but backup schedules should depend on your organization's needs.

      Answer B is incorrect also because a backup may not insure the recovery of your organization's critical data. Backup schedules depend on your organization's needs and a two-month interval may be too long.

      Answer C is incorrect because it is too late to back up a system after you have changed it. You should back up a system before you upgrade in order to recover data if the upgrade fails.

9.   Answer A is correct because running anti-virus software on the workstations, e-mail servers, file servers, and proxy servers/firewalls is the best defense against catastrophe.

      Answer B is incorrect because running anti-virus software only on the workstations will not guard your server against infection.

      Answer C is incorrect because running anti-virus software only on servers will not protect workstations from file infecting viruses.

Answer D is not correct because encryption is a security protection mechanism, and does not guard against a virus infection.

10.   Answer B is correct because anti-viruses isolate a potential virus and check it against a file of known viruses.

Answer C is correct because anti-viruses analyze viruses against a file of known virus signatures or definitions.

Answer A is incorrect because anti-viruses cannot update themselves, and you must update the virus signature definition at least once or twice a month.

Answer D is incorrect because anti-viruses can destroy only known viruses. You must update anti-viruses at least once or twice a month for the signature file to recognize the most current known viruses.

Lesson 10: Troubleshooting

This lesson introduces you to concepts for troubleshooting computer networks. The complexities of modern computer networks make it impossible to anticipate every problem that can arise in a particular network. Since it is impractical to catalog all the possibilities in the context of this course, the troubleshooting information in this lesson is intended as a model to help you develop your approach to troubleshooting computer networks.

Successful troubleshooting rests strongly upon your technical understanding of the elements and interrelationships between the client, server, and network. It also relies on logic, discernment, and the ability to isolate faults through a process of elimination. To these ends, the troubleshooting information presented in this lesson embraces a logical and systematic approach to problem resolution.

After completing this lesson, you should have a better understanding of the following topics:

- Troubleshooting Resources

- Keeping Records

- Prioritizing Problems

- Troubleshooting Methodologies

- Isolating Operator Problems

- Isolating Network Problems

- Causes of Network Problems

- Diagnostic Tools

Overview

One of the best ways to minimize computer network problems is to implement a comprehensive maintenance program, such as the one described in Lesson 9 of this study guide. However, even if you perform regular maintenance on the network, problems can still arise from the normal usage of system components over time. To effectively deal with problems, you should have in place a well thought out strategy for handling network troubles before they occur. When problems arise, you can then respond quickly and efficiently while implementing a viable recovery strategy.

If you want to create a good response and recovery agenda, first consult the appropriate information sources for your network and its components. This will help you establish a baseline understanding of your network and guide your steps in a troubleshooting scenario.

Troubleshooting Resources

The most extensive information available to support your troubleshooting efforts on hardware, software, or the operating system is vendor and manufacturer documentation.

 Note: The vendor and manufacturer are not necessarily the same company. The manufacturer is referred to as the original equipment manufacturer (OEM). The vendor is the company that supplies the equipment to the consumer, such as Dell or Compaq.

For example, a vendor may order equipment from the manufacturer, configure it with applications, and then resell the package to the consumer. The vendor normally supplies user documentation with the original equipment purchase.

Sources for the most useful information available to support computer network troubleshooting are:

■   Documentation provided by the vendor with the original product purchase

■   Vendor updates on technical changes, revisions, and resolutions to problems encountered (from their user base)

- A comprehensive database on the vendor's Web site that documents common problems and solutions
- Technical bulletins posted on the vendor's Web site that address problems encountered and fixes since the product's release
- Documentation created by the OEM
- Vendor knowledge bases such as Microsoft®TechNet or Novell®Support Connection, both available on CD-ROM

 Note: If the vendor's knowledge base is not available on CD-ROM, it will usually be provided on their Web site.

Technical Support

In the event that you cannot resolve a problem after researching all the resources listed in the previous section, call the technical support department of the vendor or manufacturer. This is the most valuable source of information available to you, although support calls can be costly (unless you have a support contract in place).

When you receive technical support, document everything for future reference. Also, maintain a log for all of your network hardware and software that specifies the manufacturer, technical support phone numbers, and support availability hours.

Reference Manuals

You should have a comprehensive collection of up-to-date professional networking manuals for reference. These can give you insights into the complexities of computer networking and thereby support your troubleshooting efforts.

Record Keeping

It is a good idea to create a ledger to keep records of problems and resolutions. The ledger can cross reference all the network hardware and software components and record other important information about them. When problems are identified and

resolved, you can update the ledger so you have a historical problem/resolution
record for future reference. This is extremely valuable if you have to fix the same
problem again. The ledger can include the following entries:

- Device or product name and model number

- Vendor or manufacturer name

- Technical support information

- Vendor/manufacturer's Web site URL

- Knowledge base Web site URL

- Directory (or other) location of diagnostic software (if applicable)

- Date of incident or problem

- Domain name or location

- Prioritization level

- Response date

- Problem identification and solution

Maintaining these records can support an efficient response plan for your
information services (IS) department. For example, knowing the problem history of
the network or workstation and establishing priorities by which IS resources are
allocated, IS professionals can better apportion their time and effort in problem
solving.

Prioritizing Problems

Since the severity of network problems can vary, it is important to classify problem
calls according to priority. With respect to establishing priorities, you should assess
how critical the problem is and how much of the network is affected. Some network
components are critical to network operation while others are not.

Problems with some components, such as shared resources, can affect large
numbers of people and compromise their productivity. Problems that receive the
highest priority are those that completely curtail productivity, including problems

with a server, router, or hub. Other problems, such as those limited to a workstation, take a lower priority. However, it should not be overlooked that workstation problems can be critical too, since workers need to access network resources for job performance.

The following priority levels can be assigned to problems so you can establish a guideline for impact on network or workstation availability:

Mission critical—Problems affecting significant portions of the network to the extent that business operations stop. These kinds of problems cause the entire network or at least a critical portion to go down.

Critical—The network is still up, but the problem affects everyone on the network to the extent that workflow is significantly impeded.

Sub-critical—The network is up, but the problem affects only certain user groups or workstations, such as when applications won't run.

Non-critical—The network, user groups, and workstations are all up, but a minor problem affects a small number of people.

Troubleshooting Methodologies

When troubleshooting a network, it is imperative that you understand all the existing devices, operating system, and computer hardware, or you will not be able to execute the diagnostics required for network problem isolation. If you understand the network architecture and component interaction, you can solve network problems very effectively.

Diagnosing a problem and its location—particularly those that are very complex—often involves first eliminating where the problem does not exist. This helps direct you to the actual location of the trouble area. Also, recognize that some problems have very simple solutions that should be tried before employing techniques that are more rigorous.

Configuration Records

Keeping records of configuration changes during the troubleshooting process is important. This allows you to return the network to its initial configuration at the time the problem occurred, should your trouble-shooting approach not yield the desired solution.

Documentation

When attempting to isolate a problem, document in a log all the steps, methods, or processes you performed. Then, if you need to consult with technical support personnel, you can accurately describe the steps taken and where you encountered difficulties. This greatly assists technical support personnel to help you solve the problem.

Troubleshooting Guidelines

The following diagram (Figure 10.1) defines the general steps you should follow when organizing your troubleshooting and problem-solving efforts:

Figure 10.1 General Troubleshooting Steps

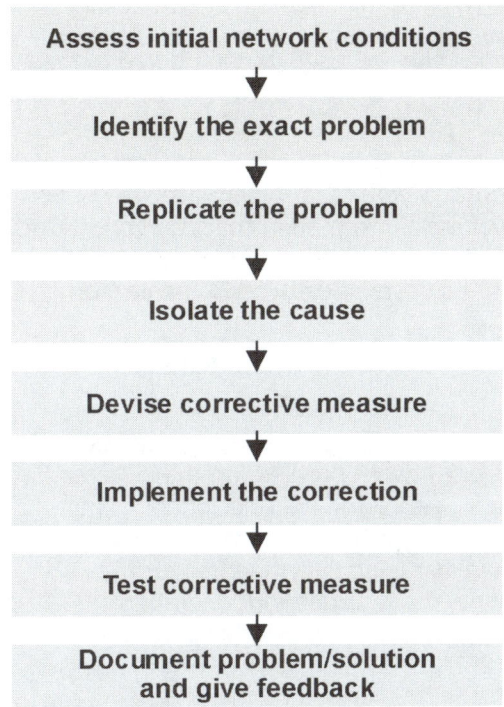

Assessing Initial Network Conditions

Before you try to identify a problem, consult the ledger you previously created for records of network hardware and software components. Your records may show a history of repeated occurrences of the same problem and how you resolved it.

Referring to this ledger can save you a lot of time and effort in problem resolution. If your ledger doesn't reveal anything useful, the next step is to establish the initial network condition at the time the problem occurred, so you have a baseline for proceeding in your troubleshooting efforts. Assess the initial conditions as follows:

- Determine whether the network was functioning normally prior to the problem occurrence or if you had any earlier indication of network trouble

- Ascertain the length of time the network was working before the problem began

- Check your records to verify whether any changes were recently incorporated in the network

- Verify if any client or administrator activities either caused or contributed to the problem when it occurred

These steps establish an initial profile of the network at the time of the problem occurrence, and subsequently, either point you to a possible problem area or eliminate some variables at the least. Other valuable information that can contribute to your initial assessment is network configuration and status. Use the Windows® NT Diagnostic utility to obtain this information, as described ahead in the section entitled "Isolating Network Problems."

Identifying the Problem

When a problem manifests in your network, frequently an error message appears. Usually, error messages point you to a device or network problem area. However, other error messages are cryptic and you might need some help deciphering them. Developers often create error messages meaningful for them, but not necessarily for others, thus making it difficult for you to know what is actually happening. In this situation, consult the technical support personnel for your network operating system, devices, or workstation, and see if they have a database that tracks error messages versus problem resolution.

Also, be aware that you can receive error messages from various network devices or you can monitor error logs and displays (logical and physical indicators) for signs of trouble. Tools that you can use to view error logs, performance statistics, and network characteristics are discussed later in "Isolating Network Problems."

Replicating the Problem

When you first become aware of network trouble, it is important to try to replicate the problem. If you can do this, then you have a good starting point to isolate the cause of trouble. If the problem can be replicated by performing certain known functions or actions, then it is not a one-time incident but a consistent error. If the problem cannot be replicated and does not display an error message, you will probably be faced with a trial-and-error approach to resolving the problem. This is primarily where to use the process of elimination.

Note: If the problem is isolated to operator error at the
                 workstation, have the operator replicate the procedures,
                 commands, or keystrokes while you observe. This could
                 reveal obvious operator errors that can be easily
                 corrected.

Isolating the Cause by Process of Elimination

Approaching network troubleshooting through the process of elimination is very
effective and sometimes the only way to isolate problems. To be proficient with this,
you should be very familiar with the functions and interaction of all network
components as well as their interface architecture. In the process of elimination,
remember to consider the most obvious sources of problems so you don't overlook
simpler solutions. Also, use the technique of expanding and contracting your focus
to eliminate the variables.

For example, if you became aware of a problem that appeared to be on a
workstation, you would not necessarily assume the problem is on the workstation.
Instead, expand your thinking about the problem up to the network level to
determine if it has an impact there. If the problem does not appear at the network
level, then you have isolated the problem to the workstation. Then, contract your
thinking down to the workstation components, to isolate the source of the problem
at that level. The following problem scenario (Figure 10.2) illustrates these
concepts.

Figure 10.2 Problem Scenario—Workstation(s) Does Not Connect to the Internet

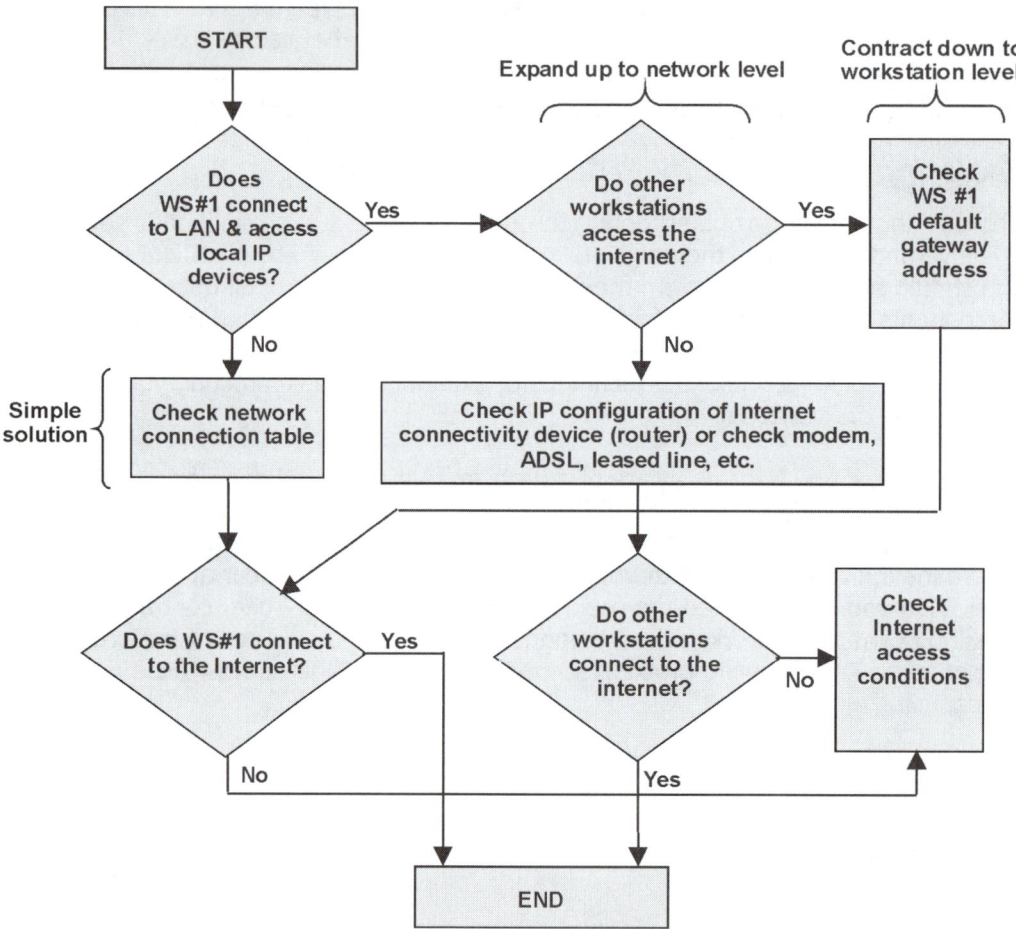

When attempting to eliminate variables to isolate a problem, change system or device configurations one at a time so you can track the system's response. If you want to isolate the problem, you must carefully isolate the changes you make. Making more than one change at a time can obscure the problem location and hinder your efforts.

Implementing Corrective Measures

How you devise and implement corrective measures is entirely dependent on the nature of the problem. Some solutions require a simple cable or device replacement or an application reinstall; others require extensive processes—like reconfiguring the entire network IP addressing scheme.

Sometimes your solution may be a product upgrade or upscaling components to better support growing system demands. Some products are designed for scalability, but remember that glitches can exist in the upgrade path provided by the vendor. It is up to you to make sure that your upgrade is compatible with other network functionalities.

Testing the Corrective Measure

After your corrections are implemented, a comprehensive testing plan should be carried out to verify that your solution is valid. What you do in your testing plan depends on the changes made and what other systems or functionalities are impacted by the change.

Documentation and Feedback

Whatever measures you take, be sure to document everything and update your ledger for future reference. It is also important to provide feedback to the user who experienced the problem to let them know the problem is resolved and how you fixed it. Alternatively, you can use a work order tracking system to inform users of status.

Isolating Operator Problems

Operator-caused problems are usually limited to operator procedures, operator applications, operator devices, the workstation, or the operator's interface with network devices. If an operator calls you with a problem, most likely they are unaware of what caused it.

To find out the cause, you should ask the operator to describe the symptoms and the actions taken when the problem occurred. If the information you obtain is insufficient to point to an obvious solution, you may need to probe further. Figure 10.3 presents a problem scenario that illustrates a method of isolating operator errors. For example, this method could be used to isolate the problem when an operator has trouble accessing a company-shared folder on the server.

Figure 10.3 Problem Scenario—Isolating Operator Errors

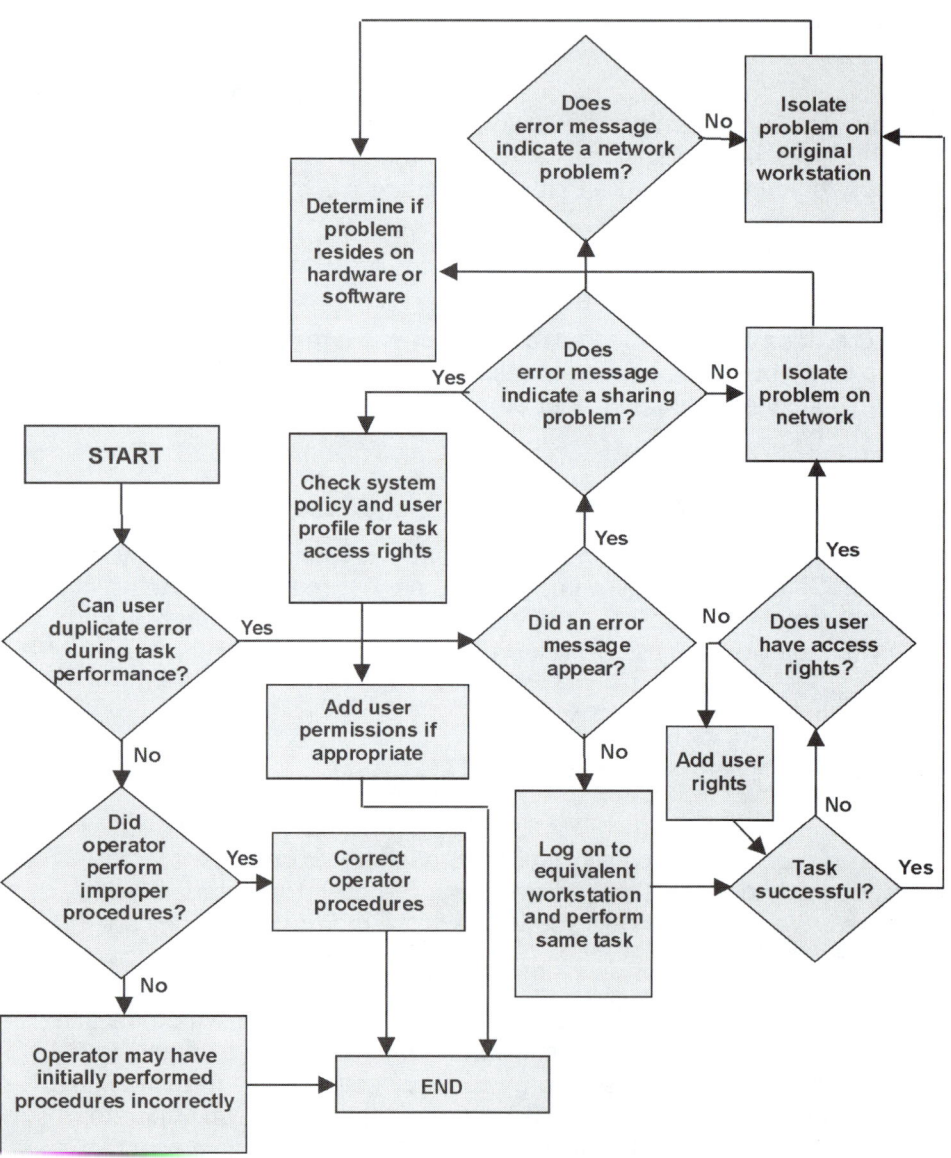

Tip: If you suspect a workstation problem other than operator error, you can use the Windows NT Performance Monitor to gather statistics and performance data on certain system objects and processes.

Isolating Network Problems

Network problems exist on two basic levels: localized and widespread. The material that follows describes what you can do to isolate and resolve localized or widespread network problems.

Localized Problems

Problems on this level are usually limited to network devices and servers. If you suspect a localized network problem, you can use the resources discussed in the following paragraphs to assist you in problem identification and location.

Windows NT Diagnostics

Knowing the baseline configuration and status of the network gives you valuable information to work with when attempting to isolate network problems. To obtain a snapshot of the network configuration, you can use the Windows NT Diagnostics utility (WINMSD). WINMSD lets you view registry values in an easy-to-read format without actually opening the registry, as shown in Figure 10.4.

Figure 10.4 Windows NT Diagnostic Utility

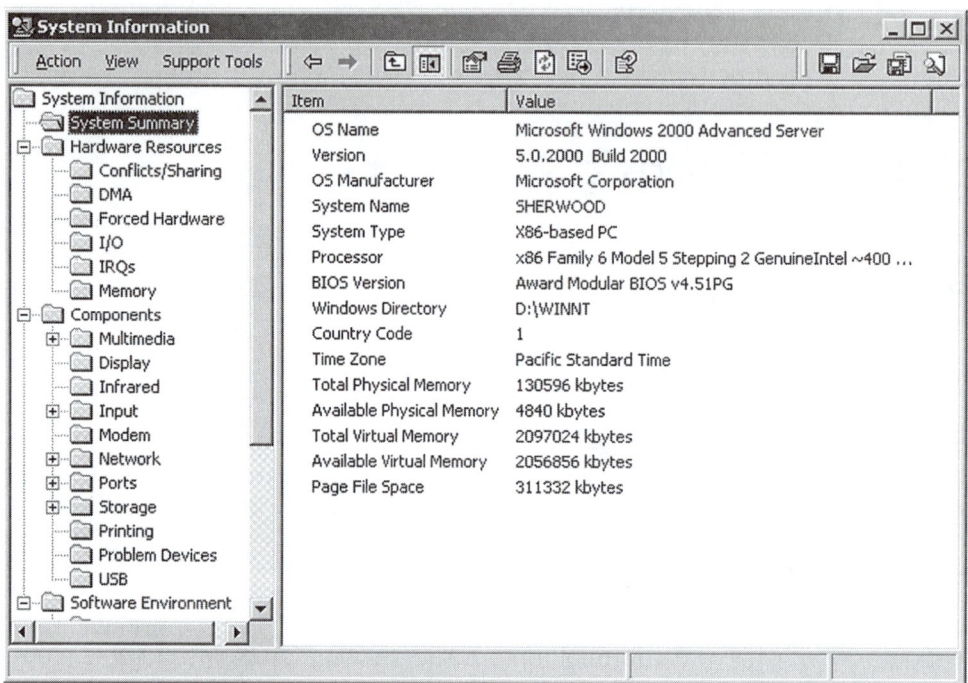

With the Windows NT diagnostic utility, you can view or generate a report, which includes the following diagnostic information:

System Summary—Displays operating system version, manufacturer, CPU, RAM, virtual memory, and BIOS information.

Hardware Resources—Contains six buttons—Conflicts/Sharing, DMA, Forced Hardware, I/O Port, IRQ, and Memory—to display hardware device status and resource usage.

Software Environment—Shows drivers, environment variables, network connections, running tasks, services, program groups, and startup programs. Other information includes installed protocols, adapter settings, and network statistics

Tip: To activate Windows NT Diagnostics, choose Run from the
             Start menu and type `WINMSD`.

Windows NT Event Viewer

The administrator's account has access to all system/event logs. Check system logs
first if you suspect a localized network problem. The Windows NT Event Viewer is a
logging facility that records errors and messages for system, security, and
application events. Typical event types follow:

System—Errors and status messages related to system hardware, drivers, and
services.

Security—Security violations and other security related events.

Application—Errors related to applications.

Note: The Windows NT Event Viewer does not record errors
              related to internal application processes.

The Windows NT Event Viewer uses the following indicators to identify the type of
error or status message recorded:

Error—A red stop sign icon indicates significant problems such as loss of data or
functions. For example, this type of error might occur if a service was not started on
the server.

Warning—A yellow triangle icon indicates events that can create future problems.
For example, a warning event might be logged if server disk space becomes low.

Information—A black circle with a "I" indicates significant events for informative
purposes.

Failure Audit—A "//" symbol indicates failed security access attempts. For
example, if a user tries to access a network drive and fails, a Failure Audit event is
recorded.

Success Audit—Icon of a key indicates a successful security access attempt. For example, if a user successfully logs on to the server, a Success Audit event is logged.

 Note: The Windows NT Event Viewer can be used in association with the Windows NT Performance Monitor. For example, the Performance Monitor can be configured to log an alert to the Event Viewer when certain critical system thresholds are exceeded.

Figure 10.5 shows an example of data displayed in a Windows NT Event Viewer log.

Figure 10.5 Windows NT Event Viewer Log Sample

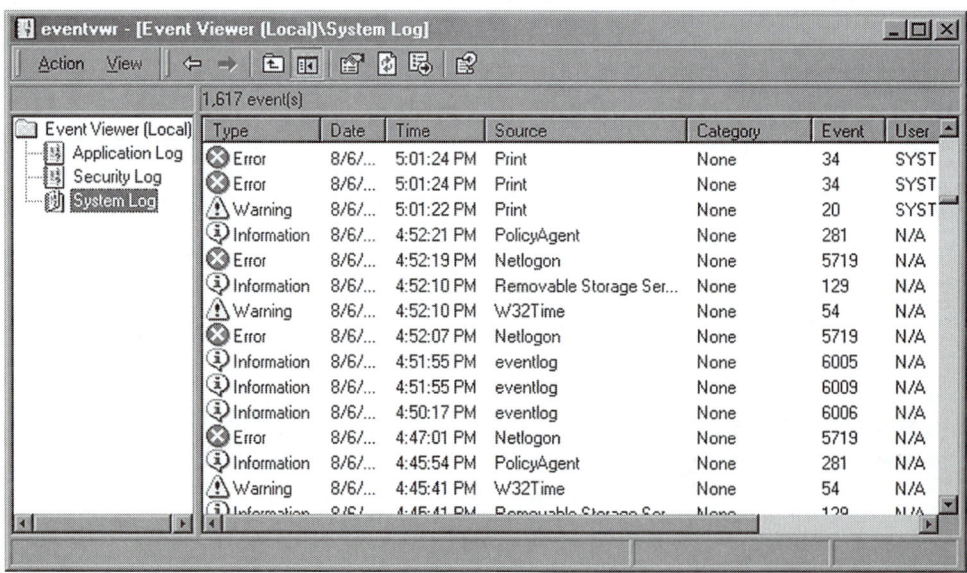

The Windows NT Event Viewer logs can often point you to the exact location of a problem if you understand the meaning of the information presented. Following are the information fields displayed in the Windows NT Event Viewer:

Date—The date the event was logged. An error message or status icon appears to the left of the date.

Time—The time the event was logged.

Source—Identifies the application, driver, or other source that logged the event.

Category—Identifies the category of the event. System and application events have a category of "none."

Event—The code that identifies the error.

User—The name of user logged on at the event occurrence.

Tip: From within the Event View, to view detailed error descriptions, double-click the event entry.

NetWare Error logs

Novell NetWare 5 has the following three error log files that can help diagnose problems on a NetWare server:

Console log file (CONSOLE.LOG)—This file keeps a history of all errors displayed on the server's console. For example, if the server started to behave oddly after a reboot, this log would be a good place to check for errors.

Abnormal end file (ABEND.LOG)—This file registers all abnormal endings (abends) on the NetWare server. This type of error can lock the server or halt operations.

For example, if the ABEND event was appended with a "page fault" or "stack" tag, this means the fault is related to a memory problem. It could be that an application or process tried to utilize memory in use with another program. In that event, the NetWare server will shut down the offending process and issue an abend.

Server log file (SYS$LOG.ERR)—This is a general log file which lists all errors occurring on a NetWare server including abends and network directory services (NDS) errors, and identifies them with a date and time stamp.

Errors recorded here incorporate a numbering code that indicates the severity, location (locus), and type (class) of error for ease of interpretation by the troubleshooter. For example, an error might read a severity of 1 (a warning condition), a locus of 17 (relates the error to the operating system), and a class of 19 (a domain type problem).

LINUX Log Files

LINUX is an implementation of the UNIX operating system. In LINUX, many programs, which run in the background, generate error logs as well as records of when a user logs in and out of a system. These programs include the Web server, e-mail, ftp, and so on. When troubleshooting in LINUX, you can check the log files for errors to point you to possible problem areas.

You can also do manual security audits by viewing records of bad logins to make sure you have no unauthorized outside traffic attempting to access services on your system. To browse the login record, use the last command, which lists the following:

- Username
- Terminal used to access the system
- IP address (if the login is remote)
- Session date and duration

Windows NT Performance Monitor

Windows NT Performance Monitor is a powerful and flexible utility that can assist you in identifying network server problems. Specifically, it allows you to do the following:

- Gather performance data for analysis and create reports
- Observe the performance of system objects in real time
- Log data over a specified interval and run alert profiles on the performance statistics gathered

- Create performance monitor alerts to warn you when operating parameters of certain system objects exceed specified levels
- Run executable commands in response to alert conditions
- Compare the performance of different systems

Performance Monitor provides over 300 counters for objects such as processor, processes and threads, disk usage, and memory cache. Network and application counters can also be obtained, depending on your network application configuration.

If you suspect a problem with a process on the network, server, or workstation, you can run the Performance Monitor to display or log specific performance statistics. This invaluable tool helps isolate problems and bottlenecks on the workstation or network since you can run multiple counters simultaneously.

You can also configure Performance Monitor to broadcast a network alert to warn you when system resource usage or performance tolerances approach a critical level. This is particularly useful to help you identify and avoid server or network problems before they have significant impact on the system. If counter thresholds are appropriately set, the alert serves as a warning for developing trends that could be detrimental to network operation.

**Tip:**   Performance Monitor can simultaneously monitor up to 25 workstations in the network.

For example, you can monitor the percent utilization of the server's processor using the Processor object and the % Processor Time counter. If processor utilization is too high, this might be an indication that you need to upgrade to a faster processor or reduce the number of processes competing for processor time.

As network clients increase the use of server resources, which can occur during peak times or as a business grows, the amount of time the server processor is utilized increases. To alert you when processor usage approaches a predefined critical limit, configure the appropriate counter threshold value and then launch Performance Monitor. Figure 10.6 shows a sample of a Performance Monitor statistics display using the % Processor Time counter.

Figure 10.6 Windows NT Performance Monitor Example

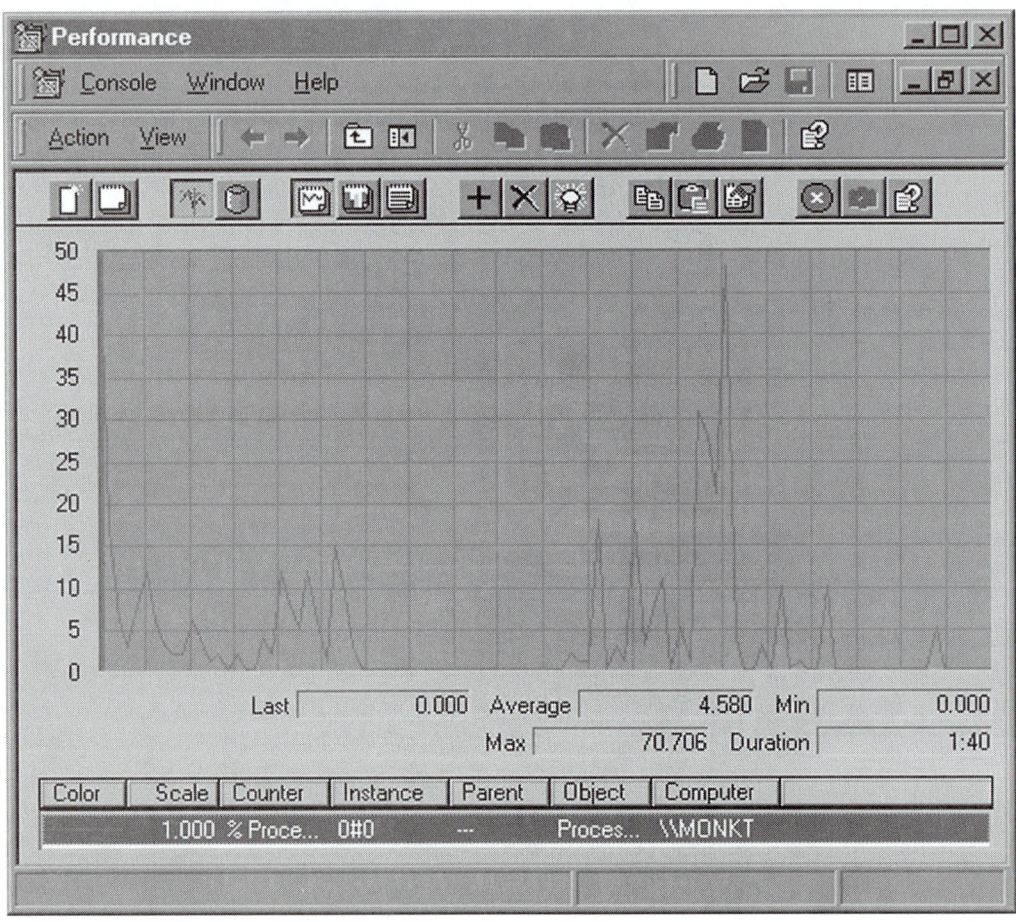

Novell Monitor

The Novell Monitor tool views system information and tracks various utilization statistics when troubleshooting a NetWare environment. For example, if applications or processes are performing slowly, observing server processor utilization can indicate if you have a bottleneck condition.

Tip: Some common causes of processor bottlenecks are CPU-intensive applications, such as e-mail and database, and excessive interrupts generated by an inadequate disk system. A remedy for this situation might be to upgrade to a faster processor or increase RAM and disk capacities. Increasing RAM makes more memory available to the applications you run and greater disk capacity can reduce excessive interrupts to the processor by improving file I/O performance.

The Monitor tool is launched at the server console with the following command—

```
LOAD [path]MONITOR [option]
```

—where path specifies the directory location of the Monitor program and `option` defines some preliminary loading options. When Monitor is launched, the main screen appears as shown in Figure 10.7.

Figure 10.7 Novell Monitor Main Screen

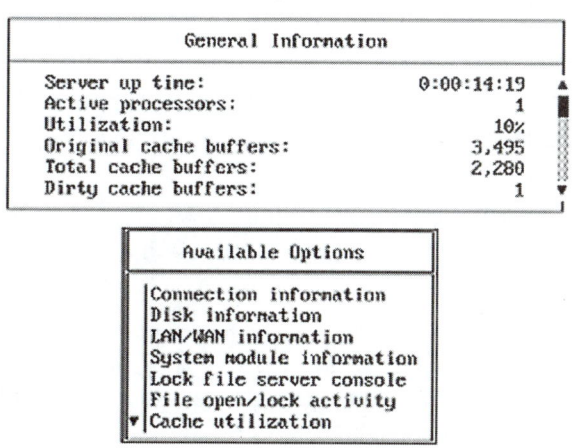

In the main Monitor screen, the general information you can view includes the following:

Server up time—Server running time since its last boot.

Active processors—The number of processors enabled on the server.

Utilization—For a uniprocessor server, the statistic shows CPU utilization. For a multiprocessor server, the statistic shows the average utilization of all active processors.

Original cache buffers—The number of blocks installed as cache memory on the server. Indicates the number of cache buffers available when the server is booted.

Total cache buffers—The number of blocks available for file caching. As modules are loaded, this value decreases.

Dirty cache buffers—The number of file blocks in memory waiting to be written to disk.

Other information that can be viewed from the Available Options menu includes the following:

Connection information—Displays lists of active connections (licensed and unlicensed) or record locks for a user, clear a connection, open files, and send a message to a workstation connection.

Disk information—Displays system hard disks and volume segments, activate or deactivate a hard disk, flash the hard disk light or change its status, and mount or dismount a removable media device and lock or unlock it.

LAN/WAN information—Displays statistics and the configuration of the LAN driver, including the protocols bound to it. Also displays node and network addresses.

System module information—Displays a list of modules loaded on the system along with name, size, and version. Also displays size of module code, data image, and resource tags for the module.

Cache utilization—Displays disk cache block request statistics such as total cache block requests, block request waiting incidents, and long and short-term cache/dirty cache hits. Long-term cache hits information can be used to assess server RAM utilization.

Processor utilization—Displays a histogram for selected active processes including the CPU usage and the number of times the CPU has serviced the process during the sample period.

Resource utilization—Displays memory usage statistics for the cache buffer pool, memory allocated in movable and non-movable memory pools, as well as code and data memory. Also displays tracked resources allocated by operating system and NetWare Load Module (NLM) programs.

Memory utilization—Displays memory information for the system or a selected system module, including memory statistics such as percentage of allocated memory in use, free blocks, and memory blocks and bytes in use.

LINUX Monitoring Commands

LINUX has several commands you can use to monitor certain system parameters that support troubleshooting in the LINUX operating system. These commands include the following:

Free—Displays available real RAM and virtual memory.

Procinfo—Displays current memory usage, when last boot occurred, recent system average load values, number of processes, system time spent in user code, system code, or idle, virtual memory and disk information, and interrupt request (IRQ) information.

Top—Displays information about the principal processes running on the computer including system up time, number of users, load averages, total number of processes and state of each, CPU state (such as CPU time spent executing user or system code and low priority processes), and resource consumption for CPU and memory.

PS—Process Status is useful for finding the ID of the process owner (PID) so you can kill or reprioritize the process.

Kill—Enables you to terminate a process if necessary.

DU—Disk Usage allows you to display a listing of how much space a directory uses in 1K (kilobyte) increments.

DF—Disk Free displays the amount of disk space available to devices on your system.

Microsoft Network Monitor

Microsoft Network Monitor is another very useful tool that makes the task of troubleshooting complex networks easier. If you suspect you have a network communication problem at the packet level, Network Monitor can help you make that assessment. The utility works by configuring the network adapter to capture traffic to and from the local host computer and then copies the network stream to memory.

This process works only if the local computer's network adapter supports the "promiscuous" mode. Most computers listen to all network traffic, but only pass packets on to their operating system specifically addressed to them. The promiscuous mode enables the network adapter to pass all the network traffic to the operating system.

Tip: To capture traffic on computers in the network other than the local host, use the Microsoft Systems Management Server (SMS) Network Monitor.

With the Network Monitor, you can do the following:

- Monitor collisions and analyze bad packets

- Evaluate short frames

- Isolate failed network adapters

- Find other faulty network communication devices

- Isolate and resolve network traffic bottlenecks

- Monitor bandwidth usage and peak traffic times

Tip: In Windows NT, the administrator account is called Administrator; in NetWare, it is called Admin; and in UNIX, it is called the Root.

Figure 10.8 Network Monitor Capture Window

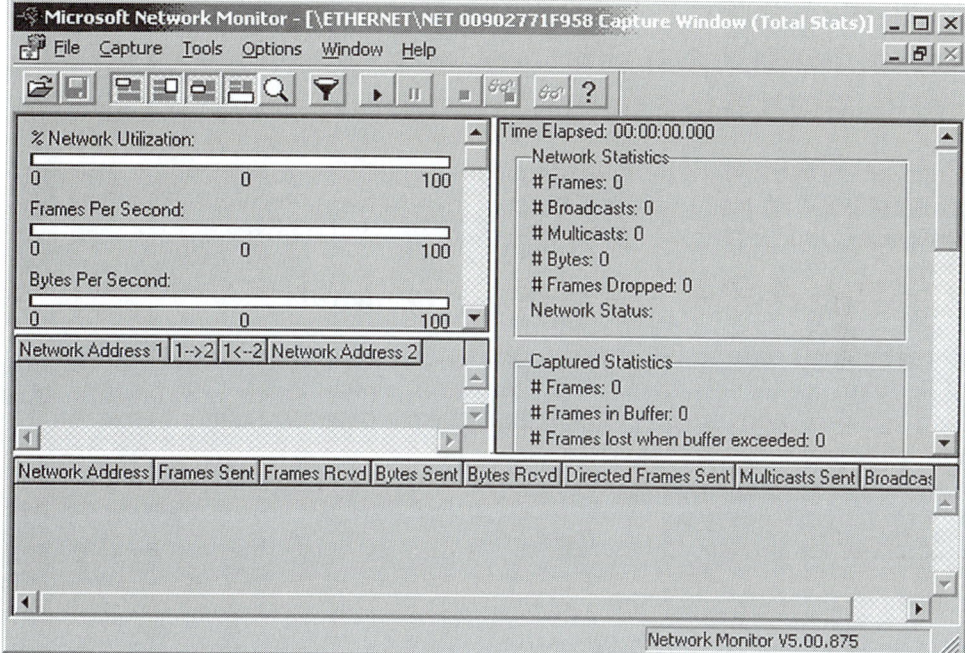

With Network Monitor, you can sample all network traffic to and from a local host or create a filter that captures only certain information. With capture filters, you can isolate and analyze specific frames. You can configure filters to isolate traffic based on the following criteria:

Data pattern matches—Limits the capture to ASCII or hexadecimal data patterns.

Protocols—Captures specific network protocols or their properties.

Source/destination addresses—Captures frames sent to and from a specific network adapter (host).

After data capture, you can view the following information in the Network Monitor's Capture window:

Graph—A graphical presentation of percent network utilization, frames per second count, and bytes per second.

Session statistics—Information on individual network sessions.

Station statistics—Information on the computer session running Network Monitor.

Total statistics—A summary of network activity.

Once a frame is captured, Network Monitor's display filtering feature can be used to isolate a problem or select a certain protocol. Then Network Monitor's dynamic link libraries parse and interpret the binary trace data to make it readable.

Performance Monitor and Network Monitor

If you want to log network performance statistics within a predefined time interval, launch Performance Monitor from within the Network Monitor application. The following counters capture performance data for each network adapter in the system.

- Percent network utilization

- Total frames per second received

- Broadcast frames received per second

- Multicast frames received per second

- Total bytes received per second

- Percent of total frames which are broadcast frames

- Percent of total frames which are multicast frames

Simple Network Management Protocol (SNMP)

This protocol has agents that monitor network devices in a TCP/IP network and report back to a network management console. In the console, control information is maintained for each device in a management information block (MIB).

The control information, which is also communicated to various other devices and servers, includes errors, packet collisions, and hardware failures. The protocol can be configured to send messages to the console when a device has exceeded specified operating thresholds, thus providing you with a tool for monitoring device status and detecting network problems.

Physical Indicators

You can also observe the following physical indicators for signs of trouble on local network devices, if they are available:

Link lights—These indicators are usually green-colored light emitting diodes (LEDs) found on network adapters, switches, routers, and hubs. When illuminated, these lights generally confirm the physical connection and logical communication path between the devices.

 Note: Link lights for some network adapters will not illuminate until the device drivers are installed.

Power lights—These indicators are provided on most computers and network devices to let you know when they are powered up. Power is required for equipment to work. This simple matter tends to be overlooked. Make sure all devices are turned on and powered up to eliminate problems related to unpowered equipment.

Error displays—Most servers have a liquid crystal display (LCD) to indicate status. Error displays for other devices include packet collision lights and other error indicator types. The color red generally signifies a problem condition.

Collision detectors—these visual indicators show when collisions occur between signals transmitted by devices or workstations on the same channel of a network. Normally, devices monitor transmission paths and wait until they are clear before sending a signal out. However, as a network grows and more traffic is generated, collisions inevitably occur.

Collisions can cause transmission failures, garbled signals, and cable throughput inefficiencies. Observing a collision detector gives you a way to monitor the collision rate on your network, which serves as an indicator of the network traffic density level.

Widespread Problems

If you suspect you have widespread network trouble (across the WAN), a Microsoft SMS version of Network Monitor may be of some help in locating the problem. Other types of network monitors are also available. With a network monitor, you can record packet collisions, network traffic levels, and network broadcasts. They can be set up to capture data in a window of time for trend analysis as described earlier. Network monitors also display errors—if you are using this type of device, check its error displays for problems.

If you are having network communication problems, check error displays on the routers that interconnect your LANs. Data in a routing table might be corrupt, disabling the router's capacity to recognize network addresses. Switch and hub operability should also be verified. In addition, you can use the TRACERT utility (discussed ahead) to display the devices along the exact path to a host, thus exposing possible problem areas.

Causes of Network Problems

Network problems can have any number of causes. How you approach solving a network problem depends on where it appears to originate. Almost all network problems can be associated with any of the following causes:

- User errors
- Viruses
- Hardware failure, degradation, or misconfiguration
- Configuration and addressing issues
- Data Communication problems

User Errors

The following are typical problem areas that can affect a user:

User credentials—Describes the user name and password.

User access rights—Ability of the user to access data and services. Permissions are controlled from the administrator account.

Figure 10.9 illustrates how to proceed if the workstation operator has problems logging on to the server.

Figure 10.9 Operator Cannot Log On to Server

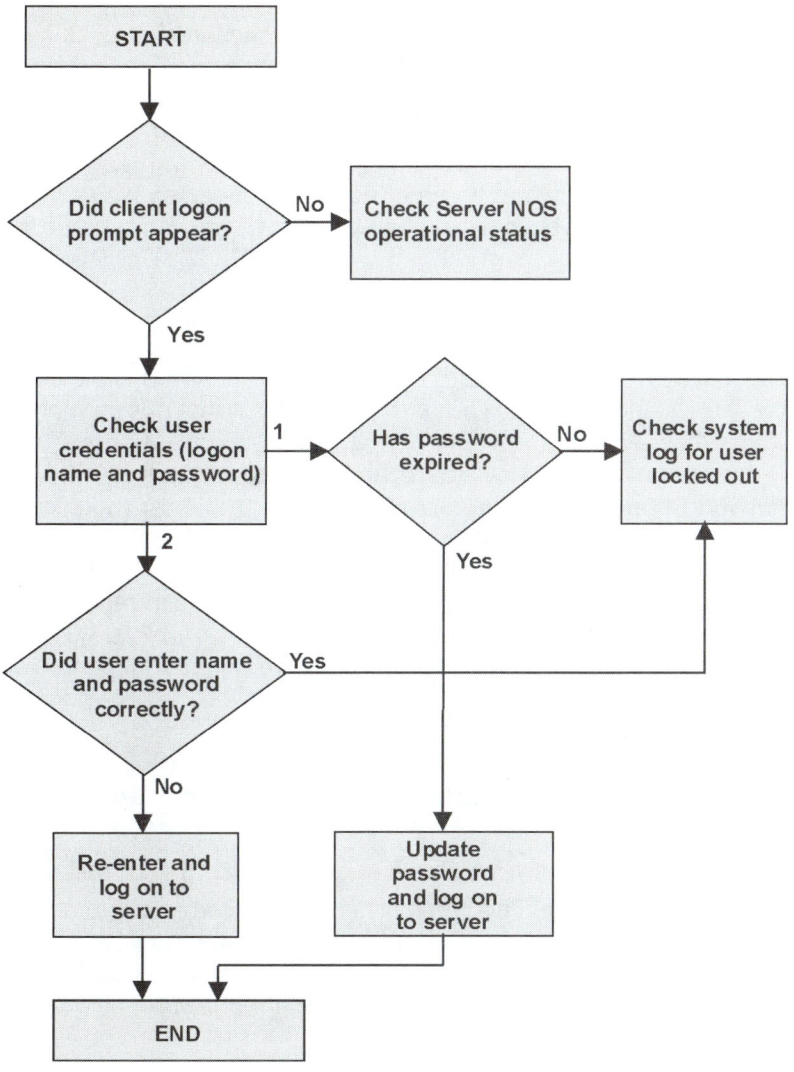

Viruses

Viruses can be very difficult to pin down since they can appear to be a software, hardware, or communication problem. Sometimes viruses find their way into your network when a new application is loaded onto the server or a workstation. If you begin to have problems shortly after new software is added, or if files with unknown macros are opened in the client environment, it might be a virus infection.

If you think this has occurred, the first thing to do is to obtain the latest virus signature update and perform a virus scan on the suspected machine. Then perform a virus scan across the entire network. This eliminates the possibility that the virus spread farther than the suspected machine.

The most damaging aspect of a virus is that it can affect the boot sector, master boot sector, and partition table. If the boot sectors are damaged, the computer will be disabled or at least unable to boot.

Hardware Failure, Degradation, or Misconfiguration

Usually, hardware components are provided with error indicators that let you know when a problem occurs. Checking these indicators requires only visual observation. If a device has indicators that allow you to verify normal operation, these should be checked. In addition, the integrity of physical interconnections to hardware components in the network should be verified. Some hardware problems, which can occur in network components, are as follows:

Server—Most network operating systems have logging facilities that can report problems with server software and hardware. In addition, they usually have management software that runs diagnostics to detect hardware failures. Repeated server reboots is an indication that you might have a server hardware problem.

Modem—Common problems with modems include interrupt request (IRQ) or COM port conflicts, incorrect drivers, I/O addresses, hardware incompatibility problems, or failure.

- **COM port conflicts**. These occur if you have another device trying to share the same COM port as the modem. This can be checked and corrected from the Modem Properties utility. For example, if another device is sharing COM2 with the modem, remove the device from COM2 (reassign it to another available COM port) and then reinstall the modem with the modem utility.

■ **Modem driver**. From the Modem utility, you can also verify the modem driver version. The correct modem drivers from the manufacturer are usually installed during modem setup. However, an incorrect driver could have been installed by mistake. To avoid any uncertainties, download the latest version of the modem driver from the manufacturer's Web site and reinstall the modem.

■ **Hardware compatibility**. Another possibility is a hardware compatibility problem. Manufacturers of most network operating systems usually maintain a Web site with a hardware compatibility list (HCL). The list describes modems, drivers, network adapters, and other hardware components compatible with the NOS. Check here to verify whether you are using a compatible modem.

■ **Failure**. Always keep the modem manufacturer documentation and technical support phone numbers available. If the modem fails and it is your only way to connect to the Internet, you will not be able to access the manufacturer's Web site.

Network adapter card—Common problems that exist with network adapters are incorrect drivers, hardware incompatibility, incorrect configuration, and failure.

■ **Driver**. An incorrect driver will disable the network adapter and prevent access to the network. If you need the correct driver, download it from the manufacturer's website and then reinstall the network adapter from the Network utility.

■ **Hardware compatibility**. These problems are similar in nature to those described above for modems. Make sure you have a network adapter compatible with the NOS by checking the manufacturer's HCL.

■ **Adapter configuration**. Some network adapters are configured with jumpers while others use software configurations. If you suspect a network adapter is the cause of a problem, you might have an IRQ conflict, incorrect memory, or an IP addressing issue. The latter issue might include an incorrect default gateway address or the wrong IP address configured on the adapter. These can be corrected from the Network utility.

■ **Failure**. A failing network adapter can flood the entire network with garbled traffic and cause it to go down. You can use Windows NT Diagnostics (WINMSD) and the Event Viewer logs to identify network adapter errors. In some cases, a faulty network adapter can transmit a large number of packets, causing a broadcast storm. You can use a protocol analyzer to detect this and isolate the broadcasting computer's location.

 Note: Verify the integrity of the physical connection between the adapter and the network. If the connection is faulty, network access will fail or connect intermittently.

Router—Routers connect multiple paths between different network locations. Routers segment traffic by subnetwork, which reduces overall network traffic. Only the traffic destined for a network segment is sent into it, while non-routable traffic is not passed through the router.

When a router receives a packet, it forwards the data to the appropriate network based on its routing table. These tables can be either static or dynamic. Static tables are created and updated by the system administrator to reflect the network configuration. Dynamic tables are automatically updated as routers communicate with each other using Routing Information Protocol (RIP).

Run diagnostics—Router problems can be hardware or software related. If you suspect a hardware problem, use a manufacturer-supplied diagnostic program to verify proper operation of the router's hardware components. Routers require programming for proper setup. If the router is already up and running, hardware (not the software) is most likely the cause of trouble.

If you have WAN segments interconnected with two routers that have stopped communicating, use the following troubleshooting steps:
1.   Verify the integrity of the connections between the routers.
2.   Verify that router power lights are illuminated and that you do not have any fault indications.
3.   Run hardware diagnostics on the two routers. If the hardware checks out, proceed to step 4.
4.   Check the software configuration, especially the routing tables and access lists. Routing tables contain the addresses of other network routers. If the address data is corrupt, the routers will not forward packets to the appropriate network. Access lists are used to restrict access to and from other routers—so make sure access is enabled between the routers.
5.   Make sure the routers you are using support the network protocols in use.

Routers are very complex devices. Many routers are designed with other capabilities beyond basic routing functions, such as dynamic host configuration protocol (DHCP) server, network address translation (NAT), and proxy server. Setting up a router correctly can be difficult and misconfigurations can result. Be sure to consult the manufacturer's documentation when configuring routers in your network.

 Note: NAT is a process where LAN-routable IP addresses are converted to Internet-routable IP addresses, and vice-versa. This enables use of a large number of IP addresses within a subdomain without depleting the limited number of available Internet IP addresses.

Gateway—A gateway is a device consisting of both hardware and software. A gateway performs protocol conversion, routing services, and information transfer. It connects networks having dissimilar communications protocols, enabling messages to be passed from one network to the other.

Gateways operate at several layers of the OSI model and act as the primary link between heterogeneous environments such as PC-based LANs and host environments such as IBM™ System Network Architecture (SNA). Sometimes gateways are used to connect mail hosts, allowing them to pass messages. If you suspect a gateway problem, do the following:

1.   Check the log files for network errors.

2.   Check the originating network or the destination host. If all users in the originating network are affected, the problem is most likely with the gateway or destination host.

3.   Restart the gateway.

4.   Be sure the gateway(s) you are using support the network protocols in use.

Switch—The main function of a switch is media access control (MAC). MAC addresses direct data within the same network, but do not send data across routers. If you suspect a switch problem, do the following:

1.   Check for illumination of switch power and fault indicators.

2.   Switches are normally programmed to set up port address groupings. If a network segment is not functioning, you can move that segment to another port and test the connection again.

3.   Reset the switch to the factory default condition as a media access control switch.

Hub—A hub is a passive or active repeating device that duplicates and directs its input signal to each of its multiple output ports. If you suspect a problem with a hub, you can do the following:

1.   Check for illumination of power and fault indicators.

2.   Replace the hub and see if that fixes the problem.

3.   Move the device attached to a port suspected of failure to another port. If communication resumes, the port is faulty and should not be used.

4.   Check the integrity of the physical connections to the hub.

Bridge—A bridge is a LAN device that connects two or more network segments and effectively makes them one logical network. The bridge drops packets if the destination is on the same segment; otherwise, it forwards traffic to the appropriate segment. A bridge can also support remote connectivity and provide remote segmentation by filtering traffic.

Bridges work at the Data Link Layer of the OSI model, thus minimizing logical configuration errors. If you suspect a bridge is not properly working, do the following:

1.   Check for illumination of power and fault indicators.

2.   Verify the integrity of interconnecting cables.

3.   Run manufacturer-supplied diagnostic programs to isolate possible problem areas.

Tip: As a temporary corrective measure, you can sometimes apply MAC address filters to bridge ports to restrict access to devices connected to those ports.

Printer Server—Printers are connected to print servers, which can be internal to the printer or an external device with a parallel port and network connection. Print queues, located in the network operating system, point to the print server. If you suspect a printer has stopped working because of a problem with a printer server, do the following:

1.   Stop and then start the print queue.

2.   Reset the printer server.

Network printers are extensively used, so errors can occur frequently. Keep in mind that a printer error can actually be occurring even though it appears like a printer server problem. Some problems that can occur on a printer are as follows:

- Printer power is off

- Printer is offline or the network cable is disconnected

- Printer is out of paper or has a paper jam

- Printer configuration is incorrect

As indicated, an incorrect printer configuration can cause printer failure. General symptoms of a configuration problem are:

- **Garbled print**. Commonly the result of incorrect printer driver selection. This can occur when printer pools are configured or if multiple print queues are attached to the same printer. In addition, bad font files, incompatible fonts, or an incorrect font cartridge can cause garbled print.

- **No output**. If this occurs, test the printer locally to make sure it is working properly. Also, make sure the printer queue is not suspended and that the printer availability times are correctly set.

- **Partial output**. This occurs when the spooler starts printing as soon as part of the print job is available. To correct this, configure the printer to start only after the entire job is spooled.

Tape Drive—Tape backup errors can be software or hardware related. If you suspect a hardware problem, do the following:

1.   Clean and service the tape drive.

2.   Run a hardware diagnostic program.

3.   Contact technical support at the software manufacturer and assess whether you
        have a software problem.

Configuration and Addressing Issues

Configuration errors frequently occur during initial setup of a computer network.
Generally, you fix these errors during the installation process to get your basic
network up and running. After setup, everything should stabilize for an initial period.
However, even if your maintenance and monitoring plans are well organized,
network problems invariably occur. This might happen because something was
overlooked, something changed, or fine-tuning is required.

It is not uncommon for misconfigurations to occur in the network IP addressing
scheme. In a large network, this is a very complex issue. You should approach it
systematically, just as any other problem described earlier in this lesson. The
following paragraphs highlight some of the areas or services where problems related
to IP addressing might occur.

Domain Name System (DNS)—This service is usually provided by an Internet
service provider (ISP) and is primarily used when connecting to the Internet. It
allows Internet hosts to resolve a domain name address, such as
www.myorganization.com to an IP address like 207.111.17.3—this permits the
computer domain to be located on the Internet. A DNS server maintains a database
that maps computer domain names to their IP addresses. Another use of DNS is to
allow client access to information across UNIX-based systems and intranets.

Some possible problems with DNS are as follows:

- **Access by DNS name**. You can access a domain by IP address but not with its
  DNS name. This could be the result of incorrect configuration of DNS zones or
  servers.

- **No access**. If you cannot access the domain by name or IP address, you might
  have a connection problem that is caused either by misconfiguration of network
  IP addresses or faulty hardware.

Windows Internet Naming Service (WINS)—WINS is a dynamic database for
registering and querying dynamic mappings of NetBIOS computer names on a
network. It allows for resolution of NetBIOS computer names to IP addresses in
routed networks that use NetBIOS over TCP/IP. WINS also provides central

management of NetBIOS name databases and reduces local IP broadcast traffic for NetBIOS name resolution.

When a client logs on to the network, the client computer name and IP address automatically register in the WINS database. The WINS database is also updated whenever there are dynamic address configuration changes. For example, when a DHCP server issues a new IP address to a WINS-enabled client, WINS information for the client is also updated.

When a client makes a request for a NetBIOS name, the client computer performs the following steps in the order shown, to resolve the address:

1. Checks for the NetBIOS name

2. Checks the local cache of remote NetBIOS names

3. Queries the WINS server for address resolution

4. Broadcasts a local name query

5. Checks the LMHOSTS file for a match to the query, if LMHOSTS support is configured on the client

6. Checks any existing HOSTS files

7. Queries DNS if it is supported

Medium-to-large networks often have several WINS servers set up to collect all computer names and their IP addresses for storage in a WINS database. Mapping entries are replicated among the WINS servers so that when a client needs to communicate with another host operating on the TCP/IP network, the mapping information is available to locate the host.

■   **WINS Client Problems.** The most common WINS client problem is failed name resolution. Do the following when attempting to correct failed name resolutions:

    1. Make sure the name used in the client's request is a NetBIOS name and not a DNS host name.

        For example, a NetBIOS name such as "PRNTR-SERV2" and a DNS name as if "prntr-serv2.network.company.com" might both refer to the same network printer server. When a client issues a NetBIOS name resolution command, WINS is used to resolve the name. If a DNS name is involved in the failure, then DNS is most likely the cause of failed name resolution.

2.   Make sure the client is using a version of Windows and applications that require WINS to resolve names.

Not all Windows computers and applications need WINS. If a failed name resolution involved a URL entered in a Web browser or an FTP program, or occurred with an Internet e-mail address request, DNS is most likely the cause of failed name resolution. Some examples of applications that require WINS for name resolution include Network Neighborhood, Map Network Drive (of Windows Explorer), and the "net" command used at the MS-DOS® prompt.

3.   Make sure the client is properly configured to use both TCP/IP and WINS.

WINS-related client settings can be configured either manually using the TCP/IP Properties dialog box or dynamically by providing the client's TCP/IP configuration with a DHCP server. With the IPCONFIG utility (discussed ahead in "Diagnostic Tools"), check the following parameters to verify a valid client IP configuration:

- IP address
- Subnet mask
- Default gateway
- Primary and secondary WINS server

 Note: If the client does not have a valid TCP/IP configuration, you can use the "IPCONFIG /RENEW" command to cause the DHCP server to renew the client's IP configuration. Reconfiguration can also be done using the TCP/IP properties.

---

4.   Make sure the client can connect with the WINS server.

To verify the client can connect to the WINS server, use the PING utility (discussed ahead in "Diagnostic Tools") to ping the WINS server with its IP address. If the WINS server responds, use the "NBSTAT –RR" command (discussed ahead in "Diagnostic Tools") at both the client and the resource server the client is trying to locate by NetBIOS name. This command causes WINS services on both computers to send name releases and refresh requests to the WINS server and re-registers their names.

 Note: If the WINS server does not respond to pinging its IP address, a network connectivity problem may exist between the client and WINS server.

■ **WINS Server Problems**. The most common WINS server problem is the inability to resolve names for clients. When this occurs, the WINS server might send an error message indicating the name was not found or it may send a response with incorrect information. Do the following when attempting to correct WINS server name resolution failures:

1.   Make sure the WINS server is started and running.

2.   At the primary or secondary WINS server (servicing the client where name resolution failed), use the WINS Manager or Event Viewer to determine whether the WINS server is currently running. If running, use the WINS Manager to search the WINS database for the name that failed resolution.

3.   If the WINS server is registering database corruption errors, the database can be restored using database recovery techniques. If the name does not appear in the database, verify correct replication configurations of WINS servers in the network.

4.   Make sure replication is occurring between WINS servers.

5.   Check the WINS servers to make sure they are configured for replication. If replication is not occurring, the resolution process is slowed down because the client is forced to query other sources for NetBIOS name mappings. Also, check the type of replication configured. Using one-way replication partnerships (push-only or pull-only) can sometimes cause the failure of name replication to all WINS servers in the network.

6.   Make sure static mapping is not an issue.

Static mapping of WINS client names can be used, but is not recommended since WINS allows for dynamic (automatic) update of client names to the WINS database. If name resolution information returned to a client is incorrect, the name entry in the WINS database may be an out-of-date static mapping. Either edit the static mapping or modify the replication configuration to overwrite it with a dynamic record to resolve the problem.

At times, static mappings are required. For example, if Windows NT is used to do IP routing between multiple subnets, the WINS database should be configured with multi-homed static mappings.

Tip: A large network with multiple subnets and domains may use browsing lists and trust relationships between WINS servers. Make sure these are correctly configured.

Dynamic Host Configuration Protocol (DHCP)—This service dynamically allocates IP addresses to a specific system for a period of time. If DHCP is not used, static (fixed) IP addresses must be configured manually for network components. DHCP automates this process.

For example, the Windows NT DHCP service can be used to assign an IP address (from its database) to a network client at logon. The client's assigned IP address is used to identify the client when fulfilling URL requests on the Internet.

Note: Normally, a DHCP server issues non-routable IP addresses to subnet clients. To facilitate the routing of client requests on the Internet, a DHCP server must also do Network Address Translation (NAT) to create an Internet-routable IP address. This can also be done by a router (or proxy server) with that capability.

Another example is when a Windows NT server is configured as a DHCP client. At startup, the system attempts to lease an IP address from the DHCP server. If this fails, an error message is displayed. If successful, the NT client receives an IP address and uses it until the lease period expires. When expiration approaches, the NT client attempts to renew the lease. Failing that, the NT client requests a new IP address.

A DHCP server provides centralized management of IP addresses, which helps prevent IP address conflicts. The DHCP server's "scope" is a term that refers to the range of non-routable IP addresses issued to subnet clients by the server. A DHCP server can also support multiple scopes, which gives it the ability to provide IP addresses and configuration parameters to clients on multiple subnetworks.

 Note: In a NetWare environment, use the Novell DHCPCFG utility at the server console to manage the DHCP service.

A DHCP problem in your network might be one of the following:

- It is possible the lease for an allocated IP address has expired and was not automatically renewed. The default period is three days. Open the DHCP Manager to view lease status.

- The DHCP server may be issuing IP addresses to the subnet address space that conflict with other network IP addresses.

- The DHCP server's IP address and any other applicable addresses to be excluded must be included in the DHCP server's exclusion range. If they are not, the DHCP server may assign its own IP address or another invalid one to a client, resulting in routing problems.

- The correct subnet mask must be used to identify the network ID and host ID portions of the IP address.

- The name that identifies the subnet must be correct.

- Some components cannot be configured as DHCP clients. For those that need a fixed address, make sure a static IP address is manually configured. These components and services include, but are not limited to DHCP servers, WINS servers, DNS servers, gateways, and routers.

- The LMHOSTS file is a static text file that locates remote computers. It maps NetBIOS machine names to their permanent, manually configured IP address. Some uses of the LMHOSTS file are to support printing, replication, remote access, domain services logon, and browsing.

  LMHOSTS address resolution is enabled from the WINS address tab of TCP/IP properties. Address resolution is performed as follows:

  1.   The name is checked to see if it is a local machine name.

2.   The local LMHOSTS cache is checked for the name.

3.   A name query is broadcast

4.   All LMHOSTS files are checked.

 Note: If LMHOSTS cannot resolve the name, an error message
is returned stating that the network name could not be
found.

If you suspect a problem with the LMHOSTS file, check the following:

1.   Make sure workstations are configured for LMHOSTS support when using
LMHOSTS file for address resolution.

2.   Check the LMHOSTS configuration file for keyword errors or other incorrect
entries.

3.   Make sure all NetBIOS machine names with corresponding static IP addresses
are entered in the LMHOSTS configuration file.

The default gateway parameter specifies an IP address assigned to a device that
acts as a gateway from one network to another. This causes traffic destined for
another network to follow the path defined by the default gateway into that
network. For example, the local IP address of the router shown in Figure 10.10 is
specified as the default gateway on the server's external network adapter to enable
routing of Internet requests.

Figure 10.10 Network Adapter Default Gateway Configuration

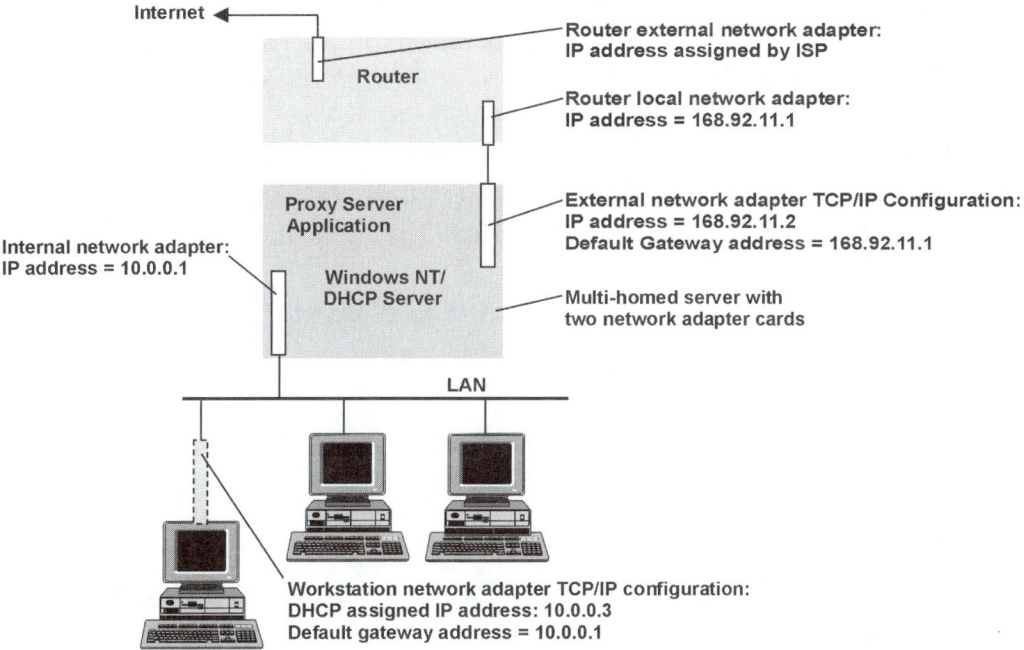

A default gateway IP address must also be configured on each workstation to define their paths out of the internal network. This is the IP address of the server's internal network adapter card. If a workstation can access local devices while everything else beyond the local subnet is inaccessible, the workstation may have an incorrect default gateway value or none specified at all.

To verify that TCP/IP protocol has binded to the host network adapter, do the following:

1.   On the workstation, open an MS-DOS window.

2.   At the prompt, type `ping 127.0.0.1`. This checks the protocol binding. To
      ping the adapter, you would ping the host's IP address.

If you do not receive a reply (the request times out), the IP configuration of the
host adapter can be corrected as follows:

1.   Configure an IP address on the network adapter in the TCP/IP properties. This
      allows the host (network adapter) to be identified by a routing device on the
      Internet.

2.   Assign a default gateway IP address to the host network adapter.

3.   Assign the correct default gateway address on all workstation network adapter
      cards. This allows each workstation to send requests out of the internal
      network.

4.   Assign the appropriate subnet mask to the host's external network adapter.

Tip: In a NetWare network, use either the BIND or INETCFG utility to link
      a communication protocol to a network board and its LAN driver.
      This enables the board to process packets.

Another default gateway configuration commonly used is the LAN router, as shown
in Figure 10.11.

Figure 10.11 LAN Router Default Gateway Configuration

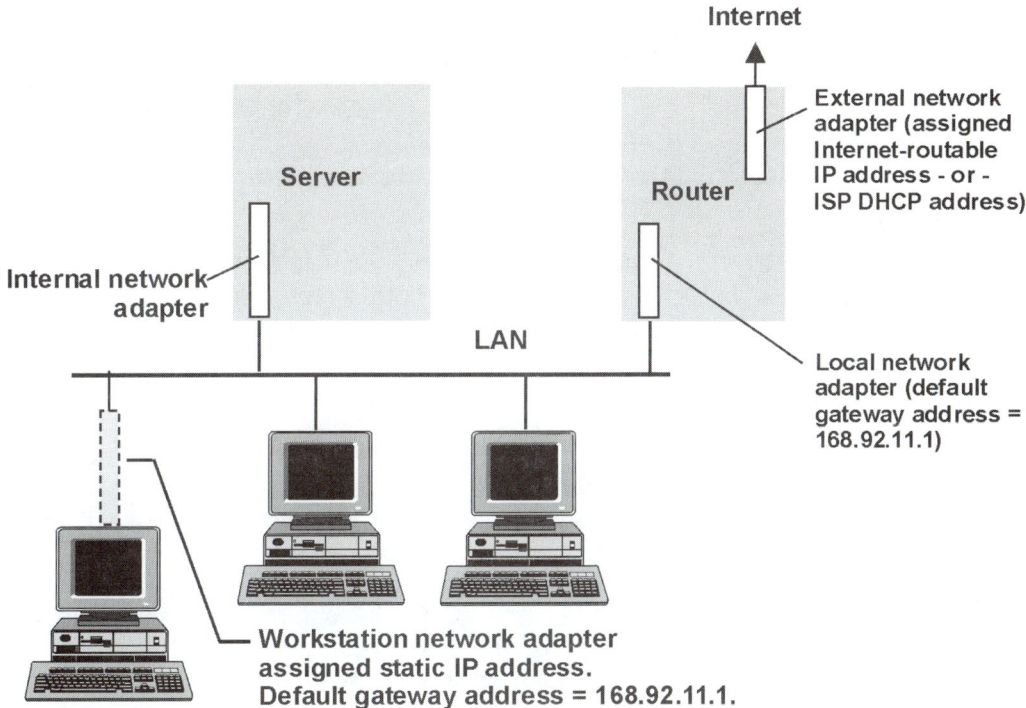

In Figure 10.11, the router is connected to the LAN for direct client Internet access. In this configuration, the router is set as the default gateway for each computer on the LAN and handles client requests outside the local address space. This requires each client on the LAN to have a static Internet IP address, unless the router does NAT and DHCP. In that case, client IP addresses are assigned by the router's DHCP server.

If you suspect the router has a problem acting as the default gateway, do the following:

1.   Verify the router has a default gateway address assigned to its local network adapter.

2.   Verify that an Internet-routable IP address is assigned to the router's external network adapter. This can be a static IP address or dynamically assigned, depending on the ISP configuration.

3.   Verify that each client has the correct default gateway IP address assigned.

4.   Verify that each client has a static IP address assigned (unless the router is a DHCP server).

Warning: Do not have two DHCP servers on the same subnet. If this occurs, then both servers can issue IP addresses to clients and serious addressing problems result.

If the router is configured as a DHCP server, Microsoft® Windows NT Server should automatically disable its own DHCP server. A built in rogue detection feature shuts off its DHCP service if it identifies another DHCP in the same subnet. Likewise, if you use the Windows NT DHCP Server, the router's DHCP server must be disabled.

To shut off the Windows NT DHCP Server, go to Control Panel, double-click Services to display the Services dialog box, and stop the DHCP service. To shut off the router's DHCP server, consult the manufacturer's documentation.

Subnetworks—Different networks connected to each other through gateways are often referred to as subnetworks, since they are a component part of an overall larger internetwork. Subnetworks are complete networks, independent of the larger network with which they interact, although comparatively smaller. With the use of subnets, network efficiency is improved by reducing network traffic and making it easier to manage routers.

If you want to communicate with a remote host on a subnet in an internetworked TCP/IP environment, the IP address of the domain, subnet mask, and a default gateway address are needed. The subnet mask is used to mask part of the IP address so the TCP/IP protocol can distinguish between the network and host ID portions of the address. The host ID portion then allows the destination host to be located once the IP packet routing reaches the remote subnetwork.

To better understand how the subnet mask works, some background is needed. Each host on a subnet has a unique IP address binded to its network adapter to identify it on the internetwork. An IP address is a 32-bit address consisting of four octets—this format is known as dotted quad decimal notation. Each octet is a decimal representation of an 8-bit binary number, as shown in the following IP address:

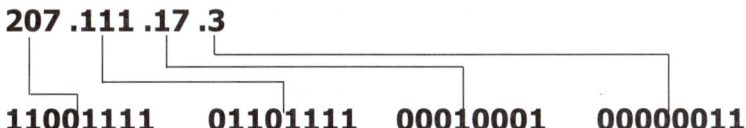

207 .111 .17 .3

11001111   01101111   00010001   00000011

The IP address is segmented into two parts that identify the network and host portions. The following IP address is a Class C address since it has three network ID octets (Class A has one; Class B has two):

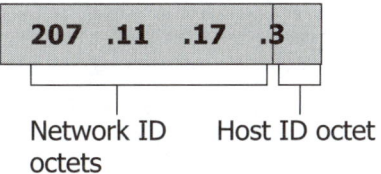

207 .11   .17   .3

Network ID   Host ID octet
octets

A standard value for a Class C subnet mask is 255.255.255.0. The first three octets in this value mask the Network ID octets of the IP address so only the Host ID octet remains—this identifies the host on the network. The subnet mask does this by preventing the TCP/IP protocol from recognizing octet values under 255 in the Network ID segment. The remaining octet of the IP address (.3 in the example) is the host address.

If you suspect a problem with the subnet mask on a network adapter, check the following:

1.   Make sure the TCP/IP protocol is binded to the network adapter. If there is no binding, a subnet mask cannot be specified. You can verify the protocol binding by pinging the network adapter (use the PING utility).

2.   Make sure a valid subnet mask was specified on the network adapter.

 Note: Without a valid subnet mask configured in TCP/IP properties,
                                     you run the risk of breaching network security. With an
                                     invalid or non-existent subnet mask, the network may be
                                     open to inbound IP address requests from unknown origins.
                                     The security threat is reduced if a proxy server protects
                                     your network, but you should correct this situation by using
                                     a valid subnet mask to isolate your LAN from the Internet.

Isolating IP Addressing Problems

Figure 10.12 presents an example of steps to take when configuring the IP
addressing scheme of a domain (LAN served by a local host). If you have trouble
with IP addressing, follow the example to expose potential problems with the IP
configurations of LAN services and components.

Tip: If you need to reconfigure internetwork addressing in a
                                     NetWare environment, use the INETCFG utility to set up the
                                     IPX, IP, or AppleTalk configuration.

INETCFG allows you to configure LANs to work with network and routing protocols
supported by NetWare. In addition, it allows for configuring boards, network
interfaces, WAN Call Directory, protocols, bindings, and node information. Launch
INETCFG using the `LOAD INETCFG` command at the server console.

In addition, you can use NWIPCFG to configure a NetWare server as a DNS client
and set it up for IP service.

Figure 10.12 Misconfigured Domain IP Addressing (NIC Default Gateway)

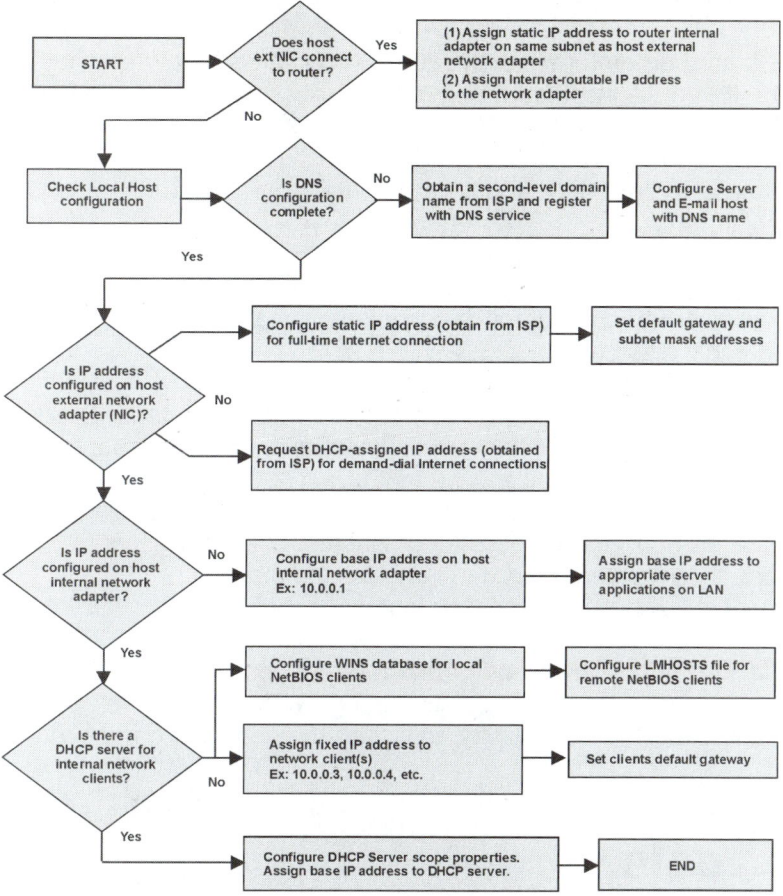

Data Communication Problems

Data communication problems can be classified in the following categories:

Intermittent—These type of problems can occur if equipment is overheated, poorly connected, or near the threshold of maximum performance. Intermittent errors are very difficult to isolate since they do not consistently occur.

Persistent—These types of problems are easier to isolate because they are usually accompanied with consistent error messages and can be replicated.

The following list describes several intermittent data communication problems you may encounter and potential resolutions:

Read/write errors—A hard disk intermittently returns error messages when reading or writing data. Run a diagnostic program on the workstation or server hard disk.

Data corruption—You receive intermittent data errors. Run a virus scan.

Interference—You have data communication interference that might be caused by intermittent connections. Check the integrity of physical network connections.
The following list describes several persistent data communication problems you may encounter and potential resolutions:

Network communication—Data communication across the network stops. Check network devices, cables and connections, and run diagnostics on network adapters.

Time-out errors—Data communication across the network is sluggish causing time-out errors or messages from non-responding processes. Check for network bottlenecks. In addition, check cable/connection integrity and make sure interconnecting network cable lengths are not too long.

Internet communication—Inbound and outbound Internet communication stops. Check router, IP addressing, and Internet connection integrity. Determine the Internet access route using the TRACERT utility, to isolate the network components where problems might exist.

Diagnostic Tools

Once a problem is isolated to a network component or device, running diagnostics on the component can further pinpoint the cause, and subsequently, tell you whether the component should be serviced or replaced. The paragraphs that follow describe several tools, which can be used to resolve network component faults.

Diagnostic Programs

Diagnostic programs check specific operating parameters, allowing you to quickly determine whether the equipment is functioning within normal operating tolerances or if a fault has occurred. If you have a device or component that does not provide diagnostics, you can usually find third-party vendors who provide them.

Cable Testers

Using a cable tester, you can verify if your network cabling is operating within specifications or at fault. Most available cable testers are sufficient to perform the task, but make sure the one you purchase supports the different types of cabling in your network.

Hardware Query Tool

If you suspect a problem with computer hardware, Windows NT has a Hardware Query Tool (HQT), which lets you identify the following internal component types, and the resources they use:

- Industry Standard Architecture (ISA) and Extended Industry Standard Architecture (EISA) devices
- Peripheral Component Interconnect (PCI) devices
- Legacy devices
- System components
- Micro Channel Architecture (MCA) devices

After the specific hardware is detected, HQT lets you display the following information:

- **System**. System processor, memory, and BIOS information.
- **Motherboard**. Built-in components such as DMA controller, I/O ports, and system timer.
- **Network**. Network adapter configuration settings.
- **Video**. Video controller information.

- **Storage**. Storage devices including floppy drives, IDE devices, and SCSI adapters. Individual SCSI devices cannot be detected with the Hardware Query Tool.

- **Others**. Devices not included in other categories, such as sound cards.

- **Compatibility**. Expansion cards are listed along with their compatibility/non-compatibility with Windows NT.

Novell CONFIG Utility

In a Novell NetWare environment, you can use the CONFIG utility to view various hardware, diagnostic, and system information to assist in your troubleshooting efforts. CONFIG allows you to display the following:

- Netware server name, internal network number, server uptime, and loaded LAN drivers

- Hardware settings, node (station) addresses, external network number of the cabling scheme, board name, frame type, and communication protocol for each network board

- Server's directory tree and bindery context

Hardware Loop-Back

By switching the transmit and receive pins of a connector, the hardware loop-back cable allows diagnostic utilities to test the transmit and receive functions of a device. For example, if you run diagnostics on a network adapter card, you need a hardware loop-back cable to test the transmit and receive functions of the adapter.

Packet Internet Groper

The Packet Internet Groper (PING) can help you determine whether a network route is available to a remote host. If you suspect there is a problem connecting to a host computer, PING can determine if the IP stack on the system is working.

The PING utility works by sending an Internet Control Message Protocol (ICMP) echo request message to a host on a TCP/IP network. The utility then registers the remote host's response time as shown in Figure 10.13.

Figure 10.13 PING Test Responses

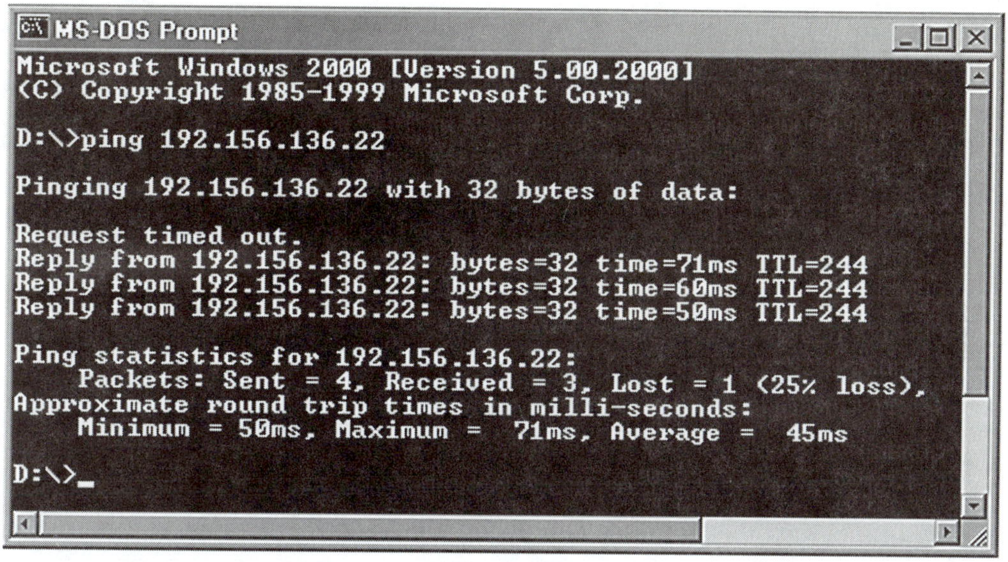

```
MS-DOS Prompt                                                    _ |□| ×|
Microsoft Windows 2000 [Version 5.00.2000]
(C) Copyright 1985-1999 Microsoft Corp.

D:\>ping 192.156.136.22

Pinging 192.156.136.22 with 32 bytes of data:

Request timed out.
Reply from 192.156.136.22: bytes=32 time=71ms TTL=244
Reply from 192.156.136.22: bytes=32 time=60ms TTL=244
Reply from 192.156.136.22: bytes=32 time=50ms TTL=244

Ping statistics for 192.156.136.22:
    Packets: Sent = 4, Received = 3, Lost = 1 (25% loss),
Approximate round trip times in milli-seconds:
    Minimum = 50ms, Maximum =  71ms, Average =  45ms

D:\>_
```

To use the PING utility to generate an echo request message, do the following:

Open the MS-DOS window.

At the prompt, type `ping 192.166.121.11`, using a valid IP address in place of the value shown here. Alternatively, you could use the DNS name as follows: **ping www.dnsname.com**. Windows® machines send four ping queries before quitting.

Observe the ping response on the MS-DOS screen. If the IP address you enter is valid and the network route to the host is functioning, a response should appear.

Tip:

The PING utility also allows you to specify several command options to define ping query and response parameters. An example is `ping -t 192.166.121.11`, where –t signifies pinging the specified host until interrupted.

Note: In a NetWare environment, use IPXPING to check
                      connectivity to devices on an IPX internetwork. IPXPING
                      is launched from the server console using the `LOAD`
                      `IPXPING` command.

IPXPING sends an IPX ping request packet to an IPX target node. When the target
node (a server or workstation) receives the request packet, it sends back a reply
packet.

Telnet

You can use the Telnet utility to connect to a TCP/IP host computer across the
Internet using its IP address. The local and remote hosts must both support Telnet
protocol to make the connection, which occurs on Port 23. You might need a user
account with credentials set up in advance of making the connection, unless you are
accessing a public account.

 Note: You must have Telnet software to make this kind of
                      connection, provided by either a third party or as shipped
                      in the Windows® operating systems. A minimum
                      functional version of Telnet is provided with
                      Windows NT.

NETSTAT

This utility allows you to display TCP or UDP port activity on a local workstation or
server. If you want this information for a machine outside the local network,
NETSTAT allows you to define the route. The NETSTAT utility has several switches
to display network statistics, as follows:

- **-a** displays all connections and listening ports

- **-e** displays Ethernet statistics

- **-n** displays addresses and port numbers in numerical form

- ■ **-p proto** displays connections for TCP or UDP protocols

- ■ **-r** displays the routing table

- ■ **-s** displays protocol statistics for TCP, UDP, and IP

For example, if you wanted to look at the routing table of a router, use "NETSTAT – r" to display the configuration as shown in Figure 10.14.

Figure 10.14 NETSTAT Routing Table Configuration

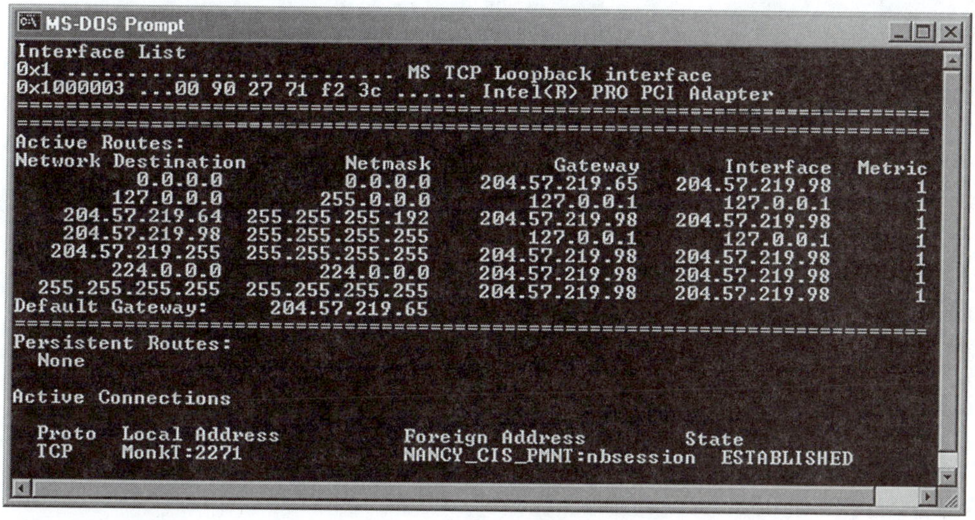

```
MS-DOS Prompt                                                      _ □ ×
Interface List
0x1 ................................ MS TCP Loopback interface
0x1000003 ...00 90 27 71 f2 3c ...... Intel(R) PRO PCI Adapter
==================================================================
Active Routes:
Network Destination        Netmask          Gateway       Interface  Metric
          0.0.0.0          0.0.0.0    204.57.219.65   204.57.219.98       1
        127.0.0.0        255.0.0.0        127.0.0.1       127.0.0.1       1
   204.57.219.64  255.255.255.192    204.57.219.98   204.57.219.98       1
   204.57.219.98  255.255.255.255        127.0.0.1       127.0.0.1       1
  204.57.219.255  255.255.255.255    204.57.219.98   204.57.219.98       1
        224.0.0.0        224.0.0.0    204.57.219.98   204.57.219.98       1
  255.255.255.255  255.255.255.255   204.57.219.98   204.57.219.98       1
Default Gateway:     204.57.219.65
==================================================================
Persistent Routes:
  None

Active Connections

  Proto  Local Address              Foreign Address         State
  TCP    MonkT:2271                 NANCY_CIS_PMNT:nbsession  ESTABLISHED
```

TRACERT

This utility traces the exact route to a specific host and specifies the packet routing time. This is useful for identifying the devices along a certain path when trying to isolate a problem. At the MS-DOS prompt, type TRACERT followed by the IP address (or domain name) of the host with which you want to connect. The utility then displays the entire route (hops) taken to reach the host, as shown in Figure 10.15.

Figure 10.15 TRACERT Display

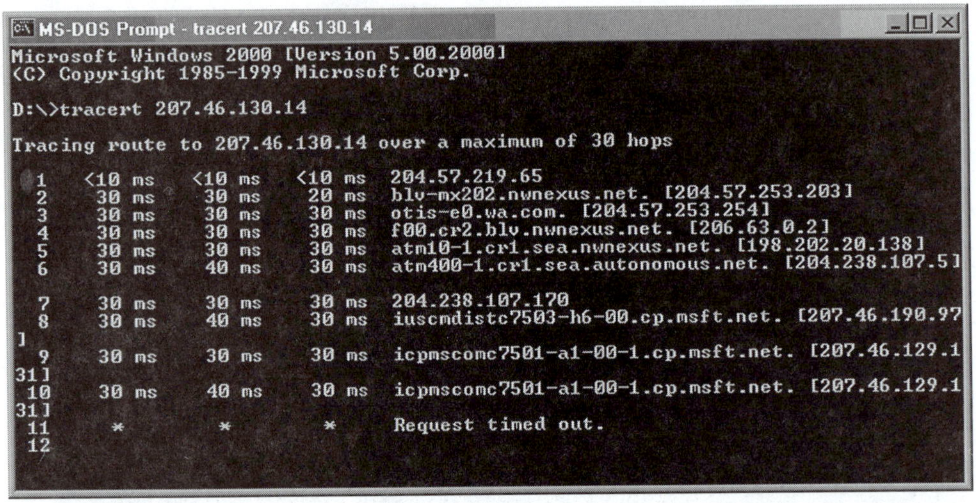

```
MS-DOS Prompt - tracert 207.46.130.14                                    _ □ x

Microsoft Windows 2000 [Version 5.00.2000]
(C) Copyright 1985-1999 Microsoft Corp.

D:\>tracert 207.46.130.14

Tracing route to 207.46.130.14 over a maximum of 30 hops

  1   <10 ms    <10 ms    <10 ms    204.57.219.65
  2    30 ms     30 ms     20 ms    blv-mx202.nwnexus.net. [204.57.253.203]
  3    30 ms     30 ms     30 ms    otis-e0.wa.com. [204.57.253.254]
  4    30 ms     30 ms     30 ms    f00.cr2.blv.nwnexus.net. [206.63.0.2]
  5    30 ms     30 ms     30 ms    atm10-1.cr1.sea.nwnexus.net. [198.202.20.138]
  6    30 ms     40 ms     30 ms    atm400-1.cr1.sea.autonomous.net. [204.238.107.5]

  7    30 ms     30 ms     30 ms    204.238.107.170
  8    30 ms     40 ms     30 ms    iuscmdistc7503-h6-00.cp.msft.net. [207.46.190.97
]
  9    30 ms     30 ms     30 ms    icpmscomc7501-a1-00-1.cp.msft.net. [207.46.129.1
31]
 10    30 ms     40 ms     30 ms    icpmscomc7501-a1-00-1.cp.msft.net. [207.46.129.1
31]
 11      *         *         *      Request timed out.
 12
```

IPCONFIG

IPCONFIG is a Windows NT-based utility that displays diagnostic information for the TCP/IP network configuration. This includes the IP address, subnet mask, and default gateway for each network adapter with a TCP/IP binding. IPCONFIG also supports several DHCP commands, which causes a system to update or release its TCP/IP configuration. This utility can help you resolve IP addressing problems on Windows NT machines.

At the MS-DOS prompt, type IPCONFIG to run the utility. In addition, several switches can be added to enhance functionality of the "IPCONFIG" command, as follows:

- **/all** shows full IP configuration information

- **/release** releases IP configuration for the specified network adapter

- **/renew** renews IP configuration for the specified network adapter

- **/flushdns** purges DNS resolver cache

- **/registerdns** refreshes DHCP leases and re-register DNS names

- **/displaydns** displays contents of DNS resolver cache

An example of an IP configuration display using the "IPCONFIG /ALL" command is shown in Figure 10.16.

Figure 10.16 IPCONFIG Utility

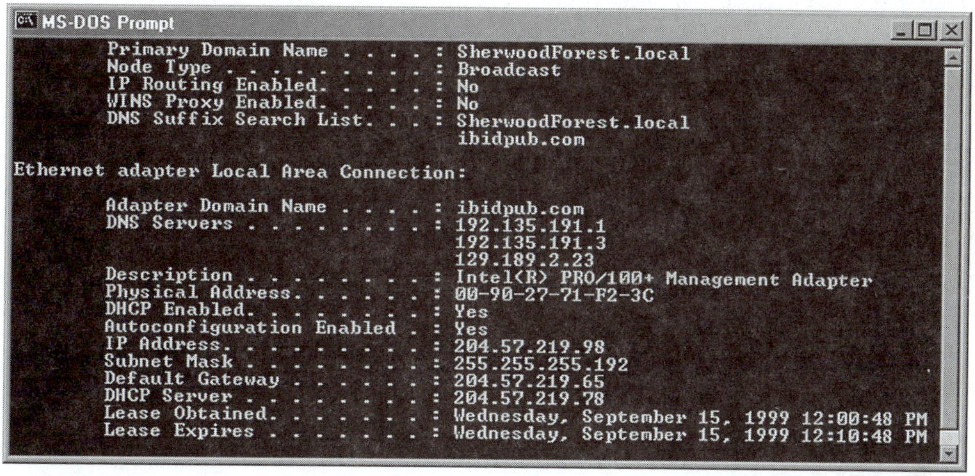

WINIPCFG

WINIPCFG is a Windows 95/98-based utility that can assist you in resolving IP addressing problems on these machines. The utility displays IP configuration information for the host and network adapter, as shown in Figure 10.17. To run WINIPCFG, type `WINIPCFG` in the Run utility.

Figure 10.17 WINIPCFG Utility (IP Configuration)

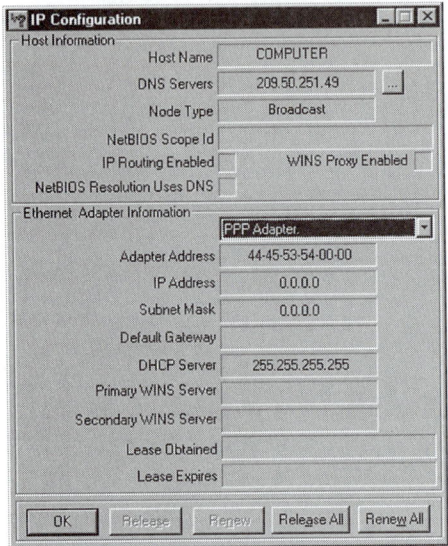

LINUX IFCONFIG

In LINUX, a network interface represents the way that networking software uses the hardware, driver, IP address, and so on. The IFCONFIG command can be used to configure or display a network interface. Parameters that can be configured or displayed include the following:

- Network interface address

- Netmask

- Broadcast address

- IP address at boot time

- Network interface parameters—including the enabling of ARP, driver- dependent debugging code, and the one-packet mode

For further information on LINUX, see http://www.linux.com.

ARP

Sending data from one computer to another is a problem if the recipient's physical address is unknown and there is no resolution system for determining the address. Each network computer may not have a list of all physical addresses of other computers or devices. In addition, it is difficult to effectively maintain an updated table with this information in each computer.

Address Resolution Protocol (ARP) is a TCP/IP protocol that solves this problem with a table that converts IP addresses to local network device physical addresses. This enables local hosts to translate IP addresses to Media Access Control (MAC) addresses that are used at the data link level to communicate with destination nodes. With ARP, applications no longer need to know physical addresses to communicate.

ARP Cache

The table holding the conversion information is known as an ARP cache, as shown in Table 10.1. Each row (entry) corresponds to one subnet device with four pieces of information related to it. As an example, some values are given in Table 10.1 for Entry 1.

Table 10.1 ARP Cache

	Interface Index (physical port)	Physical Address (Hexadecimal)	IP Address	Type
Entry 1	0x1000003	00-90-27-71-F9-58	203.56.211.77	Dynamic
Entry 2	etc.			
Entry 3				

Parameters for the "Type" column can be any of the following:

- Undefined
- Invalid entry
- Dynamic
- Static

Address Resolution Process

When a host sends out an ARP request for the MAC address of a network device or node, ARP first registers the IP address of the recipient device. Next, it searches the ARP cache for a match. If it is found, the hardware (MAC) address is returned. If it is not found, ARP broadcasts a request message containing the recipient's IP address to all local network devices. If a device recognizes this IP address as its own, the device or node replies to the computer generating the ARP request with a message containing its hexadecimal (MAC) address. The information is then placed in the ARP cache of the requesting host for future use.

In reverse, when a MAC address is resolved to an IP address, it is called Reverse Address Resolution Protocol (RARP).

 Note: When the ARP cache receives an ARP request, the cache is dynamically updated with the ARP request information. This accommodates for changes or new additions in the network without requiring separate ARP requests to be generated. Having an ARP cache improves network performance because it reduces network traffic generated by excessive ARP requests and replies.

Using ARP Utility Commands

Launch ARP from the MS-DOS screen by typing `arp` at the command prompt. When ARP first activates, a list of switches appear describing various ARP commands, as follows:

- **-a** displays current ARP cache entries

- **-g** is the same as –a

- **inet_addr** specifies an Internet address

- **-N if_addr** displays the ARP cache for the network interface specified by "if_addr"

- **-d** deletes entry for the host specified by the "inet-addr"

- **-s** adds a static entry

- **eth_addr** specifies a physical address

- **if_addr** specifies the IP address of the interface maintaining the ARP address table to be modified

For example, when you type the "arp –g" command, an ARP cache is displayed similar to the one shown in Figure 10.18.

Figure 10.18 ARP Cache

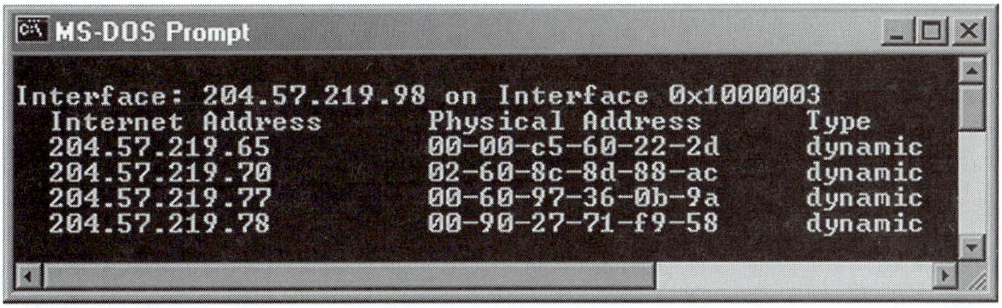

In the example of Figure 10.18, an additional static entry could be configured for a device with an IP address of 204.47.101.52 and a NIC address of 00608C0E6C6A, using the "arp –s" command as follows:

```
C:\>arp –s 204.47.101.52 00-60-8c-0e-6c-6a 204.57.219.98
```

The value 204.57.219.98 specifies the IP address of the host (interface) maintaining the ARP cache. The ARP cache displays the following after the static entry is made (new static entry is bolded):

```
C:\>arp -a
Interface: 204.57.219.98
    Internet Address   Physical Address      Type
    204.57.219.65      00-00-c5-60-22-2d     dynamic
    204.57.219.70      02-60-8c-8d-88-ac     dynamic
    204.57.219.77      00-00-c5-60-22-2d     dynamic
    204.57.219.78      02-60-8c-8d-88-ac     dynamic
    204.47.101.52      00-60-8c-0e-6c-6a     static
```

Troubleshooting with the ARP Utility

With the ARP utility, you can display the ARP cache and add or delete entries. This can assist in resolving the following types of problems:

- **Applications on two nodes cannot communicate.** Check the ARP cache on the source node to see if it contains the entry mapping the IP and MAC addresses of the destination node. Also, verify that the entry type is valid.

- **Network traffic impacted by excessive ARP requests.** Excessive ARP requests and replies can contribute to a network traffic bottleneck. Use a network monitor to detect ARP traffic levels on the network. If you detect excessive traffic, do the following:

    1.   Make sure the ARP cache contains all the IP and MAC addresses for all devices on the local network. This minimizes ARP broadcasts on the network.

    2.   Use an implementation of ARP that has an aging capability which purges expired addresses. This should reduce ARP lookup time.

- **Slow retrievals from the ARP cache.** ARP retrieves the contents of the ARP cache.

 Note: The ARP cache should automatically purge itself of expired dynamic entries.

- **IP address conflict.** If you get a system error message indicating you have an IP address conflict, two subnet devices may have the same IP address by mistake. To correct this, do the following:

1.   View the local host's ARP cache to verify whether the two subnet devices have an IP address conflict.

2.   Correct the TCP/IP configuration on the conflicting device.

 Note: ARP works at the Data Link Layer and does not contain IP headers. Since ARP requests cannot be routed on an IP network, they are limited to their logical subnet. ARP requests for devices not on the same subnet may return a router or default gateway address instead.

■   **Address resolution conflicts.** Verify that ARP is working properly by pinging a host with a known IP address. If working properly, this should add an entry to the ARP cache specifying the host's IP and physical (hexadecimal) address. If the host is remote, the ARP cache should also contain the default gateway address that PING used to locate the remote host.

Also, make sure all device or node entries in the ARP cache are assigned the correct "Type" information. For example, computers that are assigned IP addresses by a DHCP server should have "dynamic" entries. If static entries are required, add them according to the steps described earlier.

Novell IPXCON

To assist your troubleshooting efforts with routers in a NetWare environment, the Novell IPXCON utility can be used to do the following:

■   Locate, monitor, and troubleshoot active IPX routers and network segments in an IPX internetwork

■   View IPX router or network segment status

■   View all routing paths for IPX packets

■   Monitor remote IPX routers running NetWare IPX router software

Launch IPXCON using the `LOAD IPXCON` command at the server console. When loaded, the utility displays a summary of routing statistics, as shown in the partial view of Figure 10.19.

Figure 10.19 IPXCON Routing Statistics

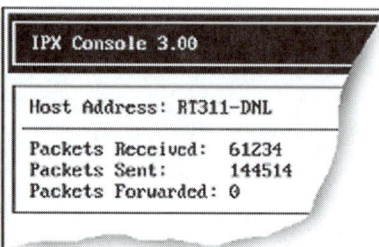

```
IPX Console 3.00

Host Address: RT311-DNL

Packets Received:  61234
Packets Sent:      144514
Packets Forwarded: 0
```

The statistics in Figure 10.19 are defined as follows:

- **Host address**. Names the router monitored.

- **Packets received**. The number of IPX packets received by the monitored router.

- **Packets sent**. The number of packets sent by the monitored router.

- **Packets forwarded**. The number of packets forwarded by the monitored router.

Additional information that can be viewed with IPXCON includes:

- **Circuits**. The number of circuits in use on the router. A circuit can be either a LAN or WAN connection.

- **Networks**. The number of networks identified by the router.

- **Services**. The number of services, which can be accessed from the router.

- **SNMP access configuration**. Monitor local or remote servers. Remote systems can be monitored through IPX or TCP/IP.

- **IPX information**. View statistics for a router's IPX packet routing.

- **IPX router information**. View general information about an IPX router.

Novell TCPCON

When troubleshooting the TCP/IP segments of a NetWare internetwork, the TCPCON utility can help you do the following:

- Monitor and gather information from the TCP/IP portion of your internetwork

- View configuration and other statistics for the following TCP/IP protocols: TCP, UDP, IP, ICMP, OSPF (Open Shortest Path First), RIP and EGP (External Gateway Protocol)

- View the IP routes known to a TCP/IP node

- View network interfaces supported by a TCP/IP node

- Use SNMP over TCP/IP or IPX to access TCP/IP information in any remote protocol stack supporting the TCP/IP Management Information Base (MIB)

Protocol Analyzers

A Protocol Analyzer, which is sometimes referred to as a network analyzer, is the tool most often used by network engineers to interactively monitor their networks. It is a very important piece of troubleshooting equipment, which performs the following key functions:

- Real-time network traffic analysis

- Packet capture, decoding, and transmission

- Internal packet structure analysis

- Network traffic statistic generation. This can verify the network configuration including cabling, software, workstations, and interface cards.

A Protocol Analyzer can help you isolate problems in the following areas:

- Protocol problems

- Traffic fluctuations and bottlenecks

- Configuration and connection errors

- Faulty network components

- Conflicting applications

Since a protocol analyzer can identify a wide scope of network behavior, you can also do the following:

- Locate the most active computer or one that is sending out bad packets

- Distinguish different protocol traffic across an internetwork segment or router by using packet filtering

- Monitor network performance over a defined time interval to identify network trends

- Generate test packets to test network cabling, components, and connections

 Note: Most protocol analyzers have a Time Domain Reflectometer, which sends out signals that can identify the location of cable breaks or faulty connections.

Protocol analyzers transmit a specific signal and measure the time it takes to get a reflection back from an anomaly such as a cable break, and then relates the travel time to distance. In this manner, the distance to a cable break or faulty connection can be approximated.

- Generate alerts when network traffic performance falls outside certain tolerances, which you set with protocol analyzer parameters

Some protocol analyzers in common use today include the Network General Sniffer, Hewlett-Packard Internet Advisor, and the Novell LANalyzer.

Vocabulary

Review the following terms in preparation for the certification exam.

Term	Description
ADSL	Asymmetric Digital Subscriber Line is a broadband device used for high-speed digital communications (including video) across twisted-pair copper phone lines
ARP	Address Resolution Protocol is a TCP/IP protocol used by local hosts to retrieve MAC addresses from local subnet devices and return it to the host's ARP cache where the device physical address is mapped to its IP address (for future reference), thus enabling device communication at the datalink layer. ARP is also a utility, accessed from the MS-DOS screen that allows the ARP cache to be displayed for diagnostic purposes.
ASCII	The American Standard Code for Information Interchange is a coding scheme that assigns 7 or 8 bit numerical values to represent as many as 256 different characters
bandwidth	The data transfer capacity of a digital communication system
binding	The process of attaching information from program elements or values to their storage location, such as an address binded to a device
BIOS	Basic Input Output System is a set of software routines on a PC that test hardware at startup, start the operating system, and support data transfer to hardware devices.
bottleneck	A heavy traffic condition that results when the processor, memory, or a disk system is overloaded by process or application requests
broadcast frame	Frames sent to all network devices

broadcast storm	A network broadcast that causes multiple hosts to simultaneously respond and overload the network
buffer	A region of memory allocated to hold data waiting to be transferred
byte	An 8-bit binary number
cache	A memory subsystem that stores frequently used data for easy future access
capture filters	A filter that defines certain parameters used when capturing network packets for analysis
collisions	When two network devices attempt to transmit data signals simultaneously, a collision can result which distorts and destroys the data
COM	A name used by MS-DOS for serial communication ports
default gateway address	A Microsoft term implicating that the IP address assigned to a device on one network acts as a default path for packet routing to other networks
DHCP	Dynamic Host Configuration Protocol is a service that automatically assigns IP addresses to clients on a network, permitting facilitation of client IP address requests
DMA	Direct Memory Access is a process of data transfer directly from memory to an intelligent peripheral device, where the processor is not involved
DNS	Domain Name System is the system by which hosts on the Internet resolve domain names to IP addresses
driver	A device driver is a program that enables the computer to work with a specific device, such as a printer, modem, or network adapter

dynamic link library	Code that runs only when it is needed by an executable program. Usually has a file extension of .dll.
errors	Faults that occur in a process, application, or device on a workstation or network
frame	A package of information formatted as a single unit and used in synchronous communications. Maintains organization, control, addressing, and error checking for transmitted data
hardware query tool	A diagnostic utility that identifies hardware internal component types and the resources they use
hexadecimal	A system that uses 16 rather than 10 as the basis to represent numerical values.
idle	Describes a state of inactivity but with readiness, such as an idle processor waiting for a command
I/O port	The channel through which data transfers to and from an input or output device and the processor
interrupt	A request for processor time that suspends the processor's current activity, saves its data, and passes control to an interrupt handler to deal with the hardware or software issue that created the interrupt
IP	Internet Protocol is the part of TCP/IP responsible for formatting data packets, routing the packets to a destination network, and reassembling the packets back into their original form at the destination
IPCONFIG	A Windows NT-based utility accessed from the MS-DOS screen that displays the TCP/IP configuration of a subnet
IRQ	Interrupt Request is a hardware interrupt identified by a number that signifies a certain handling priority

IS	Information Services is an organization's data processing department
knowledge base	A database maintained by manufacturers such as Microsoft, which provides technical support information for resolving problems and errors
LAN	Local area network is a logical grouping of computers and devices on a local network where any device can communicate with any other through a communications link
LAN router	Acts as a default gateway to connect local network clients directly to the Internet
LMHOSTS	A static text file used to map remote NetBIOS computer names to their permanent, manually-configured IP address
macro	A set of keystrokes and instructions recorded and saved under a macro name. When the macro is executed, it performs all the recorded instructions at once, thus simplifying multiple keyboard-based operations
multicast frame	The process of simultaneously sending a message to more than one destination on a network
memory block	A contiguous section of random access memory (RAM) temporarily allocated to a program or process
module	A group of routines or data structures that implement a certain task or function
NetBIOS	Network Basic Input/Output System is an application programming interface used by MS-DOS applications on a LAN to enable information transmission and sessions between nodes
NETSTAT	A utility that allows you to display TCP or UDP port activity on a local workstation or server

network adapter	A device in a computer responsible for the physical interface connection to the network, which allows information transfer at the data link/physical level.
node	A device in a LAN capable of communicating with other LAN devices
NOS	A Network Operating System provides file, print, and communication services on a network
Performance Monitor counters	Performance Monitor counters gather performance statistics on various system objects and processes, including processor, threads, disk usage, and memory cache objects
PING	The Packet Internet Groper sends echo request signals to various points in a network to determine valid communication paths and transit times
protocol	A protocol is a set of rules by which communication and transfer of information takes place in a computer system
registry	A database that keeps track of system, hardware, software, user profile, and application configurations
SNMP	Simple Network Management Protocol monitors the status of hardware devices on a network and reports it to a central management console
T1	A high-bandwidth digital telephone line that carries voice, text, and images, commonly used by corporations for Internet connectivity and is equivalent to 1.54 megabits per second
TCP/IP	Transmission Control Protocol/Internet Protocol is a set of protocols that enable communication between computers on internetworks.
Telnet	A utility that allows you to connect to a TCP/IP host computer across the Internet using its IP address

TRACERT	A utility that traces the exact route to a specific host and specifies the packet routing time
traffic	The load carried by a data communication channel
UDP	User Datagram Protocol is the connectionless protocol in TCP/IP, which corresponds to the transport layer of the OSI model. It converts data messages generated by applications to packets sent by IP
volume	A division of a hard disk which stores computer data virtual memory Memory appearing larger than actual to an application
WINIPCFG	A Windows 95/98-based utility that displays IP configuration information for the network adapter and local host
WINMSD	Windows Diagnostic utility generates reports on system configuration information
WINS	Windows Internet Naming Service is a database that dynamically maps local NetBIOS computer names to their IP addresses

In Brief

If you want to...	Then do this...
Locate troubleshooting information resources	Consult vendor and manufacturer equipment documentation
Obtain professional help and support to resolve your troubleshooting problem	Consult technical support personnel
Maintain records of problems and resolutions	Create a system ledger that tracks/documents hardware and software problems
Maintain efficiency when addressing network problems	Prioritize the problems
Keep a backup of your original system configuration when troubleshooting	Create a system configuration record
Minimize your time and effort during the initial stages of problem resolution	Consult the system ledger for past similar problems; try to replicate the problem
Isolate difficult problems	Use the process of elimination and expand and contract your focus
Eliminate possible causes of problems in a troubleshooting scenario	Change one variable at a time
Isolate a procedural problem at a workstation	Try the same procedure on an equivalent workstation

Obtain a snapshot of your system configuration	Use Windows NT Diagnostics to generate a report
Isolate widespread network problems	Use the TRACERT and PING utilities to communicate with remote hosts
View error and system messages	Use the Windows NT Event Viewer in Windows NT based systems and the SYS$LOG.ERR file in a NetWare environment
Monitor system performance and gather statistics for trend analysis	Use the Windows NT Performance Monitor
Capture network traffic for display and analysis	Use a Network Monitor or protocol analyzer
Monitor hardware device status	Use SNMP
Determine the integrity of physical network connections	Check link lights
Find simple solutions to network problems	Check power lights and error displays
Isolate user access errors	Check user credentials and permissions
Detect whether your network is infected with a virus	Run a virus scan
Correct modem problems	Check the modem COM port configuration, latest drivers installed, hardware compatibility, and device failure
Correct network adapter problems	Check the network adapter for latest drivers installed, hardware compatibility, the IP configuration, and device failure

Correct router problems	Run hardware diagnostics, check routing table, check IP configuration, check power lights, and verify physical network connection integrity
Correct gateway problems	Check log files for network errors, restart the gateway, check gateway protocol configuration
Correct switch, hub, or bridge problems	Check power indicators and network connections, check error displays, reset the component, or replace it
Correct printer problems	Reset the printer server, stop and restart the printer queue, check printer power, check printer configuration
Correct tape drive problems	Clean and service the drive, run a hardware diagnostic program, contact technical support personnel
Correct DNS problems	Check DNS zones and servers
Correct WINS problems	Check NetBIOS names, WINS server replication, browsing lists, trusts, static mappings, and client connectivity to WINS server
Correct DHCP problems	Check for lease expiration, check DHCP scope configuration, and make sure DHCP is not assigning an IP address to a component that requires a static address

Correct LMHOSTS problems	Make sure workstations are configured for LMHOSTS support, check LMHOSTS configuration for keyword errors or other incorrect entries, verify that all NetBIOS computers and static IP addresses are entered in the LMHOSTS configuration file
Correct default gateway problems	Ping the network adapter on the gateway and verify the response. If no response, check IP configuration
Correct LAN router problems	Verify the correct IP configuration for router's external and internal network adapters; also check that clients have static IP addresses and the correct default gateway address
Correct subnet mask problems	Verify TCP/IP protocol binding on the host network adapter, and check for valid subnet mask assignment
Correct cable problems	Use a cable tester to verify proper cable specifications
Correct computer hardware problems	Run the Hardware Query Tool to display hardware configuration and resource usage data
Test network adapter transmit and receive functions	Use a Hardware Loop Back cable
Test a network route to a remote host	Use the PING and TRACERT utilities
Connect to a remote TCP/IP host across the Internet	Use the Telnet utility

Display TCP or UDP port activity on a local server or workstation	Use the NETSTAT utility
Identify the route (hops) to a remote host	Use the TRACERT utility
Display diagnostic information for the TCP/IP network configuration on Windows NT computers	Use the IPCONFIG utility
Display the IP configuration on Windows 95/98 computers	Use the WINIPCFG utility
Troubleshoot routers in a NetWare environment	Use the Novell IPXCON utility
Troubleshoot the TCP/IP segments of a NetWare internetwork	Use the Novell TCPCON utility
Configure or display a LINUX network interface	Use the LINUX IFCONFIG utility
View system information and track various utilization statistics in a NetWare environment	Use the Novell Monitor utility
Configure a network adapter card	Use the Windows Network utility in Control Panel
Check IP to MAC address resolutions for subnet devices	Use the ARP utility to display the ARP cache maintained by the local host
Configure a modem	Use the Modem utility in Control Panel
Troubleshoot complex network problems	Use a protocol analyzer

Lesson 10 Activities

Complete the following activities to better prepare you for the certification exam.

1.   Describe at least four information sources you can consult when troubleshooting your computer network.

2.   Name the general steps to follow when organizing and applying your troubleshooting efforts.

3.   Given that a problem has occurred in your network, discuss how to assess the initial network conditions.

4.   Discuss how you can troubleshoot networks and workstations with the Windows NT Performance Monitor.

5.   You suspect a network communications problem and need to capture network traffic for analysis. What utility can you use and on what three parameters can you base the design of a capture filter? Describe each parameter in your discussion.

6.   Identify four typical problems that can occur with a modem.

7.   Describe the functions of a router and define the meaning of static and dynamic routing tables.

8.   You suspect a network adapter's IP configuration is at fault. How can you verify that this is the case? Give an example scenario.

9.   Describe the functions of a DHCP server and why you would not want more than one operating on the same subnet.

10.   You suspect a problem across your WAN and you want to expose the devices along the exact communication path to a remote host. Discuss how to do this.

Answers to Lesson 10 Activities

1.   The following information sources can be consulted to support troubleshooting
      activities:

      ■   Documentation provided by the vendor with the original product purchase

      ■   Vendor updates on technical changes, revisions, and resolutions to problems
           encountered (from their user base)

      ■   Troubleshooting databases and technical bulletins located at vendor Web
           sites

      ■   Documentation created by the OEM

      ■   Vendor knowledge bases

2.   The following steps are used to guide your troubleshooting efforts:
      1.   Assess the initial network conditions
      2.   Identify the exact problem
      3.   Replicate the problem
      4.   Isolate the cause through the process of elimination
      5.   Devise a corrective measure
      6.   Implement the correction
      7.   Test the corrective measure
      8.   Document the problem/solution and provide feedback to the user

3.   Before attempting to isolate a problem, consult the system ledger for hardware
      and software component records to determine if the same problem has occurred
      before. Then establish the initial network condition at the time of problem
      occurrence as follows:

      ■   Determine how long the network was functioning normally prior to the
           problem occurrence

      ■   Check records to verify recent network changes

      ■   Verify if any client or administrator activities occurring at the time of the
           problem either caused or contributed to the problem

■   Generate a report on initial network configuration and status using the Windows NT Diagnostic utility

4.   Windows NT Performance Monitor can be used to view or log performance statistics on various system objects and processes, on a server or a Windows workstation. The troubleshooter can also use this utility to show where system bottlenecks are occurring. In addition, it can be used to set up broadcast alerts, which point to detrimental system trends or warn you when operating thresholds have been exceeded.

5.   The Network Monitor can be used to sample all network traffic to and from a host computer or create a filter that captures only certain information. With capture filters, you can isolate and analyze specific frames using the following parameters to define what the filter collects:

■   **Data pattern matches.** Limits the capture to ASCII or hexadecimal data patterns.

■   **Protocols**. Captures specific network protocols or their properties.

■   **Source/destination addresses**. Capture frames sent to and from a specific network adapter (host).

6.   Four problems that typically occur with modems are incompatible hardware, incorrect drivers, COM port conflict, and device failure.

7.   Routers connect multiple paths between different network locations. They segment traffic by subnetwork so that only the packets destined for a specific subnetwork are sent there by the router. No other traffic is passed through the router.

When a router receives packets, it forwards the data to the destination address based on a routing table. A static routing table is created and updated manually by the administrator as devices and paths are added to the network. Dynamic routing tables are automatically updated as routers communicate with each other using Routing Information Protocol (RIP).

8.   Use the PING utility to send an echo request signal to the network adapter. If you receive a reply, this verifies a functional communication path to the network adapter and that the TCP/IP protocol stack is binded there. To send a ping, open

the MS-DOS window and type `ping 127.0.0.1` at the command prompt. This locates the network adapter on the computer where you launched the MS-DOS screen.

If that works, check the IP configuration of the network adapter using the Windows IPCONFIG Network utility. From the utility, verify whether the IP address, subnet mask, and default gateway address are all valid.

9.   A DHCP server dynamically and automatically allocates IP addresses to systems (a DHCP client) for a specific period of time. It allows for a centralized management of IP addresses to help avoid IP address conflicts. If DHCP is not used, static IP addresses must be configured manually for network components. It is necessary to have only one DHCP server issuing IP addresses. Otherwise, if more than one DHCP server exists on the same subnet, addressing problems result.

10.   Use the TRACERT utility to expose all the devices along the communication path to a remote host. From the MS-DOS prompt, type `TRACERT` followed by the IP address (or DNS domain name) of the remote host. TRACERT then displays the entire route (hops) taken to reach the host.

Lesson 10 Quiz

These questions test your knowledge of features, vocabulary, procedures, and syntax.

1.   When available documentation is insufficient to help, you resolve a network problem and you have exhausted all other approaches, what is the next step you should take?

   A.   Do a cold reboot on the server
   B.   Consult the help files
   C.   Call technical support
   D.   Search your system ledger for a history of similar problems

2.   In a troubleshooting scenario, what is the value of replicating a problem? Choose the answers that apply.

   A.   Demonstrates it is not a one-time incident and gives you a good starting point to isolate the problem
   B.   Saves you the trouble of searching for error logs
   C.   Pinpoints the exact location of the problem
   D.   It is better than trial-and-error methods

3.   Before attempting to isolate network problems, which utility can you use to get a profile of your basic network configuration?

   A.   TRACERT
   B.   Windows NT Performance Monitor
   C.   Windows NT Diagnostics (WINMSD)
   D.   Windows NT Event Viewer

4.   If you wanted to view a record of system errors and messages, how many of the following can you use?

   A.   Device error displays
   B.   Hardware Query Tool
   C.   Windows NT Event Viewer

D.   NetWare's SYS$LOG.ERR file

5.   To record performance statistics on certain system objects or processes, which one of the following can you use?

A.   SNMP
B.   Windows NT Diagnostics
C.   Windows NT Performance Monitor
D.   Network monitor

6.   If you isolate a problem to network communications and want to capture network traffic for bottlenecking and frame analysis, which utility would you use?

A.   IPCONFIG
B.   TRACERT
C.   PING
D.   Network Monitor

7.   Which of the following indicators confirm the physical connection and logical communication path between network components?

A.   Error displays
B.   Link lights
C.   Power lights
D.   None of the above

8.   If you want to isolate a default gateway problem, you must first know the purpose of the default gateway parameter. Which of the following best describes it?

A.   An IP address assigned to a device which acts as a gateway from one network to another
B.   An Internet-routable IP address
C.   A router's static IP address
D.   A TCP/IP protocol banded to a network adapter

9.   You have determined that a LAN router does not act as the default gateway to the Internet. From the following answers, choose all the ones that might resolve this problem.

      A.   Verify the router has a default gateway address assigned to its local network adapter
      B.   Verify the router has an Internet-routable IP address binded to its external network adapter, or is set up as an ISP DHCP client
      C.   Assign a static IP address to each client
      D.   Connect the router to a dial-up modem link

10.   If you had an incorrect subnet mask on a local host's external network adapter, or none at all, choose all the following answers that describe the steps you could take to fix it.

      A.   Ping the network adapter for an echo response to determine whether TCP/IP protocol is binded to the adapter
      B.   Bind an IP address to the network adapter if it doesn't already have one
      C.   Configure the standard subnet mask value on the network adapter
      D.   Set the subnet mask to the default gateway address value

Answers to Lesson 10 Quiz

1.   Answer C is correct. A technical support call should be your last resort to resolving a problem.

Answer A is incorrect because a server reboot will probably not solve the problem.

Answer B is incorrect because the Help files probably contain no further information beyond the manufacturer's documentation supplied with the original equipment purchase.

Answer D is incorrect since the ledger is the first place to check for similar problems you had in the past, before you look at other information sources.

2.   Answer A is correct. Replicating the problem does give you a better basis upon which to identify the cause of the problem. If you can replicate the problem, it is not a one-time incident—this makes it much easier to isolate.

Answer D is also correct because if you can replicate the problem, then its location should become more obvious. This is much easier than trying to isolate the problem with a trial-and-error method.

Answer B is incorrect. Even though you can replicate the problem, this is not an indication that you can avoid searching through error logs. This might be your next step.

Answer C is incorrect because replicating the problem does not necessarily determine its exact location.

3.   Answer C is correct. The Windows NT Diagnostic utility can provide you with a snapshot of your system configuration when assessing initial system conditions prior to troubleshooting.

Answer A is incorrect because the TRACERT utility traces routing paths to remote hosts across larger networks.

Answer B is incorrect because Windows NT Performance Monitor is used primarily to view and log performance statistics on various system objects and processes.

Answer D is incorrect because the Windows NT Event Viewer is a log file, which records errors and system messages.

4. Answer C is correct. The Windows NT Event Viewer logs all system errors and messages.

Answer D is also correct because all system errors in a NetWare server environment are recorded in the SYS$LOG.ERR file.

Answer A is incorrect since device error displays do not give you system level information—they provide error indications for the device only.

Answer B is incorrect since the Hardware Query Tool identifies internal PC hardware and the resources they use.

5. Answer C is correct. The Windows NT Performance Monitor can be used to gather performance statistics and to detect system bottlenecks.

Answer A is incorrect since Simple Network Management Protocol is used to monitor hardware devices in a TCP/IP network.

Answer B is incorrect since Windows NT Diagnostics give you a profile of your network configuration.

Answer D is incorrect since a Network Monitor is used to detect network bottlenecks and capture network traffic for analysis. It does not record server and workstation performance statistics.

6. Answer D is correct. Network Monitor allows you to create capture filters based on certain parameters, analyze collected frame data, and assess network bottlenecks.

Answer A is incorrect since IPCONFIG is used to display the TCP/IP network configuration, including the IP address, subnet mask, and default gateway for each network adapter with a TCP/IP binding.

Answer B is incorrect since the TRACERT utility is used to trace routing paths to remote hosts across larger networks.

Answer C is incorrect since the PING utility is used to determine whether a network route to a remote host is available and if the IP stack on the host is working.

7.   Answer B is correct. Link lights illuminate to indicate the integrity of the physical and logical connection between devices on a network.

Answer A is incorrect since error displays usually indicate faults with specific devices.

Answer C is incorrect since power lights only indicate that a device is receiving primary power.

8.   Answer A is correct. The default gateway address is used to direct local network IP requests destined for an outside network, to the gateway to that network.

Answer B is incorrect since a default gateway address is not Internet-routable.

Answer C is incorrect since static IP addresses assigned to routers are usually binded to their external network adapter. This enables the device to be recognized on the internetwork and has nothing to do with the default gateway address.

Answer D is incorrect because TCP/IP protocol must be binded to a network adapter before a default gateway can be specified. Otherwise, the network adapter is not recognized by the TCP/IP protocol.

9.   Answer A is correct. The LAN router needs a default gateway assigned to its local network adapter.

Answer B is correct because the router must have an Internet-routable IP address binded to its external network adapter, so it can be recognized on the Internet. The ISP can also set up the router as a DHCP client with its IP address automatically assigned when the Internet connection is made.

Answer C is correct, however, static IP addresses should only be set up on client computers if the router is not a DHCP server.

Answer D is incorrect because a dial-up modem link connects clients to the Internet and is rarely used by a server (since it is a very slow connection). In addition, if you connected a modem to a router, the two devices would not communicate.

10.   Answer A is correct. PING determines whether you have the TCP/IP protocol binded to the network adapter. A TCP/IP binding is needed before any other IP addresses are assigned to the network adapter.

Answer B is correct because if you don't have TCP/IP binded to the network adapter, you must bind it with an IP address before the subnet mask value (and default gateway) can be recognized by TCP/IP.

Answer C is correct since you must have a valid subnet mask assigned to the network adapter to locate the host computer in the subnet.

Answer D is incorrect since the default gateway address assigned to the subnet mask parameter does not provide a way for TCP/IP protocol to find the host computer. It might also cause a security breach.

Glossary

This glossary will help you understand the terms covered in the exam preparation material.

Glossary Term	Definition
100VG-AnyLAN	A 100 Mbps Ethernet standard using the demand priority network access method
10BaseT	An Ethernet standard for bus type LANs using a 10 Mbps transmission rate, baseband signals, and twisted-pair cable
386SPART.PAR	Permanent swap file in Windows 3.x
Access Control List	A list that contains information that specifies user and group permissions levels
access methods	Characterize the way computers transmit data onto the network media and standards specifications identifying network media configurations and their supporting data transmission rates
access rights	Assignment of the appropriate permission level when sharing directories or files
active hub	A hub that regenerates data signals before distributing them to attached devices
active terminator	A terminator that automatically adjusts its impedance value to the network cable where it is installed

adapter	Short for expansion board, it is the circuitry required to support a particular device such as video adapters which enable the computer to support graphics monitors, and network adapters enable a computer to attach to a network
adapter cards	Printed circuit boards that plug into a computer's expansion bus to add capabilities to your computer
address class	Address classes A, B, and C define which IP address bits are used for the network ID and which are used for the host ID
ADSL	Asymmetric Digital Subscriber Line is a broadband device used for high-speed digital communications (including video) across twisted-pair copper phone lines
algorithm	A logical sequence of steps. Ciphers work according to sophisticated computer algorithms that rearrange the data bits in digital signals
analog phone lines	Standard phone lines that transmit continuous signals
anti-virus	Software that runs scans and cleans viruses by checking the computer signature or definition files
AppleTalk	An Apple LAN for communication and resource sharing using a layered set of protocols similar to the OSI model, transferring information in frame format
Application Layer	Provides general network access, flow control, and error recovery functions for services that directly support user applications such as file transfer, database access, or e-mail communications

application server	In a client/server environment, a member server dedicated to a specific task or application such as a database
ARCNet	Attached Resource Computer Network is a network architecture loosely mapping to the 802.4 specification, used for workgroup sized token-passing bus networks having broadband cable and a 2.5-Mbps data rate
ARP	Address Resolution Protocol correlates an IP address with the Media Access Control (MAC) address
ASCII	The American Standard Code for Information Interchange is a coding scheme that assigns 7 or 8 bit numerical values to represent as many as 256 different characters
ASIC	Application specific integrated circuits are used to build switched circuits that route data on a dedicated path
asynchronous	Refers to data transmissions that rely upon start and stop bits to pace the exchange of information, instead of using a timing mechanism such as a clock
ATM	Asynchronous Transfer Mode transmits data, voice, and frame relay traffic in real time by breaking data into packets containing 53 bytes each and transmitting between 1.5 and 622 Mbps
audit trail	Software or tools that allow reviewing a list of names of those who have entered a network
AUI	Attachment Unit Interface is a DB-15 connector found on some network adapter cards that accepts a mating transceiver cable connector

backbone	The trunk or main segment of a network that carries its major traffic and to which computers, devices, or LANs attach
bandwidth	The data transfer capacity of a digital communication system
base I/O port address	A channel for information to flow between a network adapter card and the CPU, identified as an address recognized by the CPU
base memory address	Identifies a starting location in RAM for a network adapter's buffer area, used when the network adapter needs to share CPU memory space for temporary data storage
baseband	Refers to networks, such as Ethernet or token ring, that utilize a transmission media handling one message stream at a time in serial (digital) data format
baud rate	In modem communications, baud rate is the number of signal events occurring per second, not to be confused with bit rate
BDC	Backup domain controller that replicates the PDC's domain database, security policy, and logon authentication
binding	The process of attaching information from program elements or values to their storage location, such as an address binded to a device
BIOS	Basic Input Output System is a set of software routines on a PC that test hardware at startup, start the operating system, and support data transfer to hardware devices

bit rate	In modem communications, bit rate is the actual number of bits transmitted per second across a telecommunications link
BNC	British Naval Connector(s) are used to interconnect Thinnet and Thicknet coaxial cabling using barrel, T-connectors, terminator, and connector components
bottleneck	A heavy traffic condition that results when the processor, memory, or a disk system is overloaded by process or application requests
bps	Bits per second or the speed at which a modem can transfer data. The bps is not the same as baud rate
bridge	A network device that connects segments broadband Refers to the technology used in WAN communications that transmits multiple analog, high-frequency radio signals carrying multiple simultaneous messages and supporting data rates extending into the gigabit range
broadcast frame	Frames sent to all network devices
broadcast storm	A network broadcast that causes multiple hosts to simultaneously respond and overload the network
brouter	A network device that incorporates both a bridge and a router in a single unit
buffer	A region of memory allocated to hold data waiting to be transferred
bus topology	A simple network topology utilizing a single trunk or backbone segment to daisy chain devices together, such as a typical Ethernet network does

byte	An 8-bit binary number
byte stream	Continuous flow of data from beginning to end
CA	A Certificate Authority guarantees the identity of two individuals exchanging online information
cache	A memory subsystem that stores frequently used data for easy future access
capture filters	A filter that defines certain parameters used when capturing network packets for analysis
carrier frequency	An analog high-frequency signal used to carry data signals across telecommunication links using WAN technology
CD-ROM	Compact Disc—Read-Only Memory, a high-capacity, optical storage media that can hold up to 650 megabytes.
characteristic impedance	The base impedance of a coaxial cable calculated on the basis of physical cable elements, without the reactive or frequency-sensitive components included
cipher text	Text translated into a secret code that is unreadable by an unauthorized person who accesses a directory or file
client/server model	A networking model utilizing a central server to manage users, security, resource sharing, data storage, and fault tolerance where the client workstation participates in sharing the processing load

coaxial cable	A transmission media used in Thinnet and Thicknet networks consisting of a copper center conductor surrounded by a Teflon dialelectric layer and a rubber outer sheathing
collisions	When two network devices attempt to transmit data signals simultaneously, a collision can result which distorts and destroys the data
COM	A name used by MS-DOS for serial communication ports
COM port	Also known as a serial port. The port to which a modem is connected
communication protocols	Rules defining how data is formatted, transmitted, and recognized, as defined by the OSI model
communication server	A server that handles the exchange of information between the central server network and remote networks utilizing dial-up or dedicated connectivity access methods
connectionless communications	Communications or data transfers that don't require direct connection among senders and receivers
connection-oriented communications	Communication protocols that require and maintain a two-way or sending and receiving packets until a systematic release of the session is communicated
contentional access	A network access method where computers contend for transmit clearance
cps	Characters per second or the speed at which printers that are not lasers (dot-matrix or ink-jet) print. Cps can also measure the rate at which a device, such as a modem, transmits data

CPU	Central processing unit contains the "brain" of the computer. This is the device that processes and transmits data
crosstalk	Cross talk occurs when data on adjacent cables radiates across unshielded cable runs and induces baseband frequency energy onto other data paths
CSMA/CA	Carrier Sense Multiple Access with Collision Avoidance is a network access method where computers ready to send data signal their intent to transmit in advance to avoid data collisions
CSMA/CD	Carrier Sense Multiple Access with Collision Detection is a network access method used in most Ethernet networks where computers ready to send data wait for traffic to clear on the network cable before transmitting their data, avoiding data collisions
CSU/DSU	Channel Service Unit/Data Service Unit provides an interface between a LAN gateway router and a digital line, such as T1, in a WAN internetworking scenario
data bus architecture	Describes the physical and logical layout of the internal bus upon which data travels in a computer, such as ISA, EISA, or PCI architecture
Data Link Layer	Organizes data (bit stream) into structured frames (logical organized structures) to add address and error control information when sending frames to the Physical Layer
data transmission rate	The rate at which data is transmitted across a network in kilobits or megabits per second

datagram	Any IP packet or unit of information along with relevant delivery information such as the destination address, that is transferred across a network
DCE	Data Communication Equipment is an intermediate device that modifies data sent from Data Terminal Equipment (DTE) in RS-232 format, such as a modem
default gateway address	A Microsoft term implicating that the IP address assigned to a device on one network acts as a default path for packet routing to other networks
default subnet mask	An IP network address number that is subdivided, and identifies the default class networks as Class A, Class B, or Class C
demand priority	A network access method using a repeater to poll network nodes for data-send requests, where contending computers ready to send data are handled on a priority basis or with alternate processing if priority levels are equal
demodulation	The process of converting a modulated data signal back to its original form as a baseband signal, as in modem communications
DES	The Data Encryption Standard is the U.S. Government standard for encryption
deterministic access	Describes a type of network access, such as token passing, which determines that only one computer at a time in possession of the token can transmit its data

DHCP	Dynamic Host Configuration Protocol is a service that automatically assigns IP addresses to clients on a network, permitting facilitation of client IP address requests
digital certificate	A digital certificate is an attachment to an electronic message that also verifies the identification of the message sender but provides the receiver the means to encode a reply
digital phone lines	Communication lines that transfer data in intervals digital signature A digital signature is a code that attaches to an electronic message and identifies the sender
disk mirroring	Duplicates identical data from one physical disk to another to prevent data loss
distributed processing	Distributed processing is a type of mainframe computing environment where independent workstations perform some of the processing, but the central computer handles most tasks
DMA	Direct Memory Access allows a device to directly access the computer's memory without using the CPU for data transfer
DNS	A network of file servers that translates domain names into IP addresses
domain	The different levels of authority in the domain name scheme of the hierarchical structure for Internet access
driver	A device driver is a program that enables the computer to work with a specific device, such as a printer, modem, or network adapter

DVD	Digital Video Disc, a high-capacity, optical and video storage media that holds a minimum of 4.7 gigabytes which is enough for a full-length movie
dynamic link library	Code that runs only when it is needed by an executable program. Usually has a file extension of .dll
dynamic router	A type of router that automatically updates its routing table using automatic discovery of routes and link-state or distance vector algorithms, as it interacts with other internetwork routers
EISA	Expanded Industry Standard Architecture is an extended version of ISA architecture that utilizes a 32-bit expansion slot while also maintaining compatibility with ISA
electromagnetic induction	The process by which electromagnetic signals induce energy at the same frequency onto other conduction paths in close proximity
EMI	Electromagnetic interference consists of random noise, frequency bursts, and crosstalk, that in sufficient quantity, can obscure recognition of data transmitted across a network media
encryption	Translating data into secret code
errors	Faults that occur in a process, application, or device on a workstation or network
Ethernet	An IEEE 802.3 standard for networks utilizing bus or star topology, baseband signals at 10 Mbps, a contention access method such as CSMA/CD, and coaxial, fiber-optic, or twisted-pair cables

fault tolerance	Data that is replicated to a physically separate and redundant source such as a different partition or hard disk to guard against data loss
FDDI	Fiber Distributed Data Interface is a standard for 100-Mbps token-passing ring networks with fiber optic media for high end computers needing more bandwidth than 10-Mbps Ethernet or 16-Mbps token-ring can provide
fiber-optic cable	A glass cable media utilizing modulated light pulses to send baseband signals across a network and provide a very secure transmission media suitable for high-speed, high-capacity data transmissions over long distance with little attenuation
fiber-optic medium	A fiber-optic conductor having a central optical fiber (a thin cylindrical glass core) surrounded by a concentric layer of glass cladding with an outer reinforcing layer of plastic
file and print server	A member server dedicated to serving files and printers to a group of workstations
firewall	A firewall is a device designed to prevent unauthorized access to or from a private network
fix	Solution to a software problem
frame	A package of information formatted as a single unit and used in synchronous communications. Maintains organization, control, addressing, and error checking for transmitted data

frame relay cloud	In X.25 packet switching networks, a large array of switches, circuits, and routers that form spontaneous best data paths according to the need of the moment are sometimes called a frame relay cloud since the path configurations change so rapidly
free token	An uncaptured token circulating in a ring network waiting to be captured and used by a computer ready to send data
FTP	File Transfer Protocol is a connection utility application on your client computer that allows you to log onto a network and transfer files to or from an FTP enabled server and your computer
full-duplex transmission	A data transmission that communicates simultaneously in both directions among computer senders and receivers
full-duplexing	A network communication method that increases network speed by allowing transmit and response signals to travel simultaneously in both directions
gateway	A dedicated network conversion device that allows one network with different protocols, data formats, or architectures to communicate with another
gigabit per second (Gpbs)	Annotation describing very high data transmission rates, for example, 2 Gbps = 2×10^9 bits per second
groupware	Allows groups to have simultaneous access to information and each other

half-duplexing	A network communication method that allows data to flow in one direction at a time, with transmit signals sent in one direction reaching their destination before response signals can be sent on the same path in the opposite direction
handshaking	The process by which two devices interchange control signals when negotiating a communications link
hardware query tool	A diagnostic utility that identifies hardware internal component types and the resources they use
HCL	Hardware compatibility list maintained by Microsoft that lists approved operating system
HDLC	High-level Data Link Control is a bit-oriented protocol used for information transfer in synchronous transmission systems, utilizing the Data Link Layer of the OSI model to transmit frames between computers
hexadecimal	A system that uses 16 rather than 10 as the basis to represent numerical values.
hierarchical networking	Hierarchical networking isolates local LAN traffic on networks such as Ethernet or token ring while transmitting internet-work traffic over a high-speed backbone
host	A host is just another node computer.
host files	Static files that contain information to resolve names to IP addresses
host name	An assigned text identifier (called an alias) that's used to designate a specific TCP/IP host in a logical way

hot patch	A fix that doesn't require rebooting the system for activation
HTTP	Hypertext Transfer Protocol connects the Internet Web server and the Web browser on the workstation
hub	A network device providing a central location for connecting computers and devices, while serving to organize cabling and pass signals to their distribution points
I/O	Input/output refers to the process of gathering data and distributing the information. Gathering is accomplished with devices such as a mouse and disks. Output is performed through display and printing features
I/O port	The channel through which data transfers to and from an input or output device and the processor idle Describes a state of inactivity but with readiness, such as an idle processor waiting for a command
intelligent hub	A hub that contains the software intelligence to manage network access for connected nodes, senses when a faulty node exists and bypass it, or provides management features that centralize network diagnostics and traffic monitoring
Internet	The worldwide collection of networks and gateways that use the TCP/IP suite of protocols to communicate among each other

interrupt	A request for processor time that suspends the processor's current activity, saves its data, and passes control to an interrupt handler to deal with the hardware or software issue that created the interrupt
intranet	A private network based on the Internet protocols such as TCP/IP but designed for information management within a company or organization. It looks like a World Wide Web site, and could be connected to the Internet only through a proxy server outside the organization network
IP	Internet Protocol is the part of TCP/IP responsible for formatting data packets, routing the packets to a destination network, and reassembling the packets back into their original form at the destination
IP address	The Internet Protocol address is a 32-bit number which identifies the host and network address and the host node address on the network
IP default gateway	The router on the network that forwards packets to other gateways and the specified destination address
IP proxy	A network service that queries the Internet for your workstation
IPCONFIG	Internet Protocol Configuration is a diagnostic utility that allows viewing the TCP/IP configuration of DHCP, DNS, and WINS server addresses
IPX/SPX	IPX is a connection-oriented network protocol created by Novell to transfer data over compatible networks, and SPX is the connection-oriented transport protocol that guarantees data packet delivery of data

IRQ	Interrupt request is a numerical setting for the interrupt priority levels of internal computer devices, allowing the CPU to recognize the order in which processing requests are carried out
IS	Information Services is an organization's data processing department
ISA	Industry Standard Architecture refers to the size of a computer's expansion slot where cards are plugged in, having either an 8-bit or 16-bit slot size as used on IBM PCs
ISP	Internet service provider is a private concern that offers connectivity to the Internet to individuals or organizations
jumpers	Small plastic connectors placed around two pins that communicate with an expansion card to determine which circuits will be used
knowledge base	A database maintained by manufacturers such as Microsoft, which provides technical support information for resolving problems and errors
LAN	A LAN consists of a limited number of computers connected together in a common area within a limited physical space, combining hardware and software technologies that allow users to share resources such as data, programs, storage devices, printers, and other peripherals
LAN router	Acts as a default gateway to connect local network clients directly to the Internet

laser	Light amplification by stimulated emission of radiation is an existing technology utilized in fiber-optic cable transmissions for high speed, large bandwidth, low attenuation, and high security computer networks
layer	A series of software functions for data communications
LED	Light emitting diodes are used in multimode grade fiber-optic cable transmissions to generate data-modulated light pulses
LLC Sublayer	Creates and terminates communication links, control frame traffic, sequence frames, and acknowledge frames
LMHOSTS	A static text file used to map remote NetBIOS computer names to their permanent, manually-configured IP address
logon	The process that validates a user's credentials and retrieves the applicable user rights and user profile when signing onto a network
loopback	A method that sends signals to a device and receives it back to test system devices
LSA	Local Security Authority is the Windows NT security subsystem that validates local and remote logons for all types of accounts
MAC	Media Access Control is a sublayer of the Data Link Layer of the OSI model, dealing with network access and collision detection
MAC address	Media Access Control addresses identify devices on a network with unique hexadecimal numbers

macro	A set of keystrokes and instructions recorded and saved under a macro name. When the macro is executed, it performs all the recorded instructions at once, thus simplifying multiple keyboard-based operations
mail server	A member server dedicated to processing e-mail for a network
mainframe	A central computer (server) containing all programs, performing all tasks, and maintaining the databases while workstations having minimal computing power access these resources only as needed
Manchester encoding	A detection scheme for baseband signals where the positive and negative going edges of the data pulses register the signal's logic levels
media	The physical medium used to carry computer data signals across a network
megabit per second (Mbps)	Annotation describing high data transmission rates, for example, 4 Mbps = 4 million bits per second
member server	A stand-alone network server, sometimes referred to as an application server, or file/print server, dedicated to redistribute the processing load by performing specific tasks for which the network is designed
memory block	A contiguous section of random access memory (RAM) temporarily allocated to a program or process
mesh topology	A network topology using routing devices to create multiple redundant paths to interconnect WAN segments

Micro Channel Architecture	A standard IBM bus architecture incompatible with ISA buses, functioning as either a 16- or 32-bit bus
microsegmenting	A method of reducing traffic by isolating LAN devices with switched access
modem	A device used to convert (modulate) digital signals to analog and back again (demodulate) to enable computer communications across telephone links
modulation	The process of superimposing lower frequency information signals on a higher frequency signal to carry information to its destination across a medium suitable only to the high-frequency signal
module	A group of routines or data structures that implement a certain task or function
MSAU	Multi Station Access Unit is a multiport hub used in token-ring networks to sense when a ring computer's network adapter is faulty and bypasses it to keep the network up
multicast frame	The process of simultaneously sending a message to more than one destination on a network
multihomed computer	A host computer containing two network adapter cards, one for the internal network (LAN) and one connecting to an internetworking device such as a router
multiprotocol router	A router capable of independently processing data frames with multiple embedded protocols and sending them to their appropriate destination networks

NBTSTAT	NetBIOS Over TCP/IP Status is a diagnostic utility that examines the state of current NetBIOS over TCP/IP (NBT) connections.
NCP	NetWare Core Protocol that controls client and server operations by defining interactions between them
NetBIOS	Network Basic Input/Output System is an application programming interface used by MS-DOS applications on a LAN to enable information transmission and sessions between nodes
NetBIOS name	A NetBIOS name is the computer name that you gave it when you installed the operating system
NETSTAT	Network Status (NETSTAT) is a diagnostic utility that displays all TCP/IP protocol statistics and the current state of TCP/IP connections with the command: NETSTAT -a
network	A group of connected computers that can serve a large area (WAN) or a smaller area (LAN)
network access	The control method used to manage the way computers access a network, such as CSMA/CD or token passing
network adapter	A device in a computer responsible for the physical interface connection to the network, which allows information transfer at the data link/physical level.
network adapter driver	The software that interfaces with the hardware and operating system for the expansion card or device used to connect a computer to a local area network (LAN)

network architecture	The standards that define the architectural components of a network such as the topology, data transmission rate, media, signal type, cable distances, number of nodes, and so on
network interface card	NIC is the expansion board plugged into computers and servers that controls the flow of information over the network
Network Layer	Provides network addressing schemes and supports internetwork routing of Network Layer data packets
network operating system	A NOS coordinates all activities of a network and provides services from a central server to network clients
NIC	A Network Interface Card is a device that allows the network cable to connect the computer to the network
NIS	UNIX has a Network Information Service that shares a server `passwd` file containing usernames, groups, and passwords with UNIX computers
node	A computer or device on a network
non-contentional access	A network access method, such as token passing, where computers do not contend for access but are managed to reduce the chance of data collisions
non-routable protocol	A Data Link Layer protocol, such as DecNet's LAT or Microsoft's NetBEUI, that cannot be routed by a router because it contains no Network Layer addressing scheme
NOS	A Network Operating System provides file, print, and communication services on a network

ODI	Open Data-Link Interface is an industry standard developed by Apple and Novel that supports multiple protocols and allows connectivity
OSI	Open Systems Interconnection is a specification that defines the layered architecture of services and interactions for computers exchanging information across a network
OSI model	A network structure that uses layering techniques to facilitate open data communications
packet	A transmission unit of fixed maximum size containing data and a header with source and destination addresses, as well as error control information
packet forwarding	The process by which packets are sent out to multiple network segments to locate their destination
parallel data	Data traveling on a bus with binary digit values on separate conduction paths running in parallel with each other
passive hub	A hub that repeats and distributes its input signal to the output ports where computers are attached, without regenerating the signal
passive topology	A topology, such as a linear bus network, in which data transmissions are not regenerated and passed along by computers on the network
password	The unique identifier chosen by a computer user to securely access the network

patch	A group of fixes written into an executable program as a temporary remedy for program or system problems
patch cables	Cables that allow the addition of MAUs to continue the flow of data on token ring networks
patch panel	A centrally located device linking cables from existing devices to additional devices on the network
PCI	Peripheral Component Interconnect is a 32-bit bus architecture used in most Pentium and Macintosh computers, with many implementations having Plug and Play compatibility
PDC	Primary domain controller that maintains a master copy of domain information, including security policies and tracking users
peer-to-peer	A network of 10 computers or less with no central server where network security and file sharing is managed by equal peers (users)
Performance Monitor counters	Performance Monitor counters gather performance statistics on various system objects and processes, including processor, threads, disk usage, and memory cache objects
peripheral	A device connected to a computer or network permissions The ability to read, execute, write, delete, or in general access a network directory or file
Physical Layer	Controls and terminates the physical connection between communicating computers

PING	The Packet Internet Groper is a TCP/IP diagnostic utility that sends IP echo request packets to a destination host on a local or remote network to determine if a destination TCP/IP host is available, functional, or has a valid IP address
polling	A managed process by which a device searches for specific information from a group of network computers
POP3	Post Office Protocol version 3, is a mail server protocol that retrieves e-mail from an e-mail server and delivers the e-mail to your local client computer
port	An interface through which data is transferred between a computer and other devices, a network, or another computer, and appears to the processing computer as an address in memory to send and receive data
POTS	Plain old telephone service refers to normal telephone lines that connect telephones
PPP	Point-to-Point Protocol is an industry-standard set of Internet protocols that facilitates dial-up networking
PPTP	Point-to-Point-Tunneling Protocol is a newer protocol used in conjunction with TCP/IP which is the main protocol for the WWW and Internet
preemptive multitasking	A function whereby the operating system takes control of the processor without cooperation from the current task
Presentation Layer	An interface between user applications and services that handles protocol conversion, data translation, data compression, encryption, and character set modification

print queue	The list of documents or print jobs waiting to be printed
print server	A computer designated to accept print jobs generated from network computers
protocol	A protocol is a set of rules by which communication and transfer of information takes place in a computer system
protocol conversion	The translation of one protocol to another to accommodate communication between networks of differing protocols, formats, or architectures, as performed by a gateway
protocol stack	The set of protocols that work together on different levels to enable communications on a network
protocol suite	A set of protocols designed by a vendor as complementary parts of a protocol stack
proxy server	A server that is located between a client workstation and a server that filters and recreates data received
PSTN	The Public Switched Telephone Network is the standard telecommunication link used by ordinary phone calls and dial-up connections for two-way computer communications with a modem
public key encryption	The security process of using keys (an algorithm) to encrypt and decrypt messages
RAS	Remote Access Service is a Windows service that allows clients to configure Windows workstations for remote access to a LAN or WAN

reactance	A frequency-sensitive component of impedance, such as inductive or capacitive reactance, that can affect signal transmission loss in a coaxial cable segment
redirector	Routes requests to either computers or peripherals such as a network printer
refractive index	The measure of light refraction in a fiber-optic medium determining the transmission efficiency of the medium
registry	A database that keeps track of system, hardware, software, user profile, and application configurations
removable media	Tapes, floppies, disks, and zip drives
repeater	A network device that restores data signals and passes them along to other devices to ensure data signal recognition
ring topology	A network topology using computers attached in an unterminated ring and a free token to facilitate network access and communication between nodes
RJ-11	An ordinary modular telephone jack containing four wires for connecting and interfacing devices such as modems to the PSTN
RJ-45	A modular jack similar in appearance to an RJ-11 connector, but containing eight wires for connecting UTP or STP cabling to various network devices
rollout	Implementation of an upgrade from the test environment to the production environment

ROM	Read-only memory is a computer chip containing firmware that can only be read by a program and not altered
routable protocol	A protocol, such as TCP/IP or Novell's IPX, capable of being routed by a router
router	Directs packets to and from one network to another based on the destination ID
routing protocols	Protocols that decide which path data should take if the two computers are geographically separated
routing table	The table maintained within a router that is updated either manually by an administrator or automatically by discovered network addresses as the router communicates with other internetwork routers
SAM	In the Windows NT security model, the Security Account Manager manages a database which contains all user and group account information components
SAP	Service Advertising Protocol that enables servers to broadcast their available services across the network
SCSI controller	Small Computer System Interface controller is a parallel interface that connects multiple devices, such as printers and hard disks, in a daisy chain fashion
SDLC	A data transmission protocol commonly used on IBM networks with SNA architecture

security	The means of protecting an organization's network from malicious intrusion or unwanted solicitation using server security systems and appropriate transmission media
segment	A section of cable to which computers are attached that forms a separate logical network
segmentation	The process of dividing a network into smaller segments to reduce traffic on each
serial data	A data format using a single stream of baseband signals across a single cable to transmit data between computers on a network
server	A central computer on a network dedicated to handling the processing burden of specific tasks and responds to requests by the client workstations it services
Session Layer	Controls and terminates communication sessions between computers by implementing an interactive dialogue between user applications on those computers
shadow passwords	In UNIX security, shadow passwords are installed on the system by default and are important for protecting system passwords because when this file is in place, all passwords contained in the `/etc/passwd` file can only be read by root signal propagation The manner in which a signal traverses the physical network media
signal reflections	Energy at the baseband frequency that bounces off poorly terminated or unterminated cable connections and interferes with data recognition

SLIP	Serial Line Internet Protocol is an older industry-standard set of protocols that facilitates dial-up Internet and network connections
SMTP	Part of the TCP/IP protocol stack designed for message transfer between remote network computers on the Internet
SNA	System Network Architecture is an IBM communications framework that defines network functions and standards that enable computer information exchange and processing
SNMP	Simple Network Management Protocol monitors the status of hardware devices on a network and reports it to a central management console
SRM	In the Windows NT security model, the Security Reference Monitor manages a database which contains all user and group account information
stack	Another name for a set of protocols
stand-alone operating system	Interfaces and communicates between resident application programs such as a word processor and the computer's hardware
star topology	A network topology resembling a star configuration where computers are connected point-to-point with a central hub, offering centralized management and network fault tolerance
start bit	In asynchronous communications, a bit that signifies the beginning of a character
static router	A router that must have its routing table updated manually by an administrator, rendering routes more secure for that reason

stop bit	In asynchronous communications, the stop bit signifies the end of a character
STP	Shielded Twisted Pair is a network transmission media using internally twisted wire pairs with foil and braided mesh ground shielding that shunts noise and EMI to ground to prevent interference and provide a high data transmission rate over long distances
strong password	A strong password is complex and not easily guessed by another. They are preferable and should be eight or more characters and a combination of letters and numbers
subnet	A network that is a part of a larger network.
subnet mask	Allows you to split a single TCP/IP network address into multiple subnets
switch	A network device that can be used to segment (or microsegment) a network using hardware-based switched circuits, resulting in dramatic bandwidth increases
synchronous	Synchronous communication refers to data transmissions that rely upon the use of a timing mechanism such as a clock to pace the exchange of information between two digital systems
T1	A widely used digital line, also known as a T-carrier network, for point-to-point transmission over a wire pair with full duplexing at a 1.544 Mbps rate, handling voice, data, and video, and using multiplexing to place up to 24 voice channels per frame on a single cable

TCP	Transport layer protocol used to send messages across a network, and manage and deliver error free messages from the beginning of a connection to the end
TCP/IP	Transmission Control Protocol/Internet Protocol is a protocol invented by the U.S. Government as the "standard" for transmitting data over the Internet
Telnet	A connection utility application that allows running applications on a remote system. You can use Telnet if the computer you want to access has an active Telnet service, and you have a valid Telnet account on that server
terminator	A cabling component used to absorb reflected data signal energy propagating on a network to prevent interference with the data streams of transmitting computers
test documentation	Patch release notes that give you the information about the purpose of the fix
Thicknet	A rigid coaxial cable similar to Thinnet in construction (sometimes referred to as standard Ethernet) having a thickness of ½-inch, and carrying signals up to 500 meters before attenuation starts to occur
Thinnet	A thin, flexible coaxial cable belonging to the RG-58 family of cables with a thickness of ¼-inch, a 50-ohm characteristic impedance, and carrying signals up to 185 meters before attenuation starts to occur
throughput	The measure of useful information transmitted across a communication channel

token passing	A non-contentional network access method free of collisions that uses a token circulating around a ring of computers to enable individual computers to transmit only when taking control of the token
token ring	A network topology connecting computers and devices in ring configuration and utilizing a token passing method to manage network access
TokenEase Relay Set-Up Tool	A tool that resets the ports of TokenEase MSAU or TokenEase Slim MAUs once you have installed either device
token-ring media filter	An adapter that allows UTP cables in a token ring network to connect properly to the devices on the network that require traditional token ring connectors
topology	Network topology refers to the way cables, computers, and other components are physically and logically laid out and organized on a network
TRACERT	A diagnostic utility that traces the path of a packet as it travels in real time through routers (hops) from the local host to a remote host.
traffic	The load carried by a data communication channel
transceiver	A network device, such as a UART, that handles both transmission and reception of data signals and performs parallel to serial data format conversions when placing data on the network media
Transport Layer	Provides end-to-end error recovery, flow control, and necessary functions that guarantee reliable packet transmission and reception

trunk	A linear segment also referred to as a backbone which handles the major traffic of a network
twisted-pair cable	A network transmission media consisting of two insulated copper wires twisted around each other to create an out-of-phase condition for electrical noise and unwanted signals from other twisted pairs, canceling them out to minimize crosstalk and other interference
UART	A universal asynchronous receiver/transmitter is a device residing on a computer's network adapter card that handles parallel to serial data format conversions when transmitting and receiving data from the network media
UDP	A User Datagram Protocol is an Internet connectionless protocol normally used for broadcast messages
UNIX	UNIX is a multi-tasking operating system developed by AT&T. Unix manages computer hardware
upgrades	Version releases of software that incorporate patches and are released on a regular basis to keep your technology current
UPS	A device that monitor electrical power to a network to prevent data loss during a power outage or power fluctuation
URL	A Uniform Resource Locator is an Internet address that can be located through use of a browser
USB	A Universal Serial Bus is a successor to the RS-232C serial port and provides a basic mechanism for connecting peripherals to your computer

user access	Logon names and passwords provide users unique levels of system access
user accounts	The first element of network security involves the creation of user accounts and assignment of passwords that provide permission to log on to the network
user ID	Typically, users of a multi-user or securely protected network system claim a unique name, often called a user ID
user profile	A computer-based record maintained for an authorized network user that defines the user's environment, configuration options, and preference settings such as installed applications, desktop settings, color options, and so on
user rights	Refers to the range of system-level rights applied to a user, such as the right to do backups or operate a printer, as distinguished from permissions which are assigned to users for accessing different resources, such as files, directories, and printers
UTP	Unshielded Twisted Pair is a network transmission media consisting of two wires twisted around each other without shielding, attached with RJ-45 connectors, and used in networks defined by the 10BaseT specification to allow cable lengths of 100 meters before significant attenuation occurs
virtual memory	Memory appearing larger than actual to an application
virus	A potentially destructive program that spreads through a program or computer and destroys data

virus signature files	Definition files that list any existing viruses in the system and how they can be cleaned from the system
volume	A division of a hard disk which stores computer data
VPN	A Virtual Private Network is a private network on a Point-to-Point Protocol (PPP) connection that is formed upon a public system that allows its restricted users to communication through encryption
WAN	A wide area network, often referred to as an enterprise network, can serve thousands of users in different cities and states, and consists of multiple LANs interconnected by routers, channel service unit/data service unit (CSU/DSUs), and leased lines from telephone carrier service providers
weak password	A weak password is easily guessed because it has been chosen as a matter of simple association. These should be avoided
WINIPCFG	A Windows 95/98-based utility that displays IP configuration information for the network adapter and local host
WINMSD	Windows Diagnostic utility generates reports on system configuration information
WINS	Windows Internet Naming Service is a database that dynamically maps local NetBIOS computer names to their IP addresses
WWW	World Wide Web

Index